Nicholas Michaelides

GW00656577

Practical Risk Theory for Actuaries

MONOGRAPHS ON
STATISTICS AND APPLIED PROBABILITY

General Editors

D.R. Cox, D.V. Hinkley, N. Reid, D.B. Rubin
and B.W. Silverman

(Full details concerning this series are available from the Publishers.)

Practical Risk Theory
for Actuaries

C.D. DAYKIN
Government Actuary of the UK

T. PENTIKÄINEN
Retired Managing Director, Ilmarinen Insurance Co.
Helsinki, Finland

and

M. PESONEN
Chief Actuary, Industrial Insurance Co.
Helsinki, Finland

CHAPMAN & HALL

London · Glasgow · Weinheim · New York · Tokyo · Melbourne · Madras

Published by Chapman & Hall, 2-6 Boundary Row, London SE1 8HN, England

Chapman & Hall, 2-6 Boundary Row, London SE1 8HN, UK

Blackie Academic & Professional, Wester Cleddens Road, Bishopbriggs, Glasgow G64 2NZ, UK

Chapman & Hall GmbH, Pappelallee 3, 69469 Weinheim, Germany

Chapman & Hall USA, One Penn Plaza, 41st Floor, New York, NY10119, USA

Chapman & Hall Japan, ITP - Japan, Kyowa Building, 3F, 2-2-1 Hirakawacho, Chiyoda-ku, Tokyo 102, Japan

Chapman & Hall Australia, Thomas Nelson Australia, 102 Dodds Street, South Melbourne, Victoria 3205, Australia

Chapman & Hall India, R. Seshadri, 32 Second Main Road, CIT East, Madras 600 035, India

First edition 1994, Reprinted 1995 (twice)

© 1994 C.D. Daykin, T. Pentikäinen and M. Pesonen

Typeset in 10/12pt Times by Thomson Press (India) Ltd, New Dehli
Printed in Great Britain by St Edmundsbury Press Ltd, Bury St Edmunds, Suffolk

ISBN 0 412 42850 4

Apart from any fair dealing for the purposes of research or private study, or criticism or review, as permitted under the UK Copyright Designs and Patents Act, 1988, this publication may not be reproduced, stored, or transmitted, in any form or by any means, without the prior permission in writing of the publishers, or in the case of reprographic reproduction only in accordance with the terms of the licences issued by the Copyright Licensing Agency in the UK, or in accordance with the terms of licences issued by the appropriate Reproduction Rights Organization outside the UK. Enquiries concerning reproduction outside the terms stated here should be sent to the publishers at the London address printed on this page.

The publisher makes no representation, express or implied, with regard to the accuracy of the information contained in this book and cannot accept any legal responsibility or liability for any errors or omissions that may be made.

A Catalogue record for this book is available from the British Library

Library of Congress Cataloging-in-Publication Data available
Daykin, C.D. (Chris D.)
    Practical risk theory for actuaries / C.D. Daykin, T. Pentikäinen, and M. Pesonen.
        p.   cm. (Monographs on statistics and applied probability ; 53)
    Includes bibliographical references and index.
    ISBN 0-412-42850-4
    1. Insurance Mathematics. 2. Risk (Insurance) 3. Stochastic processes.
    I. Pentikäinen, Teivo.   II. Pesonen, M. (Martti)   III. Title. IV. Series.
    HG8781.D28 1993
    368'.01 dc20   93-34334
                                    CIP

∞ Printed on acid-free paper, manufactured in accordance with ANSI/NISO Z39.48-1992 and ANSI/NISO Z39.48-1984. (Permanence of Paper)

Contents

Preface

Classical actuarial mathematics was based on deterministic approaches, for example decrement tables representing the average expected outcome, such as the probability of death or survival. The investment return was represented by a single fixed rate of interest. However, uncertainty is a fundamental characteristic of the insurance business, since the number and severity of claims (payments resulting from accidents or other insured events) often appear to vary in a random manner. Investment behaviour and the impact of inflation are also very variable. The need to support classical deterministic techniques by analysis of variability was recognized almost a century ago, giving rise to the birth of risk theory, which, in modern terms, is the analysis of the stochastic features of insurance business.

Early risk theory was mainly associated with life insurance and was based on consideration of individual risk units (persons insured). The behaviour of the whole portfolio was deduced as the sum of individual outcomes. A review of the state-of-the-art at that time was given by Bohlmann (1909).

A new phase of development of risk theory began with the studies of Lundberg (1909, 1919), which, thanks, to Cramér (1926, 1930) and other Swedish authors, became known as the **collective theory of risk**. The occurrence of claims was dealt with collectively, without reference to the individual policies.

Both the individual and the collective models of risk lead to **stochastic processes**, the theory of which has become a well-developed discipline since Lundberg's pioneering days. Advances in the theory of stochastic processes have been reflected in the recent development of risk theory.

The field of risk theory has grown rapidly. There are now many papers and text-books which study the foundations of risk processes along strictly theoretical lines. On the other hand there is a need to develop the theories into forms suitable for practical purposes and to demonstrate their application.

Risk Theory by Beard *et al.* (1969, rewritten 1984) was written along these lines, as was the series of research reports issued in parallel by the British and Finnish Solvency Working Parties (Daykin *et al.*, 1984, 1987, 1989, 1990 and Pentikäinen *et al.*, 1982 and 1989) and further, for instance, by Pentikäinen, 1952; Rantala, 1984, 1992; Wilkie, 1984, 1986; Coutts and Devitt, 1988; and Ramlau-Hansen, 1988.

The avalanche of new research in recent years and the advance of information technology has already made *Risk Theory* by Beard *et al.* substantially obsolete, creating an obvious need to replace it by a new book concentrating in particular on those topics which are orientated towards practical applications. Comments and criticism received from users of the previous book suggested that parts of the book contained too much technical detail, which tended to hamper the reader in obtaining an adequate view of the fundamental principles. All these features pointed to the need for a complete recompilation of the text.

New elements include the incorporation of stochastic treatment of investments and asset risk, as well as the effect of uncertainties in claim reserving, giving rise to so-called run-off errors. The life insurance section has been extended and consideration is also now given to the stochastic analysis of pension funds.

Our aim has been practical applicability and simplicity, the more so because other monographs deal with more theoretical aspects and can, therefore, complement our presentation. On the other hand, the authors have been conscious of the danger of oversimplification: ignorance of the basic premises of the theory could lead to serious mistakes when applying the theory to practical problems.

The book is divided into two parts in order, we hope, to make it useful to a wider range of readers, catering for their differing needs. Some proofs and other mathematical considerations have been moved to appendices or covered by references to other publications, so as to keep the main text as brief as possible and focused on principles and practical applications.

Part One is intended to be used as a primary text-book. Part Two gives guidance in the solution of more complex problems and in more extended applications of the capabilities of the theory.

Because the book should also be suitable for reference, an index of topics is provided, and links between related subjects are frequently identified in the text. This is designed to help the reader who is interested only in a particular problem; for instance, the evaluation of capital at risk, net retention in reinsurance, run-off phenomena, asset risks, etc. Because consideration of the problems has necessarily been divided between several sections, in some cases split between Part One and Part Two, an indication is given of how to find a coherent coverage of each of the special topics without having to read the whole book in order to find the relevant sections.

Familiarity with the basic concepts and elementary techniques of probability calculus is assumed. However, a number of formulae are given in section 1.4, partly for revision of the subject material and partly to specify the definitions and notation. This is considered essential, because the standard text-books, on which the understanding of many readers will be based, differ greatly and may not cover all of the fundamental topics which will be employed in this book. In order to make the presentation palatable for as wide an audience as possible, we have given some of the presentations using in parallel both the expectation and the integral and summation notation, since different readers will be familiar with the one or the other. No reference has been made to the measure-theoretical foundations of probabilistic theory or the theory of stochastic processes, since this is not necessary for practically orientated considerations.

There is often perceived to be a wide gap between practical actuaries and the protagonists of risk theory. This has been exacerbated by the very theoretical nature of many presentations of risk theory. However, it is the authors' belief that practical actuaries must be competent in the analysis of uncertainty, as evidenced by the British Institute of Actuaries motto *certum ex incertis* (certainty out of uncertainty). Modern computer simulation techniques open up a wide field of practical applications for risk theory concepts, without requiring the restrictive assumptions, and sophisticated mathematics, of many traditional aspects of risk theory.

As actuaries are required to play a more and more important part in the financial management of insurance companies and pension funds, including various forms of actuarial certification, financial condition reporting and appointed actuary roles, the need to be able to explore the consequences of uncertainty is inescapable. Practical tools are needed to support such activities. Over the next few years, therefore, the material presented in this book will be essential reading for every actuary who wants to keep abreast of these developments.

Many of the topics covered are still developing rapidly, in particular the stochastic analysis of inflation and of investment behaviour. It has not always been possible, therefore, to present definitive solutions or even necessarily recommended approaches. In some cases our considerations raise questions which point the way to fruitful new fields of research.

Our thanks are due to many who have contributed to this work, including a long succession of actuaries who have taken pioneering steps in developing the theory of risk and its practical applications. A special debt is owed to the other authors of the previous *Risk Theory* book, Erkki Pesonen and the late Bobbie Beard, and to the members of the British and Finnish solvency working parties, in particular Heikki Bonsdorff, Jukka Rantala and Matti Ruohonen, whose work contributed many new aspects to risk theory and significantly extended the scope of practical applications. We acknowledge with particular gratitude the work of David Wilkie in developing stochastic models of inflation and investment return and for his assistance with the description of his model in Chapter 8. A number of these people have greatly assisted us by reading portions of the text and making suggestions for improvements. We would like also to thank Marilyn Eskrick for her patience with numerous revisions of the text.

Chris Daykin
Teivo Pentikäinen
Martti Pesonen

London and Helsinki, January 1993

Nomenclature

Conventions

Stochastic variables are denoted by bold type-face letters, e.g. X or x; the state of a stochastic process at time t is denoted by $X(t)$ or X_t.

If a variable can be either stochastic or not, according to the application, the bold type-face is generally not used, limiting this notation to cases where stochasticity is predominant and where indicating this contributes to the clarity of the presentation.

Monetary variables with a monetary unit of, say, £ or £million, are denoted by capital letters, for instance, X and B.

Ratio variables are denoted by a small letter, corresponding to the capital letter of that monetary variable from which they are derived, for instance, $x = X/B$.

Vectors and matrices are typeset in bold Roman in Appendix H.

List of symbols

$A(t)$ assets   3, 226, 344, 371, 431, 444, 448

$A_k(t)$ assets of category k 251

$A_n(k)$ Anscombe transformation   142

$A^+(t)$ amount available for new investment   264

$A^-(t)$ income from sold or matured assets   264

a autoregressive coefficient   250, 336, 337

$a(t)$ asset ratio   334

$a(d)$ auxiliary coefficients   294

a_j moments about the origin of the size of a claim   50, 78

$a_k(M)$ moments net of reinsurance   103

B gross premiums   4, 8, 155, 310, 313, 336, 346, 371, 410, 420, 441

$B(t)$ earned premiums   4, 8, 441

$B_{re}(t)$ reinsurance premium   4

$B'(t)$ premiums, written   4, 8, 371

b_v bonus fund parameter   425

X_m sum of m random variables   35
$X_p(t)$ claims paid   291
$X_r(x_t, t)$ value of pension in payment   440
$X_{re}(t)$ claim recoveries from reinsurers   4, 100
X_{tot} total aggregate claim amount   100, 191
$X'(t)$ claims paid   4, 9
x standardized claim amount   35, 123
x age   427
x_0 entry age   437
$x(t)$ claim ratio   328, 357
$x_c(t)$ combined ratio   328
x_r pensionable age   436
$Y(0, t)$ accumulated residual profit   421
$Y(t)$ profit function   411, 420
$Y(t)$ surplus or deficit, pensions   444
$Y^*(t)$ actual profit   411
Y_0 discounted capital value of future residual profit   422
Y_{tot} discounted capital value of future profits   422
$Y_u(t)$ underwriting result   327
y_0 per capita future residual profit   422
y_{tot} per capita future profit   422
$y(t)$ trading ratio   328
$y_e(t)$ equity dividend yield   243
$y_p(t)$ rental yield on property price index   249
Z claim size   55
Z credibility weight   183, 316
Z_D claim size with deductible   98
Z_{ced} cedant's share of a claim   101
Z_M cedant's share of a claim   21, 102
Z_{max} maximum claim size   92
Z_{re} reinsurer's share of a claim   101
Z_t credibility coefficient   188
Z_{tot} total size of a claim   101
$Z_{re,M}$ excess part of claim over M 102
α Pareto exponent   89, 307
α price elasticity   317
α_j coefficients for time series   479, 482
α_j moment about zero   22
β Pareto parameter   89
β_k beta coefficients   270
Γ gamma function   48

γ skewness   24

γ_X skewness of variable X 60

$\gamma(t_1, t_2)$ autocovariance   478

γ_2 kurtosis   24

δ force of interest   272

δ market share   308

$\delta(t)$ attraction force   252

$\Delta A, \Delta^* A$ change in value of assets   251, 344

ΔL_0 increase in other net liabilities   12

θ quality parameter   187

ε ruin or failure probability   16, 399

$\varepsilon(t)$ (white) noise   357

κ_j cumulant   23

Λ safety loading   155, 310

λ safety loading coefficient   15, 147, 156, 313

ν phase of cycle   280

μ mean   45

μ_X mean of variable X 60

μ_i central moments   24

ρ correlation coefficient   69

$\rho(t)$ lapse rate   437

$\rho(\tau)$ autocorrelation function   479, 484

$\rho(d)$ claim settlement probability   297

$\rho(x)$ frequency of early leaving   437

σ standard deviation   24

σ_X standard deviation of variable X 60

σ_X^2 variance of variable X 60

$\psi(s)$ cumulant generating function   23, 59

$\Psi(U)$ infinite time ruin probability   362

$\Psi_T(U)$ risk of ruin during $[0, T]$ 361

ω angular frequency   280, 389, 482

Part One

Foundations
of
Practical Risk Theory

In Part One consideration is restricted
to the short-term analysis of claims and related variables

CHAPTER 1

Some preliminary ideas

1.1 Cash flow and emerging costs

(a) Basic model. The financial operations of an insurer can be viewed in terms of a series of cash inflows and outflows. As Figure 1.1.1 illustrates, premiums and the income from investments, together with certain other items of income, are added to the reservoir of assets, while the reservoir is depleted by claim payments, expenses of running the business, taxes, dividends and other items of outgo.

The reservoir of assets could, in the simplest case, be a working balance of cash, but will more normally be invested in a variety of different forms of investment, including deposits, bonds, equities, property, subsidiary companies, loans, and so on.

An analogous process would be a water reservoir where some pipes supply water into the reservoir, whilst others drain it out. The level of the water in the reservoir goes up or down according to the balance between the inflows and outflows. The analogy is not complete as the reservoir cannot change its volume other than by flows in or out, whereas the assets of an insurer can be subject to quite significant changes of value as the market changes.

(b) Transition equation. The flows and the resulting assets can be expressed in the form of a transition equation

$$A(t) = A(t-1) + B'(t) + J(t) + X_{re}(t) + U_{new}(t) + W_{new}(t)$$
$$- X'(t) - E(t) - B_{re}(t) - D(t). \tag{1.1.1}$$

This equation, and various modifications of it, will constitute one of the basic working tools in this book. It defines the amount of assets $A(t)$ at the end of period t, when the initial amount $A(t-1)$ and the relevant variables for income and outgo are known. In most applications the effective period is an accounting year counted by an integer variable t.

INFLOW   OUTFLOW

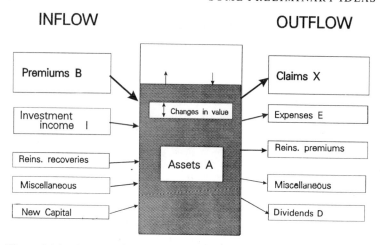

Figure 1.1.1   *Insurance business described as a balance of inflows and outflows.*

The entries in the equation are as follows:

$B'(t)$ the premium income, the prime indicating the amount of premiums **written** during the period t, as distinct from the **earned** amount to be specified in section 1.2;

$J(t)$ the return received in respect of the investments during the period t, made up of interest payments, dividends, rental income, but also from changes in value of the assets (upwards or downwards);

$X_{re}(t)$ recoveries from reinsurers during the period t;

$U_{new}(t)$ new equity capital issued and subscribed for during the period t;

$W_{new}(t)$ new debt capital issued and subscribed for during the period t and any other borrowings;

$X'(t)$ the amount of claim payments made during the period t, including payments made on account. This item is usually also taken to include expenses directly concerned with the settlement of claims;

$E(t)$ the amount of commission paid and administration and operating expenses in the period t. For brevity, taxes are also included in this term; consideration of these is deferred to Chapter 11;

$B_{re}(t)$ ceded reinsurance premium in the period t;

$D(t)$ the dividends paid to shareholders and bonuses paid to policyholders in the period t.

REMARK 1 The variables in (1.1.1) are often defined net of reinsurance, i.e. deducting the reinsurer's share from the gross amounts, in particular of $B'(t)$, $X'(t)$ and $E(t)$. Then $X_{re}(t)$ and $B_{re}(t)$ disappear from the equation, because they are broken down and absorbed into various other entries. Alternative ways of handling reinsurance are dealt with in section 14.2(a).

REMARK 2 The variables B' and X' are marked with a prime to indicate 'paid' amounts, distinguishing them from the 'earned' or 'incurred' quantities which will be specified in section 1.2. The same distinction applies in principle to several of the other variables in (1.1.1) but the effect on the stochastic behaviour of the processes is usually so insignificant that this distinction can be glossed over at this stage. This and other points of detail need to be taken into account in practical applications of risk theory models.

REMARK 3 In many applications it is helpful to treat separately the income from assets and the changes in their value. This can be done by subdividing $J(t)$ into two separate terms or by limiting $J(t)$ to income items and introducing a multiplier $[1 + g(t)]$ to be applied to $A(t-1)$ to represent asset growth (or a summation of such terms over all the different types of assets being modelled).

(c) A cash flow model concentrates on cash transactions rather than accounting quantities. However, because of the complex investment situation of an insurer and the importance of asset values for determining financial strength, and for many other applications, it is not sufficient to assume that the balance of wealth is held in cash. The **emerging costs model**, as defined here, operates on cash flows for movements in and out, but allows the insurer's reservoir of wealth, its assets, to change in value over time. In principle the value can go either up or down and may fluctuate considerably from time to time.

The important criterion in determining whether an income or outgo item should be included in the cash flow for a particular period is whether ownership of assets actually changes hands. The net excess of inflows over outflows in a period, if positive, represents an amount potentially available for investment. If there is a positive excess of outflows over inflows, this excess has to be found by the insurer, either by reducing cash balances or by sale of investments.

It will be seen that it is often necessary to keep track of other quantities, such as earned premiums and incurred claims in the accounting sense. Written premiums may not all be received in cash in a period, since part will still be in the hands of agents, or due from policyholders, at the end of the period. However, these amounts due are assets of the insurance company and the full amount of the written premium can be taken into account in the emerging costs model. It must be remembered in the modelling process that a part of the assets at any time will be amounts due, hence illiquid and non-interest-bearing.

(d) Generalization of the model. In order to restrict the scope of the subsequent development to manageable proportions, the basic emerging costs model has been stated in a firmly insurance-based context. However, essentially the same model can be applied to other financial institutions, such as pension schemes, social security schemes, banks, building societies and friendly societies. The principal feature which distinguishes all such financial institutions from other corporate entities, such as manufacturing firms, is that money is received in advance from the customer, in return for a benefit or a service at a later date. This leads to the need to hold a reservoir of wealth to meet future liabilities in respect of undertakings given or promises made. The size of this reservoir of wealth, and the importance of ensuring its adequacy in regard to liabilities already assumed, means that particular attention has to be focused on the treatment of the resulting assets.

(e) Particular cases. Although the model has been stated as referring to the whole operation of an insurer, the same principles can be applied to establish sub-models. These might concentrate on a particular part of the portfolio, or might ignore certain elements of the general model as not being of any significance in the narrow context under consideration. Application of the model to a part of a portfolio may present some problems of interpretation as to which assets are allocated to that part of the portfolio, but this does not detract from the fundamental principle that the emerging costs model can be applied to a wide variety of problems.

Another application might be in risk management assessments, where the financial consequences of accidents in an industrial plant might, for example, be considered.

(f) Deterministic or stochastic. The emerging costs model may be interpreted as either deterministic or stochastic. Many interesting and useful applications may be possible using a deterministic model. However, the true nature of almost all the items in the fundamental transition equation for an insurer is that they are subject to uncertainty. The challenge of risk theory is to model the uncertainty in a way which will provide additional insights to those who manage, supervise or appraise insurers. The development of a stochastic interpretation of the model is thus the principal objective of this book.

1.2 Accounting model

(a) Accounting concepts. In this section a parallel approach to the emerging cost and cash flow models will be presented. Financial information about an insurer is usually published in the form of accounts, which are drawn up according to certain fundamental accounting principles and concepts, and in accordance with any relevant accounting standards for insurance companies.

Fundamental accounting concepts are concerned with consistency and with prudence. However, of greater significance for our current purposes is the principle of matching (assigning items to relevant periods). Under this principle the accountant expects what is recorded in the accounts to reflect only the cash payments in or out which relate to the period in question. Thus premiums received to cover risks in a future accounting period are not treated as income until that later period. If such premiums have been received, then a **provision** must be set up in respect of amounts relating to future accounting periods. For example, if a premium which relates to a period from the middle of the accounting year to the middle of the next year is received, half of it (of the risk part) should be taken as income of the accounting year and the other half should be reserved as a provision (premium reserve) even though the total amount of the premium is included in the cash flow of the year when it is paid. This reduces the amount of premium income for which credit can be taken in the year. However, a similar provision made at the end of the previous year may be released, and this augments the premium income recognized in the year.

The accounting picture thus modifies the various quantities in the cash flow model and introduces the relevant statements of the position at the end of the accounting period, which constitute the **balance sheet**. This will show not only cash and investments as assets but also amounts due to the insurer (debtors). It will offset against these the amounts owed by the insurer (creditors). Both debtors and creditors include amounts owing to or from the previous or the next financial year.

Accounts are not always drawn up on the basis of assets shown at market value. Indeed, in many countries insurers have traditionally valued investments in their accounts at **historic cost** (the amount they paid when they purchased the investments). A more conservative approach is to use **written down historic cost**, whereby investments are held at their original purchase price, or at any lower market value which has occurred at any time since purchase.

These different features of the accounting representation, which is fundamental in terms of the recognition of profit, suggest an alternative presentation of the basic transition equation in terms of accounting quantities. The next few sections will show how the **basic accounting equation** is derived from the basic transition equation in the general insurance case once the relevant accounting quantities have been defined.

(b) Earned premiums for an accounting year are those parts of premiums written in the year, or in previous years, which relate to risks borne in that accounting year. In so far as premiums written during the accounting year provide cover for risks in the next or subsequent accounting years, the part of the premium relating to those later periods is carried forward by establishing a provision for unearned premiums. As indicated in section 1.2(a), this has the effect of reducing the amount of premiums taken into account in establishing profit in the accounting year in which the premiums are received. The provision will normally be released in the next year and become part of the earned premiums in that year, except for business where cover is provided for more than a 12-month period, in which case the provision may be released gradually over a number of years. Although an amount held on the balance sheet for a specific future liability is termed a provision by accountants, normal insurance usage in the English language (including EC and United Kingdom legislation) refers to this item as the **unearned premium reserve**. Sometimes it is more loosely referred to as the **premium reserve**.

If the unearned premium reserve at the end of the accounting period (year) t is denoted by $V(t)$, we have a simple formula for the earned premiums related to the period t

$$B(t) = B'(t) - V(t) + V(t-1) \qquad (1.2.1)$$

where $B'(t)$ is the written premium in the period, that is the premium due or receivable in that period. As mentioned above, some of the written premiums may not be received in cash form during the year but may be represented at the end of the year by amounts due from agents or brokers.

REMARK   In practice the calculation of the unearned premium reserve should take into account not only the duration and the elapsed period of cover but also the question of whether the risks being covered are evenly spread over the period. It may also be that the premiums are seen to have been inadequate for the risks covered, or there may be knowledge of particularly costly insured events that have taken place since the year end. In these cases it would normally be considered appropriate to set up an **additional reserve for unexpired risks**. The full reserve, incorporating both the unearned premium reserve and the additional amount for unexpired risks, is known as the **unexpired risk reserve**.

(c) Incurred claims in the accounting year t are defined as the total amount of claims arising from events which have occurred in the year, irrespective of when final settlement is made. It should be noted that the actual settlement (payment) of some claims may be delayed considerably beyond the year in which the event giving rise to the claim occurs and may indeed take place in stages. When accounts are drawn up at any particular point, a provision has to be established in the balance sheet for the outstanding claim payments likely to arise in respect of all previous claim events. Some of these claims may not even have been reported at the time the provision has to be set up. The provision is known as the **reserve for outstanding claims** at time t, denoted by $C(t)$, or also more loosely as the **claim reserve**. The relationship between the **incurred claims** $X(t)$ and the **paid claims** $X'(t)$ is

$$X(t) = X'(t) + C(t) - C(t-1). \qquad (1.2.2)$$

REMARK   Care is needed in ensuring that the entries in equations such as (1.2.2) are strictly consistent. For example, depending on the calculation of the claim reserve, a correction term for the yield of interest may be required in (1.2.2), as will be exemplified later in section 9.5 (see (9.5.16)).

If a claim occurs in the accounting year t and is finally settled in the same year, it is counted only in the term $X'(t)$. If payment is delayed beyond the end of the year, the estimate of the outstanding amount is booked in $C(t)$. If a claim occurred before year t but has not been (fully) paid, the estimate of the outstanding amount is carried forward in the term $C(t-1)$. If it is then settled in year t, the amount paid in year t is booked in $X'(t)$ and the claim no longer forms part of the reserve $C(t)$ carried forward. If the estimate for the outstanding amount of the claim in $C(t-1)$ were exactly correct, then the payment of the claim (in $X'(t)$) and the reserve released would precisely offset each other, having no effect on the incurred claims $X(t)$. However, if the payment and the estimate do not match exactly, the difference, if positive, adds to the incurred claims in payment year t, or if negative, reduces them. This adjustment to incurred claims in respect of claims incurred in earlier years (but inaccurately estimated) is called the **run-off error**.

REMARK The problem of estimating an appropriate outstanding claim reserve is one of the fundamental actuarial problems in general insurance. There is an extensive literature on the subject, but in spite of a wide range of possible techniques and methods the problem cannot be approached mechanically. It demands a good knowledge of the underlying business and sound professional judgement, the more so because the available relevant information might be scanty, unreliable or possibly nearly non-existent. Moreover, particular problems arise from claims which are settled in two or more parts or which are notified so late that they are not known when $C(t)$ is estimated. These issues are dealt with in section 9.5.

(d) Balance sheet. Having introduced the accounting concepts of unearned premium reserve and outstanding claim reserve, it is now useful to set these in context by giving an example of a simplified year-end balance sheet for a general insurer in Figure 1.2.1.

For the purposes of the balance sheet the assets may be shown at market value or they may, as discussed in section 1.2(a), be held at historic cost or written down historic cost, either directly or by means of an investment reserve into which all capital appreciation is transferred.

(e) Solvency margin. Having established the values of the assets and the liabilities, the balancing item in the balance sheet is the solvency margin. We introduce $L(t)$ as the total liabilities, excluding liabilities to shareholders, where

$$L(t) = V(t) + C(t) + L_0(t) \qquad (1.2.3)$$

and $L_0(t)$ incorporates liabilities other than the technical reserves, for example, sundry creditors and amounts due to reinsurers in Figure 1.2.1.

We then define the solvency margin $U(t)$ as the excess of assets over liabilities

$$U(t) = A(t) - L(t). \qquad (1.2.4)$$

BALANCE SHEET

31 December, 1992

ASSETS			LIABILITIES		
Investments			**Technical reserves**		
Bonds	200		Unearned premiums..	100	
Mortgages	300		Outstanding claims...	800	900
Loans	100		**Other liabilities**		
Equities	400	1000	due to reinsurers	60	
Bank accounts and cash		100	sundries	40	100
Other assets			**Shareholders' fund**		
Due from			Equity capital	270	
policyholders	100		Contingency reserves	100	
and agents	50		Profit of financial		
reinsurers	100		year	30	400
sundries	50	300			
Total		1400	**Total**		1400

Figure 1.2.1 *An example of the main entries of a non-life insurer's balance sheet.*

This quantity was termed the **shareholders' fund** in Figure 1.2.1, but will be called **solvency margin** or **risk reserve** in what follows, having regard to the fact that it is an indicator of the financial strength of an insurer. It is also called **surplus**, in particular in North America, but not in British usage. The term **asset margin** has been proposed in the UK (Daykin *et al.*, 1987) to get over the possible ambiguity of the term solvency margin when there is a statutory minimum solvency margin which companies must hold in order to be permitted to go on writing business. The actual solvency margin (or asset margin) is in practice usually many times the required amount. In risk theory literature the terms **risk reserve**, **equalization reserve**, **capital at risk** or **adjustment reserve** are used. In particular, these may be natural for some applications where, instead of the total business of an insurer, only a specified part of it is under consideration or when the risk theory technique is being used for the evaluation of the risks associated with industrial plants and operating arrangements, as is the case, for example, in risk management assessments. The term equalization reserve has a particular connotation in some countries,

but there is no common view on what it should include and the
term has been avoided in this text, except in section 13.2(c) where
equalization reserves are discussed briefly.

In practical applications it is crucially important to be aware of
how the assets and liabilities are valued. Depending on the national
legislation, local accounting practices and the extent to which the
management can exercise discretion, the true values of assets and
liabilities may differ considerably from their recorded values, so
giving rise to under- or overestimation of the solvency margin or to
what are often called **hidden reserves**. It will depend on the application
as to whether or not it is appropriate to include the hidden reserves
in (1.2.4). However, in general it is preferable, both for analytical
purposes, and to aid management in understanding the dynamics of
the insurance operation, that hidden reserves should be brought into
the open and realistic values are used for both assets and liabilities.

(f) Basic accounting equation. By solving for B', X' and A from
equations (1.2.1) to (1.2.4), and substituting for them into (1.1.1), the
following basic accounting equation is obtained

$$U(t) = U(t-1) + B(t) + J(t) + U_{new}(t) - \Delta L'_0(t) \\ - X(t) - E(t) - R(t) - D(t). \tag{1.2.5}$$

$\Delta L'_0(t)$ is any increase in net liabilities other than technical reserves,
excluding any increase in borrowing or debt capital or items which
are already included in other entries in (1.2.5). $R(t)$ is the balance
of dealings with the reinsurers. This term is not required if the
accounting is net of reinsurance (see Remark 1 of section 1.1(b)).

Equation (1.2.5) and its variants, in parallel with the emerging
costs equation, will be one of our key working tools.

(g) Life insurance. The basic accounting equation (1.2.5) has been
expressed in terms of the accounting principles for general insurance.
However, a similar need to establish provisions with regard to future
liabilities arises in the case of life insurance (and for other financial
systems to which a similar model might be applied). Other than for
certain types of group insurance, the majority of life insurance provisions are usually established on a policy-by-policy basis as a **premium
reserve** and a **claim reserve**, the main emphasis, contrary to general
insurance, being on premium reserves, though a distinction between
these two reserves is discretionary and varies between different
countries. The provision which needs to be set up at the end of each

year can be viewed as the excess of the value of the future outflows under the contract over the value of the future inflows.

(h) Modelling the insurance process. Having established a transition equation, whether in the emerging costs format of (1.1.1) or the accounting format of (1.2.5), the next stage is to specify sub-models for each element in the equation. The classical theory of risk concentrated on the claim process, looking first at claim numbers, then at a distribution of claim size and finally putting these two together into an aggregate claim amount process. Premiums were introduced but in a fairly simple way. More recent work on modelling the insurance process has emphasized the importance of considering the volatility of the assets and the impact of inflation, which is also very variable and normally affects both investment behaviour and claim settlement. Other important elements include the problems of reserve estimation, premium rating and profitability, expenses, reinsurance, catastrophic claims, interaction with the rest of the insurance market and the impact of formal or informal dynamic feedback mechanisms.

(i) The structure of the book. The presentation in subsequent chapters is organized so that the elements of the basic equations introduced in the preceding sections are introduced and analysed one by one, beginning with claims and proceeding to premiums, investments, market effects, dynamic control, etc., eventually arriving at comprehensive models in Chapter 14 and particular applications for life insurance (Chapter 15) and pension funds (Chapter 16). But before that a brief review of classical risk theory is given in section 1.3.

1.3 Some features of the classical theory

(a) The main features of the classical risk theory are outlined in this section in order to illustrate how the idea of stochasticity can be introduced into the insurance environment. Even though this approach, as it stands, is inadequate to describe real-world phenomena, it will be an element in the more advanced modelling and it is therefore useful to present it as a prelude to later sections.

(b) Claims as a stochastic process. The claim process can be illustrated graphically as in Figure 1.3.1. Every occurrence which gives rise to a claim is represented by a vertical step, the height of the step

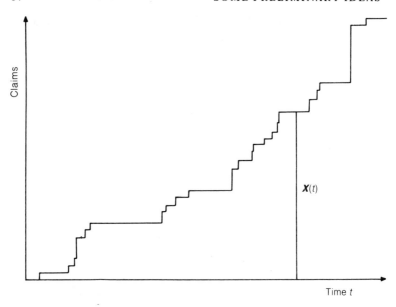

Figure 1.3.1   *A sample path of a claim process.*

indicating the amount of the claim. Time is measured to the right along the horizontal axis and the altitude X of the stepped line at time t shows the total amount of claims during the time interval $(0, t]$. The process is, in fact, a compound stochastic process in the sense that the time of occurrence and the size of each claim are both random phenomena. Any particular realization, consisting of an observed flow like that in Figure 1.3.1, is called a sample path or a realization of the process.

Note that in this section the time variable t is continuous whereas in section 1.2 and in many other contexts it is discrete, indicating the current accounting period, which usually coincides with the calendar year.

If the observation time t is fixed, then the corresponding outcome, the **state** $X(t)$, of the claim amount process $X(t)$, in our example representing the accumulated amount of claims in the interval $(0, t]$, is a **random variable** having a distribution function (abbreviated d.f.)

$$F_t(X) = \text{Prob}\,\{X(t) \leqslant X\}. \tag{1.3.1}$$

For every time point t, given the stochastic process X, the d.f. (distribution function) F_t is uniquely determined. On the other hand, unlike the case of a single random variable, mere definition of the F_t's, even if valid at every t, is not sufficient to determine the probability structure of a stochastic process. Transition rules are also needed to describe how the values of the states $X(t)$ at different times t are correlated. Hence some care is necessary when the state d.f.'s F_t are used to define the behaviour of a stochastic process. The notation used in this book, and the use of different type-faces, is explained in the Nomenclature section.

REMARK It is helpful to notice that we mostly need the **state** or **cross-sectional variable**, the amount of claims associated with the accounting period, usually year. The periods are counted by integers, for instance, $t = 1980, 1981, \ldots$. The related claim amounts are denoted by $X(t)$, i.e. using the same notation which was introduced above and which is commonly used in the literature to indicate a stochastic process. A logical procedure would, of course, be to distinguish these two concepts with different symbols. However, finding a handy different symbol for the cross-sectional variable would lead to some inconvenience, because it is one of the key variables in the book and is needed frequently. We will, therefore, use the symbols $X(t)$ and X for both the cross-sectional variable and the stochastic process, the latter being needed only on a few occasions. The context will indicate the meaning. This notation convention also concerns some other variables, for example the solvency margin $U(t)$. Similarly, the variable t, time, will mostly be taken as discrete, the counting number of periods, but in some contexts, as above in this section, as a continuous variable.

(c) Insurance process. As the next step in building up towards more advanced models, the premium income $P(t)$ is introduced. Premiums are assumed to flow continuously into a risk reserve $U(t)$, from which the claims $X(t)$ are paid out. Claim payment delays are ignored. In the classical risk theory the premium income in time period $(0, t]$ is defined to be $P(t) = (1 + \lambda) \cdot P \cdot t$, where $P = E(X(1))$ is the **pure risk premium** per unit time and λ is a **safety loading coefficient**. This definition of premium income differs from the earned premium $B(t)$ which was introduced in section 1.2(b), in that no allowance is made in $P(t)$ for expenses. In other words, when using the loaded risk premiums $P(t)$ in the classical risk theory it is assumed that the loading for expenses and the actual expenses offset each other and can therefore be dropped from the analysis of stochasticity. $B(t)$, on the other hand, is the commercial premium charged.

The process is illustrated in Figure 1.3.2. The premium income is assumed to be receivable continuously and is represented by a line sloping upwards. The claims, which can be regarded as negative

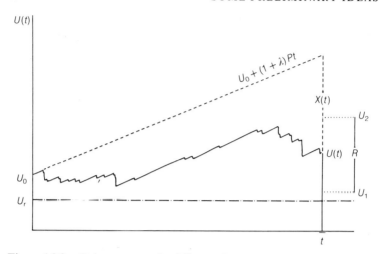

Figure 1.3.2   *Risk reserve as the difference between incoming premiums and outgoing claims.*

income, are shown as discrete downwards steps. If U_0 is the initial reserve, the difference

$$U(t) - U_0 = (1 + \lambda) \cdot P \cdot t - X(t) \qquad (1.3.2)$$

gives the net gain (which may be positive or negative) arising during the time period $(0, t]$.

REMARK $U(t)$ is a stochastic process and the risk reserve at time point t is the state of the process. However, for the sake of convenience, we do not distinguish this in the notation, as was explained in the remark in section 1.3(b).

An important problem is to evaluate the variation range R, indicated by lower and upper confidence limits U_1 and U_2, inside which the ultimate risk reserve $U(t)$ will fall with a given probability.

Of interest is the probability ε that the ultimate risk reserve $U(t)$ will, at the end of the observation period $(0, t]$, be less than some specified limit U_1, subject to the condition that the initial value is U_0

$$\varepsilon = \text{Prob} \{ U(t) < U_1 | U(0) = U_0 \}. \qquad (1.3.3)$$

A useful way to analyse the situation is to consider the capital at risk. It can be defined to be equal to $U_0 - U_1$, i.e., the capital which can at most be lost with $1 - \varepsilon$ confidence. Another way to pose the problem is to introduce the concept of the probability of ruin, i.e., the probability that $U(t)$ will fall below a limit U_r, which can be described as the **ruin barrier**. Such a limit is laid down by regulators in most countries (if not, it could be taken as equal to 0). In this context U_0 is to be understood to comprise the whole solvency margin as was defined in (1.2.4). Ruin in the risk theory terminology means a mandatory discontinuation of the company. If $U(t)$ has become negative, bankruptcy is at hand; otherwise various types of winding-up or run-off procedures are applied, as specified in national insurance laws (more discussion in section 14.6).

(d) Long-term approaches. From the point of view of the insurer, long-term survival is crucial. Therefore, the time horizon has to be extended from a single period (usually one year) to a sequence of periods, as is illustrated in Figure 1.3.3.

If the financial position (represented by $U(t)$) is assumed to be checked only at (equi-distant) time points $t = 1, 2, ..., T$, a discrete version of the model is needed. The observation points may be the ends of calendar years.

The ruin problem is now transformed into estimating the probability that the solvency margin $U(t)$ will fall below the ruin barrier at any one of the time points $t = 1, 2, ..., T$. In addition, the distribution of outcomes at the points t is of interest in many applications. Finding techniques for the calculation of these distributions is one of the central issues of the theory.

In parallel with the discrete ruin concept, a **continuous** definition is also applied, counting as a ruin any outcome where $U(t)$ falls below the ruin barrier at **any** point of the observation period $(0, T]$, including those situated between the discrete observation points, which are normally the end points of consecutive calendar years.

REMARK An approach frequently adopted in the classical theory was to let the time horizon T grow to infinity and to use the resulting 'infinite-time'

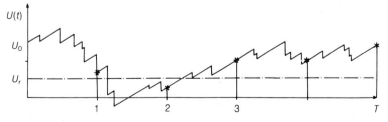

Figure 1.3.3 *Checking at points $t = 1, 2, ... T$ during the observation period.*

ruin probability as a measure of instability. Sometimes when speaking about 'ruin probability' without specifying the horizon, this particular case was implied. This procedure leads to an interesting and elegant theory. However, as will be briefly discussed in section 13.1(d) it requires restrictive assumptions and features which make it unsuitable for practical models. Therefore, we will restrict our consideration to finite time.

Note further that the common practice with the infinite-time models was to describe the various 'long-term' models as 'finite-time', so distinguishing them from the 'normal' infinite-time version.

(e) Validity of models. Finally, let us recall the well-known fact that models, including the advanced ones which will be outlined in later chapters of this book, are only an idealization of the real world. As well as the methods of mathematical statistics, common sense is also required in deciding whether and to what extent the proposed models are competent to describe the phenomena of concern.

It is useful to recognize that models are in principle subject to three types of uncertainty.

Model errors arise from the fact that no model can in practice fully represent the real world. This results in unavoidable bias or error in estimated outcomes.

Parameter error arises from the difficulty of making good estimates for any parameters which are required to run the model, given the need to use statistical inference on the observed data.

Stochastic error arises because the actual target quantities such as claim payments, return on investment, etc. are subject to random fluctuations, even in a (theoretical) situation where the model and its parameters are strictly correct.

Evaluating and analysing each of these error types may require the use of different techniques.

REMARK   Different terms may be used for the above concepts in mathematical statistics, for instance specification error, estimation error and statistical error (Taylor, G. C., 1986).

1.4 Notation and some concepts from probability theory

The purpose of this section is to set out a number of concepts of probability calculus in order to facilitate reference to them in later sections and, in particular, to introduce the notation which will be followed throughout the book. More notational conventions and an overall summary of notation is given in the Nomenclature section.

A recapitulation of some techniques and important results will be of assistance to the reader, since this material is presented in a variety of ways in standard text-books. Readers who are accustomed to the axiomatic discipline of probability calculus and stochastic processes should appreciate that consideration will be restricted to rather simple types of variables and processes, which should, however, be sufficient for practical purposes. Details of mathematical theorems and proofs have for the most part been omitted, or included only in exercises and appendices, since the emphasis of the presentation is on practical applications of the ideas.

(a) Random variables and their distribution functions. Random variables and stochastic processes are denoted by bold-face letters; for example X or k (see discussion on these concepts in section 1.3(b)). Random variables are also called **stochastic variables** or **variates** in the literature.

The probability distribution of a random variable X is uniquely determined by its (cumulative) distribution function (d.f.) F,

$$F(X) = \text{Prob}\{X \leqslant X\} \quad (-\infty < X < \infty), \tag{1.4.1}$$

i.e., $F(X)$ denotes the probability that X assumes a value that is less than or equal to X.

A number of different probability distributions will be used. Some will be continuous, others discrete, whilst yet others will be of mixed type (Figure 1.4.1). These are defined as follows:

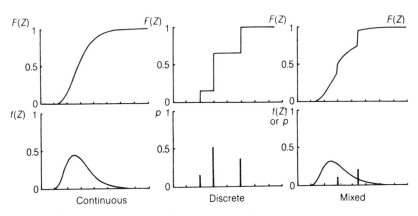

Figure 1.4.1 *Three types of distribution.*

The distribution function F of a continuously distributed random variable is a continuous function and has a derivative $f(X) = F'(X)$ everywhere, apart from a finite (or countable) set of exception points. The derivative is called the **density**. The continuous d.f. is often used to represent the size of individual claims in general insurance, where all amounts from zero up to a very large amount may occur, allowing for a variety of different sized risks and the possibility of partial damages being paid.

A random variable X with a discrete distribution has only a finite (or countable) number of values $X_1, X_2, \ldots$ The respective probabilities are usually denoted by $p_1, p_2, \ldots$ The claim number variable is of this type. Another example could be insurance where the benefits are standardized, with an option only as to how many units of cover are taken out, as is the case, for example, in some travel policies.

A distribution function F of mixed type is a combination of the discrete and continuous types, being continuous and differentiable apart from at a finite or countable set of exception points $X_1, X_2, \ldots$, which are associated with probabilities $p_1, p_2, \ldots$, causing jumps in the distribution function at these points. This type of distribution function can arise from a continuous basic distribution if there are reinsurance arrangements which have the effect of cutting off the top layer of the original risks. Similar discrete jumps can result from legal or contractual upper limits of indemnity or from policyholder deductibles. A mixed type of distribution function also results when a policy comprises benefits of several types, for example, fixed sums in the case of death or permanent invalidity and indemnity of actual costs for medical care.

Using the notation given above, an expression for the d.f. of X is

$$F(X) = \int_{-\infty}^{X} f(X)\,dX + \sum_{X_i \leqslant X} p_i. \qquad (1.4.2)$$

This applies to all three types of distribution function. In the continuous case the second term is zero, in the discrete case the first term is zero, and in the mixed case both terms are used.

For example, the mean value $E(X)$ of X can be written as follows using this notation

$$E(X) = \int_{-\infty}^{\infty} Xf(X)\,dX + \sum_{i} X_i p_i \qquad (1.4.3)$$

provided that both the integral and the sum converge.

(b) A convention for notation.   There is a frequent need for a generalized form of (1.4.3), to give expected values of type $E(g(X))$, where g is any real-valued function. We can write

$$E(g(X)) = \int_{-\infty}^{\infty} g(X)f(X)\,dX + \sum_i g(X_i)p_i \tag{1.4.4}$$

or we can use the convenient notation of the Stieltjes integral

$$E(g(X)) = \int_{-\infty}^{\infty} g(X)\,dF(X). \tag{1.4.5}$$

Readers who are not familiar with the Stieltjes integral can interpret an expression in the form (1.4.5) as an abbreviation for the double term expression in (1.4.4).

EXAMPLE   Let the gross size Z of a claim be a random variable distributed with exponential density

$$f(Z) = ae^{-aZ} \qquad (Z \geqslant 0;\ a \text{ being a positive constant}). \tag{1.4.6}$$

Suppose that the liability of the insurer has an upper limit M, so that he pays the whole amount Z only when $Z \leqslant M$ and the sum M if Z exceeds this limit. (Such a limit may arise, for example, from a contractual limitation or as a result of having ceded the top layer to reinsurers.) Hence, the truncated claims Z_M for which the insurer is responsible are defined as

$$Z_M = \min(M, Z) = \begin{cases} Z & \text{if } Z \leqslant M \\ M & \text{if } Z > M. \end{cases}$$

What is the mean value of Z_M?

By a straightforward application of formula (1.4.3), we have

$$E(Z_M) = \int_0^M Z \cdot ae^{-aZ}\,dZ + M \cdot p_M \tag{1.4.7}$$

where p_M is the probability

$$p_M = \text{Prob}\{Z_M = M\} = \text{Prob}\{Z \geqslant M\} = \int_M^{\infty} ae^{-aZ}\,dZ = e^{-aM}. \tag{1.4.8}$$

Substituting (1.4.8) into (1.4.7) and calculating the integral on the right-hand side we obtain

$$E(Z_M) = \frac{1}{a}[-aMe^{-aM} - e^{-aM} + 1] + Me^{-aM} = \frac{1}{a}[1 - e^{-aM}]. \tag{1.4.9}$$

Note, for verification, that $E(Z_M) \to 0$ when $M \to 0$ and $E(Z_M) \to 1/a = E(Z)$ when $M \to \infty$.

(c) Moments and the moment generating function.   The jth moment α_j about zero of a random variable X is defined as follows, using the notation of (1.4.5)

$$\alpha_j = \mathrm{E}(X^j) = \int_{-\infty}^{\infty} X^j \mathrm{d}F(X). \qquad (1.4.10)$$

A straightforward calculation of the moments often turns out to be difficult or impossible. A method of calculating the moments α_j, as well as facilitating many other tasks, is provided by the moment generating function (m.g.f.) M of the random variable X, which, using the notation (1.4.5), is defined by

$$M(s) = \mathrm{E}(e^{sX}) = \int_{-\infty}^{\infty} e^{sX} \mathrm{d}F(X). \qquad (1.4.11)$$

Here and in what follows the existence and convergence of the integrals and expansions is assumed. This technical condition will not cause any problems in what follows (apart from exceptional cases such as the Pareto distribution (section 3.3.7) unless it is censored or truncated). A sufficient condition for the m.g.f. $M(s)$ is that it is finite in some neighbourhood $|s| < \varepsilon$ of the origin.

The m.g.f. has the following important properties, which are derived in standard text-books and are given here without proof (see Exercise 1.4.2).

The moment α_j is equal to the jth derivative of M evaluated at the origin $s = 0$, that is

$$\alpha_j = M^{(j)}(0). \qquad (1.4.12)$$

When it is necessary to indicate which random variable is meant, the longer notation $\alpha_j(X)$ will be used.

The moments appear as coefficients in the Taylor expansion of the m.g.f.

$$M(s) = \sum_{j=0}^{\infty} \alpha_j \frac{s^j}{j!}. \qquad (1.4.13)$$

The probability distribution of a random variable is uniquely defined by its m.g.f. $M(s)$.

If M_1 and M_2 are the m.g.f.s of *independent* random variables X_1 and X_2, then the m.g.f. M of the sum $X_1 + X_2$ is obtained as the product of the m.g.f.s M_1 and M_2.

$$M(s) = M_1(s) \cdot M_2(s). \qquad (1.4.14)$$

The linear transformation $Y = aX + b$ of a random variable X

changes the m.g.f. into the form

$$M_Y(s) = e^{bs} M_X(as). \tag{1.4.15}$$

EXAMPLE   The m.g.f. of the exponential distribution (1.4.6) is

$$M(s) = \int_0^\infty e^{sX} a e^{-aX} \, dX = \frac{a}{a-s} \quad (s < a)$$

$$= 1 + \frac{s}{a} + \left(\frac{s}{a}\right)^2 + \left(\frac{s}{a}\right)^3 + \cdots. \tag{1.4.16}$$

Comparing the coefficients of this expansion with (1.4.13), the moments are obtained

$$\alpha_j = \frac{j!}{a^j}. \tag{1.4.17}$$

(d) Cumulants and the cumulant generating function.   A disadvantage of the moments α_j about zero, although they are conveniently defined in terms of the m.g.f., is that they are not additive for independent random variables (other than the mean α_1). It is useful, therefore, to introduce, in parallel with the α-moments, primary characteristics which have a simple additivity feature. The set of characteristics with this feature is known as the **cumulants**. They are obtained by means of the **cumulant generating function** (c.g.f.) ψ of X, which is defined as the natural logarithm of the m.g.f. M of X

$$\psi(s) = \ln M(s). \tag{1.4.18}$$

The cumulant κ_j of X is equal to the jth derivative of ψ evaluated at the origin (cf. 1.4.12)

$$\kappa_j = \psi^{(j)}(0). \tag{1.4.19}$$

When it is necessary to indicate which random variable is meant, the longer notation $\kappa_j(X)$ will be used. If ψ_1 and ψ_2 are the c.g.f.s of independent random variables X_1 and X_2, the c.g.f. ψ of the sum $X_1 + X_2$ is obtained by taking logarithms in (1.4.14)

$$\psi(s) = \psi_1(s) + \psi_2(s). \tag{1.4.20}$$

The implies the additivity of cumulants for independent variables, i.e. if X_1 and X_2 are independent, then

$$\kappa_j(X_1 + X_2) = \kappa_j(X_1) + \kappa_j(X_2) \tag{1.4.21}$$

as immediately follows from (1.4.19) and (1.4.20).

Since the c.g.f. uniquely defines the moment generating function, and the probability distribution of a random variable is uniquely defined by its m.g.f., it follows that the probability distribution of a random variable is also uniquely determined by its c.g.f.

The cumulants κ_j can be expressed in terms of the moments α_j, and conversely. For example

$$
\begin{aligned}
\kappa_1 &= \alpha_1 \\
\kappa_2 &= \alpha_2 - \alpha_1^2 \\
\kappa_3 &= \alpha_3 - 3 \cdot \alpha_1 \cdot \alpha_2 + 2 \cdot \alpha_1^3 \\
\kappa_4 &= \alpha_4 - 4 \cdot \alpha_1 \cdot \alpha_3 - 3 \cdot \alpha_2^2 + 12 \cdot \alpha_1^2 \cdot \alpha_2 - 6 \cdot \alpha_1^4.
\end{aligned} \tag{1.4.22}
$$

EXAMPLE   The c.g.f. of a normally distributed (briefly $N(\mu, \sigma)$-distributed) random variable with mean μ and standard deviation σ and d.f.

$$
F(X) = \frac{1}{\sigma\sqrt{2\pi}} \int_{-\infty}^{X} \exp\left(-\frac{1}{2}\left(\frac{x-\mu}{\sigma}\right)^2\right) dx \tag{1.4.23}
$$

is $\psi(s) = \mu \cdot s + \frac{1}{2} \cdot s^2 \cdot \sigma^2$ (Exercise 1.4.3). Hence,

$$
\kappa_1 = \mu, \; \kappa_2 = \sigma^2, \quad \text{and} \quad \kappa_j = 0 \quad \text{for} \quad j \geqslant 3. \tag{1.4.24}
$$

REMARK   Instead of the cumulants many standard text-books use the central moments $\mu_j = E((X - E(X))^j)$, which equate to the cumulants for $j = 1, 2$ and 3 but not for higher j, for example $\mu_4 = \kappa_4 + 3\kappa_2^2$. Additivity and some other features mean that the cumulants are much to be preferred to central moments.

(e) The conventional basic characteristics of a random variable X can readily be given in terms of the cumulants

$$
\begin{aligned}
E(X) &= \kappa_1 && \text{(mean)} \\
\text{Var}(X) &= \sigma^2 = \kappa_2 && \text{(variance, } \sigma \text{ being the standard} \\
& && \text{deviation)} \\
\gamma &= \kappa_3/\sigma^3 && \text{(skewness)} \\
\gamma_2 &= \kappa_4/\sigma^4 && \text{(kurtosis)}
\end{aligned} \tag{1.4.25}
$$

As is well-known from the probability calculus, the mean gives the position of the centre of the probability mass concerned and the standard deviation (or the variance) measures its spread. The skewness is an indicator of deviation from a fully symmetric shape. Kurtosis is the degree to which the distribution is peaked. If γ_2 is negative, then the density is more sharply peaked (leptokurtic), whereas positive γ_2 implies a more flat-topped (platykurtic) curve.

The skewness and kurtosis of a normally distributed random variable are equal to zero, as (1.4.24) immediately implies.

It follows from the additivity of the cumulants that the skewness and kurtosis of the sum of independent random variables can easily be derived from the characteristics of the summands. For example, if X and Y are independent, then the skewness of the sum $X + Y$ can be calculated from the formula (see Exercises 1.4.11 and 1.4.12)

$$\gamma_{X+Y} = \frac{\sigma_X^3 \gamma_X + \sigma_Y^3 \gamma_Y}{\sigma_{X+Y}^3}. \tag{1.4.26}$$

(f) Convolution. If X_1 and X_2 are two independent random variables, for example the aggregate amount of claims arising from two independent lines of business, with d.f.s F_1 and F_2, then the d.f. of their sum $X = X_1 + X_2$ is obtained from the **convolution** formula

$$F(X) = F_1 * F_2(X) = \int_{-\infty}^{\infty} F_1(X - X_2) \, dF_2(X_2). \tag{1.4.27}$$

In the particular case where at least one of the summands, for example X_2, is continuously distributed, f_2 denoting its density, we have

$$F(X) = F_1 * F_2(X) = \int_{-\infty}^{\infty} F_1(X - X_2) f_2(X_2) \, dX_2 \tag{1.4.28}$$

from which it follows that the sum X is continuously distributed.

(g) Conditionality. There are important applications where the d.f. of a variable X may depend on some background factor or controlling phenomenon. For example the claim expenditure in fire insurance is often dependent on the weather conditions. Then X is said to be conditional on this background factor. If the latter is described by a variable Y (for example the average temperature during the observation period) and the relevant dependencies can be quantified, i.e., the joint distribution of X and Y is known, then the conditional d.f. of variable X can be constructed.

The **conditional** d.f. of X subject to Y having the fixed value Y is denoted

$$F(X \mid Y = Y) = \text{Prob}\{X \leqslant X \mid Y = Y\}. \tag{1.4.29}$$

The conditional expectation of X, subject to Y being equal to Y, is denoted by

$$E(X \mid Y = Y) = \int_{-\infty}^{\infty} X \, dF(X \mid Y = Y). \tag{1.4.30}$$

When X does not depend on Y, this expectation is naturally equal to the (unconditional) expectation $E(X)$ for any value Y of Y, whereas if X does depend on Y, $E(X| Y = Y)$ takes different values for different Y. If the value Y of the variable Y is not fixed, but random, then the conditional expectation is a **random variable**, denoted by

$$E(X| Y). \qquad (1.4.31)$$

If Y is fixed and equal to Y, the value of this variable is equal to (1.4.30).

A few basic formulae of conditional expectations, which will be sufficient for our purposes, will be given here. For the convenience of those readers who are accustomed to expressing conditional expectations as integrals and sums, in the continuous and discrete case respectively, these traditional expressions for conditional expectations will be given in parallel in parts of the later development.

A useful basic formula for conditional expectations describes their **iterativity**

$$E(X) = E(E(X| Y)), \qquad (1.4.32)$$

which means that the expectation of a random variable can be calculated by first taking the expectation with respect to the condition variable. In the case when the condition variable Y is discrete, the formula can be reduced to the elementary form

$$E(X) = \sum_i \text{Prob}\{Y = Y_i\} \cdot E(X| Y = Y_i), \qquad (1.4.33)$$

where Y_i runs through all possible values of Y. The corresponding expansion for continuously distributed variable Y is

$$E(X) = \int_{-\infty}^{\infty} g(Y) \left[\int_{-\infty}^{\infty} X \, dF(X| Y = Y) \right] dY$$

$$= \int_{-\infty}^{\infty} g(Y) E(X| Y = Y) \, dY \qquad (1.4.34)$$

where g is the density of Y.

EXAMPLE   Consider a game where a die is first thrown once, the outcome being k, and thereafter as many dice are thrown as k indicates, resulting in a score $X = X_1 + \cdots + X_k$. What is the expected value of the sum of the scores?

Formula (1.4.32) can be applied in a straightforward way, putting k now in the role of Y. Since

$$E(X|k = k) = E(X_1) + \cdots + E(X_k) = k \cdot E(X_i) = k \cdot 3.5,$$

or $E(X|k) = k \cdot 3.5$, it follows that:

$$E(X) = E(E(X|k)) = E(k \cdot 3.5) = 3.5 \cdot E(k) = (3.5)^2 = 12.25.$$

Later on, in Chapter 3, a strict analogy will be found between the structure of the above example and the model of aggregate claims. The outcome k will have the role of the number of claims, the X_i's will be the individual claim sizes and X will be the aggregate claim amount.

In addition to formula (1.4.32), the following three rules are sufficient in order to master the calculus of conditional expectations to the extent needed in this book. Firstly

$$E(X|Y) \equiv E(X) \text{ if } X \text{ and } Y \text{ are independent,} \qquad (1.4.35)$$

i.e. if it is known that the value of Y does not give any additional information in regard to X, the conditional expectation of X is equal to the unconditional expectation $E(X)$. Secondly, the conditional expectation, like the unconditional one, is naturally additive

$$E(a \cdot X_1 + b \cdot X_2 | Y) = a \cdot E(X_1 | Y) + b \cdot E(X_2 | Y). \qquad (1.4.36)$$

The third property states that any factors that are uniquely defined by the condition variable Y can be treated as constants with respect to the conditional expectation

$$E(g(Y) \cdot X | Y) = g(Y) \cdot E(X | Y). \qquad (1.4.37)$$

This is obvious, since under the condition that the value of Y is known to be equal to Y, $g(Y)$ has the constant value $g(Y)$.

The variance of a random variable can also often be calculated easily using conditioning, by means of the formula

$$\text{Var}(X) = \text{Var}[E(X|Y)] + E[\text{Var}(X|Y)], \qquad (1.4.38)$$

where $\text{Var}(X|Y)$ is the conditional variance $E\{[X - E(X|Y)]^2 | Y\} = E(X^2|Y) - E(X|Y)^2$ (see Exercise 1.4.6).

The covariance $\text{Cov}(X_1, X_2) = E[(X_1 - EX_1) \cdot (X_2 - EX_2)]$ of the random variables X_1 and X_2 has a representation in terms of conditional covariance analogous to (1.4.38); Exercise 1.4.7.

It follows immediately from (1.4.32) that the conditional probability, the conditional d.f., and the conditional m.g.f. satisfy the corresponding formulae

$$\text{Prob}(A) = \text{E}(\text{Prob}(A \mid Y)) = \int_{-\infty}^{\infty} \text{Prob}(A \mid Y = Y) \, dG(Y)$$

$$F(x) = \text{E}(F(x \mid Y)) = \int_{-\infty}^{\infty} F(x \mid Y = Y) \, dG(Y)$$

$$M(s) = \text{E}(M(s \mid Y)) = \int_{-\infty}^{\infty} M(s \mid Y = Y) \, dG(Y), \qquad (1.4.39)$$

where G is the d.f. of Y (Exercise 1.4.8).

All formulae for conditional quantities hold true if the conditioning variable is replaced by several conditioning variables.

Exercise 1.4.1 Calculate the standard deviation of the (truncated) random variable Z_M of the example in section 1.4(b).

Exercise 1.4.2 Derive equations $(1.4.12), \ldots, (1.4.15)$.

Exercise 1.4.3 Derive the c.g.f. of the normal distribution.

Exercise 1.4.4 Calculate the four lowest cumulants of the exponential distribution (1.4.6) (a) by using the formulae (1.4.22), and (b) by using the c.g.f.

Exercise 1.4.5 Show that the cumulants κ_j ($j = 1, 2, 3, 4$) satisfy formulae (1.4.22).

Exercise 1.4.6 Prove (1.4.38).

Exercise 1.4.7 Prove the formula

$$\text{Cov}(X, Y) = \text{Cov}[\text{E}(X \mid Z), \text{E}(Y \mid Z)] + \text{E}[\text{Cov}(X, Y \mid Z)] \quad (1.4.40)$$

for conditional covariances.

Exercise 1.4.8 Prove rules (1.4.39).

Exercise 1.4.9 Show that a linear transformation $aX + b$ ($a > 0$) does not change the skewness of a random variable X. What happens if a is negative?

Exercise 1.4.10 Show that a linear transformation $aX + b$ $(a \neq 0)$ does not change the kurtosis of a random variable X.

Exercise 1.4.11 Derive formula (1.4.26) and a corresponding formula for the kurtosis.

Exercise 1.4.12 Give a formula for the skewness and kurtosis of a sum of k independent, identically distributed random variables. What is their limit when $k \to \infty$?

CHAPTER 2

The number of claims

2.1 Introduction

(a) Claim numbers. Both the number of claims and the size of each claim are generally stochastic in practical applications and our final model will be constructed accordingly. In order to introduce the relevant concepts gradually, however, this chapter is restricted to consideration of the number of claims k arising from a risk collective during a time period. The general case, where consideration is also given to the stochastic nature of individual claim sizes, will be dealt with in Chapter 3.

The behaviour of the claim number variable k can be described in terms of its probability distribution, which is determined by the probabilities

$$p_k = \text{Prob}\{k = k\} \quad (k = 0, 1, 2, \dots) \qquad (2.1.1)$$

that exactly k claims occur in the given time period.

The risk collective consists of individual risk units, such as houses, buildings and factories in fire insurance, or insured persons in life insurance. The primary events are the accidents which impinge randomly on the units, giving rise to claims. In individual risk theory the modelling of the risk process is based on consideration of these units as separate entities. The probability that one or more claims will occur in a time period is determined for each unit, as well as the distribution of the claim sizes. The claim number variable k of the whole collective is then obtained as the sum of the claim number variables of the individual risk units and, correspondingly, the aggregate claim amount is the sum of the aggregate claim amounts of the risk units.

Historically, the individual theory came as a first phase in the development of risk theory. In practice, however, it is inconvenient for handling large risk collectives and is therefore usually replaced by a collective approach. A model is developed directly for the claim number variable k and the aggregate claim amount for the whole collective, without any regard to the individual risk units. This method is nowadays commonly recognized as more satisfactory in practice and is adopted here, unless it is explicitly stated otherwise.

The individual and the collective approaches are not necessarily mutually exclusive. In sections 3.3.2 and Appendix A, section A1 it will be shown briefly that, on quite natural assumptions, the individual model leads to the collective model. It is a matter of taste whether, in introducing the foundations of risk theory, it is preferable to begin with the individual risk units and the events which randomly impinge on them or move straight to the collective description of the process.

(b) Poisson and other laws. In section 2.2 it is shown how, under certain ideal and restricted conditions, the claim number variable k is Poisson-distributed. The properties of Poisson-distributed variables are discussed in section 2.3.

A generalization of the Poisson distribution, the so-called mixed Poisson distribution, will be introduced in section 2.4. In practice the mixed Poisson distribution provides a more realistic model for the number of claims than the pure Poisson distribution. A special class of mixed Poisson distributions with useful technical properties, based on the negative binomial distribution law, is introduced in section 2.5.

2.2 The Poisson distribution

(a) The Poisson law. Insurance claims occur as a sequence of events in such a random way that it is not possible to forecast the exact time of occurrence of any one of them, nor the exact total number.

If it can be assumed that claims occur independently of each other, then the number of claims in a given time period is Poisson-distributed.

These rather loosely described characteristics need to be expressed in a rigorous form which is suitable for mathematical development. For this purpose let us consider the accumulated number $k(t)$ of claims occurring during a time period from 0 to t as a function of time t. Then $k(t)$ is a stochastic process. It is postulated that this claim number process satisfies the following three conditions.

(1) The numbers of claims occurring in any two disjoint time intervals are independent (**independence of increments**).
(2) No more than one claim may arise from the same event (**exclusion of multiple claims**).
(3) The probability that a claim occurs at a given fixed time point is equal to zero (**exclusion of special time points**).

Then the number of claims occurring in any fixed time interval is **Poisson-distributed**. The proof is given in Appendix A. Condition (3) can be replaced by the equivalent condition:

3(a) The expected number of claims $n(t) = E(k(t))$ is a continuous function of t.

If k is a *Poisson*-distributed variable, then the probabilities (2.1.1) corresponding to different values of k are obtained from the well-known formula

$$p_k = p_k(n) = e^{-n} \cdot \frac{n^k}{k!} \qquad (2.2.1)$$

where n is a positive real number, known as the Poisson parameter, being equal to the expected value of k, i.e. $n = E(k)$.

The Poisson distribution law (2.2.1) defines a class of distributions characterized by the parameter n, denoted Poisson(n).

(b) Discussion of condition (1). The condition of independence is usually only approximately satisfied in practice. There are often background factors which may cause claims in different sub-periods to be correlated. Typical factors of this nature might be the weather and economic conditions.

EXAMPLE 1   If there is an unusually large number of forest fires in the first week of July, this might be the result of a particularly dry summer, which greatly increases the risk propensity. In this case it is quite likely that the number of forest fires in the second week of July will also be rather high. Conversely, if there are no forest fires in the first week, there will probably not be many in the second week either.

EXAMPLE 2   In sickness insurance and life insurance contagious diseases and epidemics can cause correlation between claims in different time periods.

The presence of background factors such as those described above need not prevent the Poisson model from being applicable, provided that it is suitably generalized. The influence of background factors can often be quantified by introducing an auxiliary variable (or process) which controls the changes in risk propensity. In this way we arrive at the so-called mixed Poisson distribution, which will be dealt with in section 2.4.

(c) Discussion of condition (2). In some insurance lines it is possible for more than one claim to arise from the same event, for example, in a collision between two cars. In such a case the two claims are closely related and the independence condition (2) is violated. A way to circumvent the problem of multiple claims is to consider all the claims arising from the same event as forming part of a single claim. This technical convention should also be taken into account when the distribution function of the size of claims is constructed, i.e. when the relevant data is derived from the actual claim files the components of multiple claims must first be aggregated.

(d) Discussion of condition (3). It is a characteristic feature of claims that they occur randomly in such a way that their exact times of occurrence are unpredictable. Therefore, condition (3), as a rule, is automatically satisfied.

Note, however, that in some special forms of insurance there may be events which conflict with condition (3), for example, in endowment insurance a sum is paid at the terminal age. Special cases of this type should be considered separately.

2.3 Properties of Poisson variables

Poisson-distributed random variables have many useful properties. Some of these features will be considered in this section. Proofs are given in Appendix A and in the exercises.

(a) Additivity. The sum of independent Poisson variables is Poisson-distributed. More precisely, if $k_1, k_2, ..., k_m$ are independent Poisson-distributed random variables, then their sum

$$k = k_1 + k_2 + \cdots + k_m \qquad (2.3.1)$$

is also Poisson-distributed, with Poisson parameter n equal to the sum of the Poisson parameters n_i of the summands k_i

$$n = n_1 + n_2 + \cdots + n_m. \qquad (2.3.2)$$

This result can be deduced directly from the conditions of section 2.2(a). If two or more independent risk collectives are combined, for example lines of business or sections of a portfolio, and, if claims occur independently in each of these collectives, then naturally the same holds true for the combined risk collective.

A simple technical proof of additivity can be obtained by using the m.g.f. $M(s) = \exp(n \cdot (e^s - 1))$ (see Exercise 2.3.1) of a Poisson(n) variable, or its c.g.f. (see section 1.4(d))

$$\psi(s) = \ln M(s) = n \cdot (e^s - 1). \tag{2.3.3}$$

Having regard to the additivity of the c.g.f.s of independent variables (see (1.4.20)) we obtain for the sum (2.3.1) the c.g.f.

$$\psi(s) = n_1 \cdot (e^s - 1) + n_2 \cdot (e^s - 1) + \cdots + n_m \cdot (e^s - 1) = n \cdot (e^s - 1), \tag{2.3.4}$$

which is the c.g.f. of a Poisson(n) variable. The proof follows from the one-to-one correspondence of the d.f. and c.g.f. (cf. section 1.4(d)).

(b) Characteristics. The cumulants of a Poisson(n) variable are obtained as the derivatives of the c.g.f. (see (1.4.19))

$$\kappa_j = \psi^{(j)}(0) = n \tag{2.3.5}$$

for all j. Hence, the mean, standard deviation, skewness and kurtosis of a Poisson(n) variable are (see (1.4.25) and Exercise 2.3.2):

$$
\begin{aligned}
\alpha_1 &= \kappa_1 = n \quad \text{(mean)} \\
\sigma &= \kappa_2^{1/2} = \sqrt{n} \quad \text{(standard deviation)} \\
\gamma &= \kappa_3/\sigma^3 = 1/\sqrt{n} \quad \text{(skewness)} \\
\gamma_2 &= \kappa_4/\sigma^4 = 1/n \quad \text{(kurtosis)}
\end{aligned}
\tag{2.3.6}
$$

Note that the mean and variance of a Poisson variable are equal; this fact can be used in applications to test the Poisson hypothesis against the claim number data (see section 2.6).

(c) Recursion formula. The easiest way to calculate the Poisson probabilities p_k is often to use the following recursion formula

$$p_k = \frac{n}{k} \cdot p_{k-1} \tag{2.3.7}$$

with the initial value $p_0 = e^{-n}$. (Owing to overflow problems in computer applications the calculation may require rescaling or the use of logarithms when n is large.) This recursion rule is a key property when recursion algorithms for claim distributions are developed (see section 4.1).

(d) **Numerical calculation and approximations.** Let F denote the (cumulative) distribution function of a Poisson(n) variable k. Then

$$F(k) = F_n(k) = \text{Prob}\{k \leqslant k\} = \sum_{i=0}^{k} e^{-n} \frac{n^i}{i!}. \qquad (2.3.8)$$

For large k the computation of $F(k)$ is laborious. Approximate formulae are therefore required for many applications to give improved speed with acceptable accuracy.

(e) **Normal approximation.** According to the well-known **central limit theorem** of probability calculus, the distribution function F_m of a random variable X_m which is a sum of m independent and identically distributed variables, with mean μ and standard deviation σ ($< \infty$), tends asymptotically to the normal d.f. N as m grows to infinity. Applied to the claim number variable we have

$$F_m(X) = \text{Prob}\{x \leqslant x\} \to N(x) = \frac{1}{\sqrt{2\pi}} \int_{-\infty}^{x} e^{-u^2/2}\, du \quad \text{when} \quad m \to \infty,$$

$$(2.3.9)$$

where x and x are the standardized quantities

$$x = (X_m - \mu_m)/\sigma_m, \quad x = (X - \mu_m)/\sigma_m. \qquad (2.3.10)$$

Here $\mu_m = m \cdot \mu$ is the mean of X_m and $\sigma_m = \sigma \cdot \sqrt{m}$ its standard deviation (see Kendall and Stuart, 1977). This feature is also seen in Figure 2.3.1.

The formulae for computation of N and its inverse are given in Appendix D.

If X_m is asymptotically normally distributed, it can be expected that $N(x)$ can serve as an approximation for $F_m(X)$ if m is large. Since (for integer n) a Poisson(n) distribution is, by additivity, the distribution of a sum of n independent Poisson(1) variables, the premises of the central limit theorem hold. This suggests the approximation

$$F(k) \approx N\left(\frac{k-n}{\sqrt{n}}\right) \qquad (2.3.11)$$

which is obtained from (2.3.9) and (2.3.10) by substituting the relevant characteristics (2.3.6).

$p_k(n)$

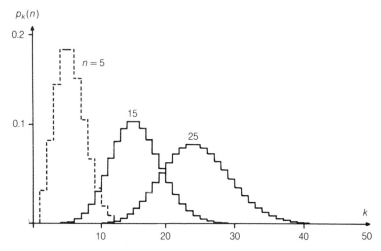

Figure 2.3.1   *Poisson probabilities (so that the distributions can be visualized, the discrete probability values $p_k(n)$ are linked as step curves).*

Table 2.3.1   *Comparison of approximations for Poisson distribution. The function shown is $F(k)$ for $k \leqslant n$ and $1 - F(k)$ for $k > n$*

n	k	Exact	Normal	Anscombe	Peizer and Pratt
10	0	0.000045	0.000783	0.000034	0.000044
	2	0.002769	0.005706	0.002672	0.002763
	8	0.332820	0.263545	0.332775	0.332833
	10	0.583040	0.500000	0.582704	0.583059
	12	0.208444	0.263545	0.208786	0.208432
	18	0.007187	0.000252	0.007137	0.007187
	23	0.000120	0.000020	0.000115	0.000120
100	80	0.022649	0.022750	0.022643	0.022649
	90	0.171385	0.158655	0.171405	0.171386
	100	0.526562	0.500000	0.526551	0.526563
	110	0.147137	0.158655	0.147161	0.147137
	120	0.022669	0.022750	0.022665	0.022669
	130	0.001707	0.001350	0.001703	0.001707
	140	0.000064	0.000032	0.000064	0.000064
	145	0.000010	0.000003	0.000010	0.000010
1000	905	0.001215	0.001332	0.001214	0.001215
	937	0.023172	0.023173	0.023172	0.023172
	968	0.159596	0.155786	0.159599	0.159596
	1032	0.152095	0.155786	0.152097	0.152095
	1063	0.023155	0.023173	0.023155	0.023155
	1095	0.001446	0.001332	0.001446	0.001446

Unfortunately, the accuracy of this approximation is poor, except for large values of n, as will be seen in Table 2.3.1.

(f) Anscombe approximation (Anscombe, 1960). This also makes use of the normal d.f. N but the argument is first suitably transformed, as in the following expression

$$F(k) \approx N\left(\frac{3}{2}\left(k + \frac{5}{8}\right)^{2/3} \cdot n^{-1/6} - \frac{3}{2}\sqrt{n} + \frac{1}{24\sqrt{n}}\right). \qquad (2.3.12)$$

As can be seen from Table 2.3.1, this formula provides a very accurate approximation.

(g) Peizer and Pratt approximation (Molenaar, 1974). This is another approximation which is based on a transformed normal d.f., and is even more accurate than the Anscombe approximation, although a little more cumbersome:

$$F(k) \approx N\left(\left[\frac{k-n}{\sqrt{n}} + \frac{1}{\sqrt{n}}\left(\frac{2}{3} + \frac{0.022}{k+1}\right)\right] \cdot \sqrt{1 + T(z)}\right), \qquad (2.3.13)$$

where
$$z = \frac{k + 0.5}{n}; \quad T(z) = \frac{1 - z^2 + 2z\ln(z)}{1 - z^2}, \quad \text{and} \quad T(1) = 0.$$

A special merit of this formula is that it is also valid for quite small values of n.

Note that (2.3.13) has the same variable as the simple normal (2.3.11) as the main term within the square brackets, but this is adjusted by both linear and multiplicative correction terms.

(h) Comparisons. Table 2.3.1 shows some illustrative results obtained using the three approximate methods referred to above.

Table 2.3.1 indicates the excellent accuracy of the Anscombe and the Peizer and Pratt formulae, whereas the normal approximation gives a poor fit unless the Poisson parameter is rather large, say of order of magnitude of at least 1000. This is to be expected, as the normal distribution is symmetric and cannot, therefore, approximate satisfactorily to any skew distribution.

Numerical tests show that the maximum absolute error of the Anscombe approximation is less than 10^{-4} for $n \geqslant 35$, the absolute error being largest in the vicinity of the mean. The corresponding limit for the Peizer and Pratt formula is $n \geqslant 6$.

A merit of the Anscombe approximation over that of Peizer and Pratt is that it is better suited to random number simulation, since the inverse transformation of the argument of N can easily be solved explicitly (section 5.3(b) for details).

REMARK   If an approximation formula does not meet the required accuracy where n is small, the exact expansion (2.3.8) has to be used, which is not too inconvenient for small values of n.

Exercise 2.3.1   Derive the m.g.f. and the c.g.f. of the Poisson(n)-distributed random variable.

Exercise 2.3.2   Prove (2.3.5) and (2.3.6).

Exercise 2.3.3   Show that the Poisson probabilities p_k satisfy the recursion formula (2.3.7).

Exercise 2.3.4   Find the mode of the Poisson(n), i.e. the value of k at which p_k reaches a maximum.

Exercise 2.3.5   Let k be a Poisson(n)-distributed variable. Show that

$$E[k \cdot (k-1) \cdot \ldots \cdot (k-i)] = n^{i+1} \quad (i = 1, 2, 3, \ldots).$$

Calculate the skewness of k directly using these expressions.

Exercise 2.3.6   The observed numbers of claims arising from a portfolio in two consecutive years were 9025 and 10 131. Do these numbers support the assumption that the Poisson parameter has remained the same?

2.4   Mixed Poisson claim number variable

(a) Volatility of risk propensity.   In section 2.2(b) it was stated that the premises of the standard Poisson law are often not valid in practice, because external background factors, such as weather, economic conditions, etc., have an impact on the underlying claim intensity. This important feature is illustrated by the examples exhibited in Figures 2.4.1 and 2.4.2.

Graph (a) of Figure 2.4.1 demonstrates seasonal variation in the

(a)

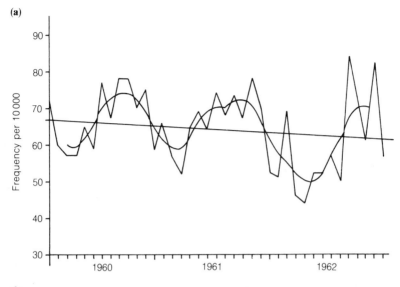

(b)

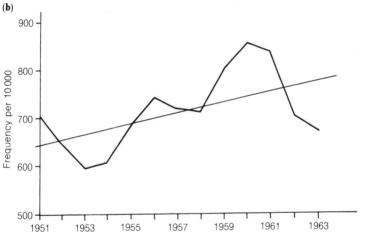

Figure 2.4.1 *Four-weekly (a) and annual claim frequencies (b) for motor cycles. The former data are smoothed by using a moving average with weights (1, 2, 3, 2, 1). UK statistics provided by Beard for Risk Theory (Beard et al., 1984).*

risk propensities, superimposed onto a downwards trend. Graph (b) shows a long-term cycle combined with an upwards trend.

The workers' compensation time series in Figure 2.4.2 shows similar behaviour.

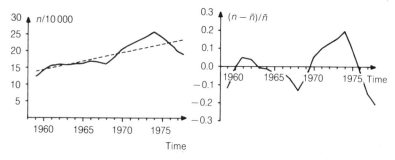

Figure 2.4.2   *The number of accidents in workers' compensation insurance; joint data of Finnish insurers. The right-hand graph displays the relative deviations from the trend line $\bar{n}$ of the left-hand graph.*

The cycles shown in Figure 2.4.2 are strongly correlated with general economic booms and recessions. During a boom, industry runs at full capacity and overtime working is frequent. The number of accidents can be expected to increase accordingly. On the other hand, during a recession working hours are reduced, which is reflected in the claim frequencies.

As the above figures suggest, the claim process can be composed of (1) **trends**, (2) periodic (non-regular) variations of risk propensity, often called **cycles**, (3) **short-term oscillation** of the risk propensity and (4) **pure random fluctuation**.

(1) Trends imply slow-moving change of the claim probabilities. They are caused, for example, by changes in building construction methods affecting fire frequencies, by changes in traffic conditions affecting accident propensity or by improvement in mortality rates in life insurance.

(2) Cycles in claim propensities are brought about, for example, by national or international economic cycles. Note the difference between claim cycles and underwriting cycles, which will be discussed in section 12.3. The latter is a broader concept, being caused also by changes and biases in premiums and, in particular, by the common feature that changes in claim propensity can only be taken into account in premium rates after a time lag. Consideration of cycles will be confined to claim cycles in Chapters 2 to 6. The more general treatment will be introduced in Chapter 12, when the time horizon is extended to cover several years.

(3) Short-term oscillations are seasonal variations or other similar effects of short duration.

(4) The pure random fluctuation is of the type which was described by the Poisson law in preceding sections.

(b) Mixing variable. The above examples show that the risk propensity, i.e., the intensity of claim-causing events, can be subject to major variations.

In so far as these variations are deterministic, such as those relating to daytime and night-time, conditions (1)–(3) of section 2.2(a) are usually still satisfied and the Poisson law applies.

When the variation of claim intensity is random, then the independence condition (1) is violated, as was seen for instance in the examples of section 2.2(b). In such a situation the stochastic variation of the claim intensity can often be interpreted as random changes of the Poisson parameter from its expected level n. The change will be described by a multiplicative factor q such that $E(q) = 1$. It is defined so that, had the claim intensity during the time period been at the expected level, the value q of q would have been equal to 1. If $q > 1$, the intensity is higher than expected and vice versa if $0 < q < 1$. Note that if the value of q is fixed and equal to q, condition (1) is satisfied and the conditional claim number d.f. $F(k|q = q)$ is Poisson($n \cdot q$).

The variable q ($q > 0$) is called a **mixing variable** and the corresponding claim number variable k a **mixed Poisson variable**.

The d.f. of the mixing variable, i.e. the mixing d.f., will be denoted by

$$H(q) = \text{Prob}\{q \leqslant q\}. \tag{2.4.1}$$

Let k be a mixed Poisson claim number variable, and let q be the corresponding mixing variable. Then, by definition, the conditional distribution $F(\cdot|q = q)$ of k is Poisson($n \cdot q$)-distributed for any value q of the mixing variable q. The claim number probabilities p_k of k can be obtained from the conditional Poisson probabilities $p_k(n \cdot q)$, given by (2.2.1), by means of the conditional probabilities formula (1.4.39)

$$p_k = E(p_k(n \cdot q)) = \int_0^\infty e^{-nq} \frac{(nq)^k}{k!} \, dH(q). \tag{2.4.2}$$

This formula has a straightforward explanation: the Poisson outcome (the integrand) is evaluated for all possible alternative states of the background factors, i.e. for all values of the mixing variable q, and then a weighted average is taken, using the mixing d.f. for weighting. The d.f. F of the mixed Poisson variable is written down as a similar weighted average (1.4.39) of the simple Poisson d.f.s

$$F(k) = E(F(k|q)) = \int_0^\infty F_{nq}(k)\, dH(q). \qquad (2.4.3)$$

Recall that due to the scaling convention the mean of q is assumed to be equal to

$$E(q) = 1. \qquad (2.4.4)$$

Since the conditional mean of the mixed Poisson variable k, given that $q = q$, is $n \cdot q$, the conditional distribution being Poisson($n \cdot q$), we have, according to (1.4.32),

$$E(k) = E(E(k|q)) = E(n \cdot q) = n \cdot E(q) = n \cdot 1 = n. \qquad (2.4.5)$$

We have checked that the mean of a mixed Poisson variable k is equal to the mean of the randomly varying Poisson parameter.

Note that the definition above, and the following considerations, are sufficiently general that there is no need to specify the causes of the mixing phenomenon. They may be cycles or short-term variations or combinations of these. However, the relevant parameters specifying the mixing distribution may depend to a great extent on the background phenomenon.

There are two ways of considering a potential cycle. On the one hand a deterministic forecast can be made of the amplitude and phase of the cycle and the Poisson parameter n be evaluated accordingly. The mixing variable q then describes only the short-term oscillation. Alternatively, q may represent the joint effect of cycles and short-term oscillation at the targeted time point t. In the latter case, of course, the variation range of q may be much greater.

REMARK   The derivation of (2.4.2) and (2.4.3) has a direct analogy with the well-known urn model of probability calculus. An urn contains lottery tickets, each having a value for the mixing variable q (it is easiest to think of q as having discrete values for a finite number of outcomes). Tickets for each value of q are included in proportion to the increments $dH(q)$ of the assumed d.f. A ticket is drawn from the urn and is assigned to the value of q which the ticket shows. The probabilities of $\{k = k\}$ and $\{k \leqslant k\}$ can be found using the rules for the multiplication and addition of probabilities. This approach is illustrated in the example in section 2.4(c) below.

(c) An example. Consider the number of forest fires in July. This is assumed to be mixed Poisson-distributed with the long-term average value of $n = 100$. To characterize variations in the weather the long-term average claim number n can be multiplied by a mixing variable q. For simplicity, let us assume that possible weather conditions in July are classified into 5 classes, $i = 1, \ldots, 5$, and, as shown in Table 2.4.1, let h_i be the respective probabilities of each type of weather condition. (Verify that $\Sigma q_i h_i = 1$ as (2.4.4) requires.) By applying (2.4.3) we have the d.f. for the number of fires

$$F(k) = \sum_i F_{nq_i}(k) \cdot h_i. \tag{2.4.6}$$

The mixed Poisson d.f. F in Table 2.4.2 can be obtained by simple calculations. For comparison the corresponding Poisson values are also given.

Table 2.4.1 *Variation of the value of the mixing variable according to the weather*

Weather type i	q_i	h_i
1 Very dry	3.00	0.05
2 Dry	1.75	0.20
3 Normal	0.80	0.40
4 Wet	0.60	0.25
5 Very wet	0.30	0.10

Table 2.4.2 *Claim distribution*

k	F(k)	F(k|q = 1)
50	0.13	0.00
70	0.47	0.01
100	0.74	0.53
150	0.76	1.00
200	0.94	1.00
300	0.98	1.00

The standard deviation increases from $\sigma_{k|q=1} = 10$ to $\sigma_k = 65.4$ due to mixing. This can be calculated by using the formula for the moments about zero

$$\alpha_j = \sum_i \alpha_j(nq_i) \cdot h_i \qquad (2.4.7)$$

which is obtained in the same straightforward way as (2.4.6) (see (1.4.10), (1.4.22) and (1.4.25)).

(d) The moment generating function of a mixed Poisson variable k can be obtained (see (1.4.39)) by taking a weighted average of the Poisson m.g.f.s:

$$M(s) = E(M(s|q)) = \int_0^\infty e^{nq(e^s - 1)} \, dH(q). \qquad (2.4.8)$$

Using conditional expectations (cf. section 1.4(g)) the following expressions are obtained

$$\begin{aligned} M(s) = E[M(s|q)] &= E[\exp(nq \cdot (e^s - 1))] \\ &= E[\exp(q \cdot (n(e^s - 1)))] = M_q(n(e^s - 1)) \\ &= M_q(\varphi(s)) \end{aligned} \qquad (2.4.9)$$

where $\varphi(s) = n \cdot (e^s - 1)$ is the c.g.f. (2.3.3) of the Poisson(n) variable.

By taking logarithms the c.g.f. ψ of the mixed Poisson variable can be written in the form

$$\psi(s) = \ln M_q(\varphi(s)) = \psi_q(\varphi(s)), \qquad (2.4.10)$$

ψ_q being the c.g.f. of the mixing variable.

(e) Characteristics. The cumulants $\kappa_j(k)$ of a mixed Poisson variable k, expressed in terms of the cumulants of the mixing variable q, are obtained from the derivatives of the c.g.f. $\psi(s) = \psi_q(\varphi(s))$ (see (2.4.10)). Note that $\psi_q'(0) = E(q) = 1$, $\varphi^{(j)}(0) = n$ (2.3.5), and $\varphi(0) = 0$. Then, for example, $\kappa_1(k) = E(k) = \psi_q'(\varphi(0)) \cdot \varphi'(0) = 1 \cdot n = n$ in accordance with (2.4.5). Similarly, the higher derivatives of the c.g.f. give (Exercise 2.4.3)

$$\begin{aligned} \kappa_2(k) &= n + n^2 \cdot \kappa_2(q) \\ \kappa_3(k) &= n + 3 \cdot n^2 \cdot \kappa_2(q) + n^3 \cdot \kappa_3(q). \end{aligned} \qquad (2.4.11)$$

By (1.4.25) the key characteristics of the mixed Poisson variable k are

Mean $\mu_k = n$

Variance $\sigma_k^2 = n + n^2 \cdot \sigma_q^2$

$$\text{Skewness } \gamma_k = \frac{n + 3 \cdot n^2 \cdot \sigma_q^2 + n^3 \cdot \gamma_q \cdot \sigma_q^3}{\sigma_k^3}. \tag{2.4.12}$$

For the kurtosis, see Exercise 2.4.6.

(f) How to get the mixing d.f. The choice of the mixing d.f. H must, of course, be based on experience (or expectation) relative to the environment under consideration. There are three commonly used techniques to introduce the mixing effect.

(1) The d.f. H is expressed in an **analytic form**. This is exemplified in section 2.5.

(2) The d.f. H is given in **tabular form**, dividing the relevant range of q into intervals. The example of section 2.4(c) above demonstrates this approach.

(3) The exact formula of the d.f. H is not specified, but only the key characteristics, in particular the standard deviation σ_q and the skewness γ_q, are given. This **method of moments** is a possible, and in fact natural, approach when approximate methods are used for the computation of claim numbers and claim amounts, as will be the standard method in the chapters which follow. The relevant formulae are based on the characteristics of the distributions, such as the mean, standard deviation, skewness, etc. (cf. also (2.4.11) and (2.4.12)), but do not require knowledge of the d.f. H.

In practice information concerning the mixing phenomenon is often sparse, sometimes allowing an estimate to be made only of the variation range. Then the analytic method (1) may be appropriate, in particular if there is experience from similar situations to support the choice of the function. If the statistical data are more abundant, then the tabular method (2) may be preferable. If the data are scanty, but there is some suggestion of skewness, then the moment method (3) may be advisable, particularly if the skewness is assumed to be large, as it often is, for instance, in the lines which are sensitive to weather conditions (storms, snow, flooding) or to booms and recessions of the national economy (credit insurance).

Note that the use of an analytic mixing function restricts the shape of the distribution. For example, in the Pólya case, which is introduced in section 2.5, the mixing d.f. has only one parameter available. The other alternatives are, therefore, often preferable. A merit of the moment method is that the necessary characteristics can be derived directly from the empirical data and from earlier experience. Construction of an analytic mixing function is then unnecessary.

A graphical presentation such as that in Figure 2.4.2 may assist the construction of the mixing d.f. If a cross-sectional distribution and its characteristics at a given time point t are required, the deviations from the trend line are relevant, as demonstrated by the right-hand graph of Figure 2.4.2. If a long-term structure is needed, regard should also be had to the trend.

Empirical data on the mixing effect were presented by Pentikäinen and Rantala (1982, section 2.3). For most classes of insurance the standard deviation σ_q of the short-term fluctuation was found to be of the order of magnitude of 0.02 to 0.08. However, for credit and forest insurance much larger values were obtained.

Note again, as was stated in section 2.4(a), that, if the fluctuation at a given time point t is of concern, both the short-term variation and the long-term cycles may simultaneously affect the total variation range and the standard deviation of 'mixing' may be much larger, as can well be seen, for example, from Figure 2.4.2. This extended problem will be considered in Chapter 9.

(g) Additivity of mixed Poisson variables.   There are two different situations where the question of additivity of mixed Poisson variables arises. Firstly, we may have two or more insurance classes, which are assumed to have mixed Poisson claim number variables, or, secondly, we may have a situation where subsequent time periods are considered, and where the number of claims occurring in each time period is mixed Poisson. Note that we are speaking about the same mathematical problem in each case.

The question is whether the sum k of mixed Poisson variables k_i is mixed Poisson-distributed and, if this is the case, what is the mixing variable? Note that the mixing variables may often correlate with each other. For example, if we think about consecutive time periods, then the circumstances of the neighbouring time periods are often correlated, especially when the mixing is caused by relatively slowly changing background factors.

The answer is 'yes', provided that the mixed Poisson(n_i)-distributed summands k_i are either mutually independent or, more generally, depend on each other only through their mixing variables q_i. Note that the mixing variables may depend on each other. To prove this additivity of mixed Poisson variables, the case of two variables k_i ($i = 1, 2$) will now be considered, but it is easy to see that the proof can be generalized for the case of more than two k_i.

So, let k_1 and k_2 depend on each other only through the mixing variables. By conditional independence and by (1.4.20) and (2.3.3) the conditional c.g.f. ψ_k of k is

$$\psi_k(s \,|\, q_1 = q_1, q_2 = q_2) = (n_1 \cdot q_1 + n_2 \cdot q_2) \cdot (e^s - 1), \qquad (2.4.13)$$

the conditional claim number d.f. of k_i being Poisson $(n_i \cdot q_i)$. By denoting $n = n_1 + n_2 = \mathrm{E}(n_1 \cdot q_1 + n_2 \cdot q_2)$ and

$$q = \frac{n_1 \cdot q_1 + n_2 \cdot q_2}{n} \qquad (2.4.14)$$

we get

$$\psi_k(s \,|\, q) = n \cdot q \cdot (e^s - 1). \qquad (2.4.15)$$

Hence, the sum $k = k_1 + k_2$ is a mixed Poisson variable with the weighted average (2.4.14) of the separate mixing variables as mixing variable.

REMARK   Even though the sum of (conditionally) independent mixed Poisson (claim number) variables is also mixed Poisson, the corresponding result is not always true for the compound mixed Poisson claim amount variables, as will be seen in section 3.2(d).

Exercise 2.4.1   Calculate the standard deviation of the mixed Poisson distribution given in the example of section 2.4(c).

Exercise 2.4.2   Suppose that the number of claims is a mixed Poisson variable with mean $n = 10\,000$ and standard deviation $\sigma = 1000$. Find the standard deviation of the mixing variable.

Exercise 2.4.3   Calculate the cumulants κ_j, $j = 1, 2, 3$, of a mixed Poisson variable k when the cumulants of the mixing variable are given.

Exercise 2.4.4   Derive the characteristics (2.4.12).

Exercise 2.4.5 Show that the standard deviation of a mixed Poisson variable, with characteristics n, $\sigma_q > 0$, is always larger than the standard deviation of the Poisson(n) distribution. Is the same true for the skewness?

Exercise 2.4.6 Show that the kurtosis of a mixed Poisson variable is

$$\gamma_2 = \frac{n + 7n^2 \cdot \sigma_q^2 + 6n^3 \cdot \gamma_q \cdot \sigma_q^3 + n^4 \cdot \gamma_{2,q} \cdot \sigma_q^4}{\sigma_k^4}.$$

2.5 The Pólya case: negative binomial distribution

(a) Gamma-distributed mixing variable. The gamma distribution Gamma(r, a) is often used as the distribution of the mixing variable q (2.4.1). Its density is

$$f(x) = \frac{a^r}{\Gamma(r)} e^{-ax} x^{r-1} \quad (x \geqslant 0) \tag{2.5.1}$$

where r and a are positive constants and

$$\Gamma(r) = \int_0^\infty e^{-u} u^{r-1} \, du \tag{2.5.2}$$

is Euler's Γ-function.

The mean value of the Gamma(r, a) variable is r/a. According to the scaling convention (2.4.4) it should be equal to 1 when used for the mixing variable, i.e. the parameters r and a should be equal. Denoting their common value by h, the d.f. H of the mixing variable q is Gamma(h, h), i.e.

$$H(q) = \frac{1}{\Gamma(h)} \int_0^{hq} e^{-z} z^{h-1} \, dz, \tag{2.5.3}$$

and its main characteristics (Appendix B, section B.1(a)) are

$$E(q) = 1; \quad \sigma_q = \frac{1}{\sqrt{h}}; \quad \gamma_q = \frac{2}{\sqrt{h}}; \quad \gamma_{2,q} = \frac{6}{h}. \tag{2.5.4}$$

Figure 2.5.1 gives some examples of the shape of the Gamma-densities H'. As the parameter h increases, the probability mass becomes more concentrated. As $h \to \infty$ this mixed Poisson distribution approaches the corresponding simple Poisson distribution. In

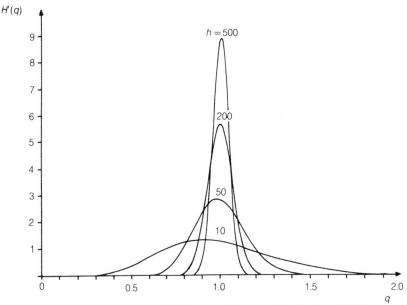

$H'(q)$

Figure 2.5.1   *Densities H' of the normed gamma distribution (2.5.3).*

fact this is an obvious outcome, because, if the mixing distribution is compressed to a point probability at $q = 1$, the mixing effect is removed. A proof is given in Appendix B.

Note that, if σ_q is given, the parameter h can be solved from (2.5.4) as $1/\sigma_q^2$. This is a useful relationship when an estimate for h is required, because σ_q can often be evaluated, at least approximately, from available data.

A three-parameter gamma distribution will be considered in section 3.3.5.

(b) The negative binomial distribution. When using the normed gamma d.f. (2.5.3) as a mixing distribution, the point probabilities (2.4.2) of the mixed compound Poisson distribution can be obtained in the form (proof in Appendix B, section B.1(b))

$$p_k = \binom{h + k - 1}{k} \cdot p^h \cdot (1 - p)^k, \qquad (2.5.5)$$

where

$$p = \frac{h}{n + h}$$

and

$$\binom{r}{s} = \frac{(r+s)!}{r!\cdot s!} = \frac{\Gamma(r+s+1)}{\Gamma(r+1)\cdot\Gamma(s+1)} \tag{2.5.6}$$

is the binomial coefficient (generalized to non-integer variable values through the Euler's Γ-function (2.5.2)).

The probability distribution defined by (2.5.5) is, in fact, well-known in the probability calculus as the **negative binomial**, also known as the **Pólya distribution**.

The m.g.f. of a Pólya-distributed claim number variable k is

$$M(s) = \left(\frac{h}{h+n-n\cdot e^s}\right)^h \tag{2.5.7}$$

and the key characteristics (Appendix B, section B.1(d)) are

$$\mu_k = E(k) = n$$

$$\sigma_k^2 = n + \frac{n^2}{h}$$

$$\gamma_k = \frac{n + \frac{3n^2}{h} + \frac{2n^3}{h^2}}{\sigma_k^3}. \tag{2.5.8}$$

The negative binomial probabilities can most easily be calculated by means of the following recursion formula

$$p_k = \left(a + \frac{b}{k}\right)\cdot p_{k-1}, \quad k = 1, 2, \ldots, \tag{2.5.9}$$

which is immediately derived from (2.5.5) (see Appendix B), where

$$p_0 = p^h; \quad p = \frac{h}{n+h}; \quad a = 1 - p; \quad b = (h-1)\cdot a. \tag{2.5.10}$$

Recall that the Poisson(n) distribution satisfies a similar recursion formula (2.3.7) with $a = 0$ and $b = n$. One advantage of the recursion formula (2.5.9) is that it leads to a recursion formula for the calculation of the aggregate claim amount distribution, as will be seen in section 4.1.

(c) Discussion. The negative binomial distribution is fairly convenient for applications, which has made it very popular. On the other hand, because only one free parameter, h, is available for the mixing distribution, the fit is not always good. It is not possible to adjust the gamma-distributed mixing distribution so that, for example, both the standard deviation and skewness assume some desired values. This is a handicap, particularly if the 'actual' mixing distribution is very skew.

The approach using the negative binomial claim number distribution is, particularly in the context of risk theory considerations, associated with the name of Pólya. He, together with Eggenberger (1923), derived it for the contamination model. The models based on the Poisson law and the negative binomial are often referred to briefly as the **Poisson case** and the **Pólya case**.

Exercise 2.5.1 Calculate and plot on the same diagram the Poisson probability function p_k and the corresponding Pólya probabilities p_k for $n = 5$ and $h = 10$.

Exercise 2.5.2 For which value of k does p_k, as given by (2.5.5), achieve its maximum?

Exercise 2.5.3 The aggregate claim numbers of two stochastically independent portfolios are Pólya-distributed. Prove that, if the portfolios are merged, the joint distribution is again of Pólya type, providing the parameters n and h are the same for both the original portfolios.

Exercise 2.5.4 Show that if k is mixed Poisson with exponential mixing distribution, then k is geometrically distributed.

2.6 Variation of risk propensity within the portfolio

(a) Individual risk propensity. Mixed Poisson distributions have applications in environments other than those just described. An example is the situation where the risk propensity of the individual risk units of the insured portfolio is considered, e.g. the problem may be to find the d.f. of the number of claims arising from a single motor-car policy. The physical process may justify the assumption of a Poisson law for the accidents, but the risk parameter n, the expected number of claims per car, can be expected to vary for different cars depending on the type, use, exposure time, mileage etc., of the car and the skill of the driver. It can be assumed that each risk unit i, or car in this example, has claim frequency parameter $n_i = nq_i$, which is the expected number of claims for this unit. Here n is an average value and q_i a coefficient indicating the deviation per unit from n. Let H be the d.f. which describes the variation of the q_i values (such a d.f. can be assumed to exist even though it may be unknown or only roughly estimated in practice). This function characterizes the distribution of risk inside the portfolio (or inside some particular part of the portfolio under consideration, e.g. some class of motor-car).

The distribution of the claim number variable k of an individual unit which is selected at random from the portfolio can be obtained by first taking the probability that the risk parameter q is in the interval $(q, q + dq)$, assuming the Poisson law for the parameter value nq, and then integrating with respect to q. The derivation of the probabilities p_k is analogous to that applied in the derivation of (2.4.2). Only the physical environment is different in the variation of the Poisson parameter n from one time unit to the next; in the present case it is the variation from one risk unit to the next. Hence (2.4.2) is readily applicable

$$p_k = \text{Prob}\{k = k\} = \int_0^\infty p_k(nq)\,dH(q), \qquad (2.6.1)$$

where $p_k(nq)$ is the standard Poisson probability (2.2.1). In this connection $H(q)$ is generally called a **structure function** (see Ammeter, 1948; Bühlmann, 1970). This is an important concept in many applications, including credibility theory (section 6.5) in particular and rate-making in general.

(b) Example. For an illustration consider the example given in Table 2.6.1. The statistics are taken from Johnson and Hey (1971) and relate to claims under UK comprehensive motor policies in

Table 2.6.1 *Comprehensive motor policies according to the number of claims*

k	Observed	Poisson	Negative binomial	Two Poissons
0	370 412	369 246	370 460	370 460
1	46 545	48 644	46 411	46 418
2	3935	3204	4045	4036
3	317	141	301	306
4	28	5	21	20
5	3	–	1	1

1968. The 421 240 policies were classified according to the number of claims in the year 1968, the average number of claims per policy being 0.13174 and the variance 0.13852. The column headed 'Poisson' sets out the distribution which would have resulted if the occurrence of claims had followed the Poisson law with $n = 0.13174$, i.e. the expected number of claims per policy in one year. As will be apparent, the Poisson distribution has a shorter tail than the data, an observation confirmed by the χ^2-test. In other words, the hypothesis that the risk propensity is the same for all policies is rejected.

The insufficiency of the Poisson law could also be anticipated from the fact that the variance is greater than the mean, whereas they should be equal if the Poisson law were valid (2.3.6).

The column headed 'Negative binomial' sets out the distribution according to this law, with parameters $n = 0.13174$ and $h = 2.555$, the latter being found by the method of maximum likelihood. The value of χ^2 is 6.9 which gives a probability of 0.14 for 4 degrees of freedom, so that the representation is acceptable. There is some indication that the negative binomial distribution may be under-representing the tail and for some applications it might be desirable to improve the model. For applications which have no significant degree of skewness the model may safely be used.

REMARK   It might be worthwhile commenting that the differences in the structure of the tails of the above distributions, even though significant, concern only a very small number of accidents and do not necessarily have any major effect on the finances of the business, which is determined by the large bulk of cases at the top of Table 2.6.1.

Another approach is to approximate the structure function $H(q)$ by a discrete d.f., assuming values $q_1, q_2, \ldots, q_r$ with probabilities $h_1, h_2, \ldots, h_r$. This means, in fact, that the d.f. is composed of r Poisson terms. The greater the number of free parameters, the better the possibility of achieving a reasonable fit, even for heterogeneous portfolios.

In the present case a two-term distribution gives quite a satisfactory result. The parameter values $q_1 = 0.65341$ and $q_2 = 2.1293$, with probabilities $h = 0.76519$ and $1 - h = 0.23481$, can be found from the equation

$$h \cdot p_k(n \cdot q_1) + (1 - h) \cdot p_k(n \cdot q_2) = p_k, \qquad (2.6.2)$$

equating the mean and standard deviation. The final column of Table 2.6.1 shows this distribution.

(c) Discussion. For most of the applications to be dealt with in this book the inner variation in the collective is not relevant. The risk behaviour will be considered in the aggregate for a given period, or broken down into consecutive periods, usually calendar years. The heterogeneity is taken care of by a proper assignment of the values for the relevant distributions and model parameters. When these are derived from actual data, this happens automatically.

Note that our notation and most of the formulae presented in section 2.4 for mixed Poisson variables are formally exactly the same as those used by many authors for structure variable considerations. This is natural, because the real structures of the processes are similar, even though the physical interpretations are different. Hence, some care is required when reading risk theory literature, the more so because a clear distinction is not always made in the terminology. In fact, owing to the similar structures the term **structure variable** has been used both in the context of a time-related mixing effect and for genuine structure heterogeneity. Unfortunately, this was the case also in *Risk Theory* by Beard *et al.* (1984), the forerunner of this book. The terminology has now been changed.

(d) References. There is a substantial literature about structure functions. The problem of finding the parameters q_i and h_i for the discrete approximation was dealt with by D'Hooge and Goovaerts (1976). Gossiaux and Lemaire (1981) studied the fit of the above methods and applied them to motor-car accident statistics. Loimaranta *et al.* (1980) have presented a cluster analysis approach as a solution of the same problem.

CHAPTER 3

The amount of claims

We will now consider the situation where the claim amounts can vary. The claim amount is the sum which the insurer has to pay on the occurrence of a fire, an accident, death or some other insured event. The sum of the individual claims constitutes the aggregate claim amount, which is one of the key concerns both in the practical management of an insurance company and in theoretical considerations.

In terms of probability calculus we construct a doubly stochastic aggregate claim amount model, where both the number of claims and the size of each claim are stochastic variables. This model applies particularly to general insurance classes. In life insurance and annuity business the amount of the claim is usually fixed or well-defined, rather than random.

3.1 Compound aggregate claim amount model

(a) Aggregate claim amount. The claim process is now generalized to include consideration of claim amounts. Let k denote the number of claims for an insurance portfolio (or any group of risks) in a certain time period, for example, one year. The aggregate claim amount X during that time period is

$$X = \sum_{i=1}^{k} Z_i \qquad (3.1.1)$$

where Z_i is the claim size of the ith claim occurring during the time period. If there are no claims, then $k = 0$ and $X = 0$.

Variables of the form (3.1.1) are known as random sums since the number k of the summands is a random number as well as the individual values of the summands.

The objective is to find an expression for the probability distribution of the aggregate claim amount X in terms of the claim number probabilities p_k (section 2.1) and the distribution of claim size. The event $\{X \leqslant X\}$ can occur in the following alternative ways

$k = 0$ (provided that X is not negative),

$k = 1$ and $Z_1 \leqslant X$,

$k = 2$ and $Z_1 + Z_2 \leqslant X$,

$k = 3$ and $Z_1 + Z_2 + Z_3 \leqslant X$,

etc.

Assuming the individual claim sizes Z_i to be independent of the claim number variable k, and applying the addition and multiplication rules of probabilities, the d.f. F of X can now be written as

$$F(X) = \text{Prob}\{X \leqslant X\} = \sum_{k=0}^{\infty} p_k \cdot \text{Prob}\left\{\sum_{i=1}^{k} Z_i \leqslant X\right\}, \qquad (3.1.2)$$

where $p_k = \text{Prob}\{k = k\}$ is the probability that exactly k claims occur.

For this result to hold it is necessary for the claim size variables to be independent of k. Otherwise the formula is quite general, with no specification of the claim number distribution, nor that for claim size, being required.

(b) Compound aggregate claim amount variable. It is now further assumed that, in addition to being independent of the number of claims k, the claim sizes Z_i making up the aggregate claim amount variable X are also mutually independent and identically distributed, each having the same d.f. S,

$$S(Z) = \text{Prob}\{Z_i \leqslant Z\}. \qquad (3.1.3)$$

An aggregate claim amount variable X satisfying these assumptions is called a **compound variable**, and its distribution is called a **compound distribution**. When the claim number variable k is (mixed) Poisson, the distribution of X is **compound (mixed) Poisson**, and when k is Pólya-distributed, the distribution of X is **compound Pólya**. Analogous terms are used for the respective aggregate claim amount processes when the accumulation of claim amounts is assumed to take place continuously over time.

The mutual independence of the claim size variables Z_i and the claim number variable k means that the probability of an individual claim being of a particular size is not affected by the number of claims that have occurred, nor by the sizes of other claims. Consequently, the d.f. F of a compound variable X is completely determined by the claim number distribution and the claim size d.f. S. More precisely, for a compound variable X the formula (3.1.2) becomes

$$F(X) = \sum_{k=0}^{\infty} p_k \cdot S^{k*}(X), \tag{3.1.4}$$

where

$$S^{k*}(X) = \text{Prob}\left\{ \sum_{i=1}^{k} Z_i \leqslant X \right\} \tag{3.1.5}$$

is the kth convolution S^{k*} of S evaluated at the point X (section 1.4(f)). In particular, $S^{0*}(X) = 0$, if $X < 0$, and $S^{0*}(X) = 1$, if $X \geqslant 0$. Note the recurrence formula $S^{k*}(X) = S*S^{(k-1)*}(X)$, whenever $k > 0$.

(c) Discussion on applicability. When the compound model is applied in practice, the insurance portfolio should be divided into sections according to the type of business, and each such section should be handled separately. Then the assumption of the existence of a claim size d.f. S for each section, which satisfactorily represents the distribution of actual claim sizes within the section, is a reasonable one and accords with general experience, at least for moderate time periods and provided that the effect of changes in monetary values is eliminated.

However, in some cases the assumption of mutual independence between the number of claims k and the claim sizes Z_i may conflict with reality. If the number of claims increases dramatically from time to time, for example as a result of a wind storm, then, besides the appearance of very large claims, small claims may predominate to a greater extent than in normal circumstances, as Ramlau-Hansen (1988) has observed. This could theoretically be dealt with by making the d.f. S depend on the claim number through the mixing variable.

The claim sizes Z_i indicate the total amount which has to be paid in order to settle the claim. Settlement is often delayed and may consist of two or more payments. The outstanding claim payments have to be estimated for reporting purposes. Inaccuracies associated with the estimation procedure need attention and will be considered in section 9.5.

A common practice in risk theory, in particular with regard to general insurance, is to restrict the claim sizes to non-negative values. Unless otherwise stated in particular cases, we have not applied this restriction and negative claim amounts are allowed. It is convenient to use the concept of a negative claim amount in cases where an event gives rise to an increase in the wealth or the reported profit of the insurer, as contrasted with the normal cases which reduce the profit (the basic process is outlined in section 1.3(c) where the claims cause downward steps in the wealth of the insurer).

Following the collective approach outlined in section 2.1(a), no regard is paid to the risk unit (or policy) from which the claim has arisen. The d.f. S describes the variability of claim size in the collective flow of claims. The connection between the individual risk units and the collective d.f. S is dealt with briefly in section 3.3.2.

3.2 Properties of compound distributions

(a) Moment and cumulant generating functions. The purpose of this section is to find expressions for the cumulant generating and moment generating functions of a compound variable X in terms of the generating functions of the claim number and the claim size distributions. These are then utilized to calculate the characteristics of X in section 3.2(b).

If X is a compound random variable with claim number variable k, then, under the condition that $k = k$, the conditional m.g.f. of X is

$$M(s|k = k) = M_{Z_1 + Z_2 + \cdots + Z_k}(s) = M_Z(s)^k \qquad (3.2.1)$$

where M_Z denotes the m.g.f. of the claim size distribution. Formula (3.2.1) follows from the multiplicativity (1.4.14) of the m.g.f.s of independent variables and from the assumption that the individual claim sizes Z_i are identically distributed. The m.g.f. M_X of X is obtained as a weighted average of the corresponding conditional m.g.f.s (1.4.39) as follows

$$M_X(s) = \sum_{k=0}^{\infty} p_k \cdot M(s|k = k) = \sum_{k=0}^{\infty} p_k \cdot M_Z(s)^k \qquad (3.2.2)$$

where $p_k = \mathrm{Prob}\{k = k\}$. This can be written more economically as

$$M_X(s) = \mathrm{E}[M(s|k)] = \mathrm{E}[M_Z(s)^k] \qquad (3.2.3)$$

and further, using the identity $t^k = e^{k \cdot \ln t}$

$$M_X(s) = \mathrm{E}[M_Z(s)^k] = \mathrm{E}[e^{k \cdot \ln M_Z(s)}] = M_k(\ln M_Z(s)) = M_k(\psi_Z(s))$$
$$(3.2.4)$$

where $\psi_Z = \ln M_Z$ is the c.g.f. of the size of one claim ((2.3.3) and (2.4.10)). The latter expressions are obtained by substituting $u = \ln M_Z(s)$ in $E(e^{k \cdot u}) = M_k(u)$. Taking logarithms, the following general formula for the c.g.f. ψ_X of the compound variable X is obtained

$$\psi_X(s) = \psi_k(\psi_Z(s)). \tag{3.2.5}$$

When the claim number variable is Poisson-distributed we have, by (2.3.3)

$$\psi_X(s) = n \cdot (e^{\psi_Z(s)} - 1) = n \cdot M_Z(s) - n \quad \text{(compound Poisson } X) \tag{3.2.6}$$

which is the c.g.f. for the compound Poisson distribution, with n being the Poisson parameter.

In the case where the number of claims is mixed Poisson-distributed this formula can be generalized (Exercise 3.2.1) as

$$\psi_X(s) = \psi_q(n \cdot M_Z(s) - n) \quad \text{(compound mixed Poisson } X). \tag{3.2.7}$$

For the compound Pólya case it is more convenient to use the m.g.f.

$$M_X(s) = \left[1 - \frac{n}{h} \cdot (M_Z(s) - 1) \right]^{-h} \quad \text{(compound Pólya } X). \tag{3.2.8}$$

(Exercise 3.2.2) instead of the c.g.f. Note that the choice here between the m.g.f. and the c.g.f. is based on the simplicity of the expressions obtained, bearing in mind that the characteristics of compound variables are derived directly from these expressions.

(b) Basic characteristics of compound distributions. We will now introduce the following standard notation, to be used throughout the book for the moments α_j about the origin of the claim size distribution and for the expected number of claims:

$$n = E(k); \quad m = a_1 = E(Z); \quad a_j = \alpha_j(Z) = E(Z^j), \tag{3.2.9}$$

where Z is distributed according to the claim size d.f. S and k is the variable for the number of claims.

The expected value of any compound aggregate claim amount variable X is simply the product of the expected number of claims n and the mean claim size m. Since $E(X|k = k) = E(Z_1 + Z_2 + \cdots + Z_k) = k \cdot E(Z_i)$, we have

$$E(X) = \sum_{k=0}^{\infty} p_k \cdot E(X|k = k) = \sum_{k=0}^{\infty} p_k \cdot k \cdot E(Z_i) = E(k) \cdot E(Z_i). \tag{3.2.10}$$

We will introduce another notation, P, for the expected value μ_X

of a claim amount variable X. P is a symbol for the risk premium, i.e. the premium required to cover the expected cost of claims, without allowance for expenses, profit or adverse deviations, which will be considered later. Using this notation, we have

$$P = \mu_X = E(X) = n \cdot m, \qquad (3.2.11)$$

whenever X is a compound variable.

From the c.g.f. of a compound aggregate claim amount variable X we obtain its cumulants. The higher order characteristics can then be obtained immediately using (1.4.25).

In the **compound Poisson case** the cumulants κ_j are obtained directly from (3.2.6):

$$\kappa_j = \psi_X^{(j)}(0) = n \cdot M_Z^{(j)}(0) = n \cdot a_j. \qquad (3.2.12)$$

Since the moment a_j about the origin is the jth derivative of the m.g.f. M_Z evaluated at zero (1.4.12). Consequently, the variance and the skewness of a compound Poisson variable X are as follows

$$\sigma_X^2 = \kappa_2 = n \cdot a_2$$

$$\gamma_X = \frac{\kappa_3}{\sigma_X^3} = \frac{n \cdot a_3}{(n \cdot a_2)^{3/2}} = \frac{a_3}{a_2^{3/2} \cdot \sqrt{n}}. \qquad (3.2.13)$$

In the **compound mixed Poisson case** these characteristics take the following forms (Exercise 3.2.3), which will be required frequently in subsequent sections and chapters

$$\sigma_X^2 = n \cdot a_2 + n^2 \cdot m^2 \cdot \sigma_q^2$$

$$\gamma_X = \frac{n \cdot a_3 + 3 \cdot n^2 \cdot m \cdot a_2 \cdot \sigma_q^2 + n^3 \cdot m^3 \cdot \gamma_q \cdot \sigma_q^3}{\sigma_X^3}, \qquad (3.2.14)$$

where q is the mixing variable. The kurtosis can be found in Exercise 3.2.4. These formulae will be the fundamental building blocks in numerous applications. Sometimes it is more convenient to use the following variant of (3.2.14), which is obtained directly by substitution:

$$\sigma_X = n \cdot m \cdot \sqrt{\frac{r_2}{n} + \sigma_q^2}$$

$$\gamma_X = \frac{r_3/n^2 + 3 r_2 \cdot \sigma_q^2/n + \gamma_q \cdot \sigma_q^3}{(r_2/n + \sigma_q^2)^{3/2}}, \qquad (3.2.15)$$

where

$$r_2 = a_2/m^2, \quad r_3 = a_3/m^3. \tag{3.2.16}$$

These ratios are called the **risk indices**, and they serve as indices of the degree of riskiness of the claim size distribution S. A rule of thumb might be to characterize the riskiness as moderate or slight if r_2 is < 30. Then typically the tail of the distribution is short. On the other hand, if r_2 exceeds, say, 200, the distribution is risky. Table 3.4.1 will illustrate the behaviour of the risk index r_2.

In the **compound Pólya case** the formulae (3.2.14) take (see (2.5.4)) the form

$$\sigma_X^2 = n \cdot a_2 + n^2 \cdot m^2/h$$
$$\gamma_X = \frac{n \cdot a_3 + 3 \cdot n^2 \cdot m \cdot a_2/h + 2 \cdot n^3 \cdot m^3/h^2}{\sigma_X^3}. \tag{3.2.17}$$

The characteristics of X above were expressed in terms of the moments a_j about the origin of the claim size distribution, since this leads to more convenient formulae in the compound mixed Poisson case. However, for other compound variables, or when the standard deviation and skewness of the claim size d.f. S are more easily available than the moments about zero, one can use the general formulae

$$\sigma_X^2 = n \cdot \sigma_Z^2 + \sigma_k^2 \cdot m^2$$
$$\gamma_X = \frac{n \cdot \sigma_Z^3 \cdot \gamma_Z + 3 \cdot \sigma_k^2 \cdot m \cdot \sigma_Z^2 + \sigma_k^3 \cdot \gamma_k \cdot m^3}{\sigma_X^3}, \tag{3.2.18}$$

which apply (Exercises 3.2.5 and 3.2.12) for any compound variable X.

(c) Asymptotic behaviour of compound mixed distributions. The above formulae for the characteristics of a compound variable X permit some useful conclusions to be drawn about the limit properties of the compound mixed Poisson distributions as the size of the portfolio grows very large. For this purpose the variance in (3.2.14) is written in the form

$$\sigma_X^2 = \sigma_0^2 + P^2 \cdot \sigma_q^2, \tag{3.2.19}$$

where

$$\sigma_0^2 = n \cdot a_2 \tag{3.2.20}$$

is the variance in the compound Poisson case.

Then the relative standard deviation of X (the coefficient of variation) is

$$\frac{\sigma_X}{E(X)} = \frac{\sigma_X}{P} = \sqrt{\frac{\sigma_0^2}{P^2} + \sigma_q^2} = \sqrt{\frac{r_2}{n} + \sigma_q^2}. \qquad (3.2.21)$$

The first term inside the square root arises from the compound Poisson fluctuation, whereas the second term introduces the additional effect of the mixing variable q. The former decreases when the volume parameter n increases but the latter is independent of n. This implies that in small collectives the pure Poisson random variation, together with the random variation of the individual claim sizes, has a more significant effect on the fluctuation of the aggregate claim amount than the mixing variable, whereas in large collectives the effect of the mixing variable predominates. This property is illustrated in Table 3.2.1, where the standard deviation of a compound mixed Poisson distribution is compared with the corresponding compound (non-mixed) Poisson case and with two compound Pólya cases. The risk index r_2 is equal to 44 in these examples.

A further decomposition of the variance (3.2.14) of a compound mixed Poisson variable is of interest in understanding the nature of the claim fluctuation.

$$\text{Var}(X) = \sigma_X^2 = m^2 \cdot n + n \cdot (a_2 - m^2) + n^2 \cdot m^2 \cdot \sigma_q^2$$
$$= m^2 \cdot \text{Var}(\text{Poisson}(n)) + n \cdot \text{Var}(Z) + n^2 \cdot m^2 \cdot \text{Var}(q). \qquad (3.2.22)$$

Table 3.2.1 *Comparison of the standard deviations of compound Poisson and compound mixed Poisson variables with two different mixing variables. The value of the risk index* r_2 *is equal to 44*

Expected number of claims n	σ_0/P	Case: $\sigma_q = 0.038$		Case: $\sigma_q = 0.100$	
		σ_X/P	σ_X/σ_0	σ_X/P	σ_X/σ_0
10	2.098	2.098	1.00	2.100	1.00
100	0.663	0.664	1.00	0.671	1.01
1000	0.210	0.213	1.02	0.232	1.11
10 000	0.066	0.076	1.15	0.120	1.81
100 000	0.021	0.043	2.07	0.102	4.87
1 000 000	0.007	0.039	5.82	0.100	15.11

σ_0 is the standard deviation of the compound Poisson distribution,
σ_X is the standard deviation of the compound mixed Poisson distribution, and
σ_q is the standard deviation of the respective mixing distribution.

Table 3.2.2 *Relative sizes of the three components of* $Var(X) = V_1 + V_2 + V_3$ *in the decomposition (3.2.22) of the variance of a compound mixed Poisson variable. Assumptions:* $\sigma_q = 0.038$ *and* $r_2 = 44$

Expected number of claims n	$V_1/Var(X)$ (%)	$V_2/Var(X)$ (%)	$V_3/Var(X)$ (%)
10	2	98	0
100	2	97	0
1000	2	95	3
10 000	2	74	24
100 000	1	23	76
1 000 000	0	3	97

$V_1 = m^2 \cdot n$, $V_2 = n \cdot \mathrm{Var}(Z)$, $V_3 = m^2 \cdot n^2 \cdot \mathrm{Var}(q)$.

Here the first term represents the variance in the case in which only the claim number is stochastic, but without any mixing, and where the claim size variation is ignored. The second term represents the contribution caused by the variation of individual claim sizes, and the third term represents the contribution arising from introducing the mixing variable q. In Table 3.2.2 an example is given of the influence of the three components in (3.2.22) which control the variation range of the aggregate claim amount. The data employed in the table are the same as in the case $\sigma_q = 0.038$ in Table 3.2.1.

A glance at Table 3.2.2 demonstrates how the structure of the claim fluctuation differs between small and large collectives. The effect of the pure Poisson claim number variation is clearly slight for all values of n. For small collectives the claim size variation is predominant and, as was already seen above, for large collectives the variation arising from the mixing variable predominates. In the compound Poisson (i.e. non-mixed) case there is no random variation in the claim frequency and the variation of the aggregate claim amount is relatively small if the collective is very large. On the other hand, in the mixed case, where the claim frequency is not stable but depends on changing conditions, such as weather variations or economic cycles, the mixing contributes the same amount σ_q^2 to the variance of the relative variable X/P independently of the size of the portfolio. These features accord with experience.

Formula (3.2.21) also permits important conclusions to be drawn about the asymptotic behaviour of the aggregate claim amount distribution. Note that (3.2.21) represents the standard deviation of the loss ratio

$$\frac{X}{E(X)} = \frac{X}{P}. \tag{3.2.23}$$

Letting $n \to \infty$, this standard deviation has the limit

$$\sigma_{X/P} \to \sigma_q. \tag{3.2.24}$$

In particular, in the compound Poisson case $\sigma_q = 0$, so $\sigma_{X/P} \to 0$, as $n \to \infty$, i.e. the relative deviation of a compound Poisson variable X, as a fraction of its mean, approaches zero. So, in the compound Poisson case the impact of random fluctuation becomes insignificant when the aggregate claim amount is derived from a very large collective of risks. This is the law of large numbers, which has been applied in rate-making and in many other practical problems in the insurance industry for as long as the industry has been in existence, in its basic form since the times of Hammurab and the ancient Romans. Burning rates and similar characteristics have been derived from mean values, obtained from as large a body of claim statistics as possible, relying on the proposition that the mean values provide a satisfactory basis for setting the actual premium rates.

Moreover, in the compound Poisson case the central limit theorem can also be applied and the d.f. F of X is therefore asymptotically normal (Exercise 3.2.8) as the size of the collective increases. This means that the normal approximation is asymptotically accurate for the d.f. F of X in the compound Poisson case. However, especially when the risk index r_2 is large, a satisfactory approximation is provided only for very large values of the Poisson parameter n. The approximation of the d.f. F will be considered in Chapter 4.

A crucial observation is that the limit of (3.2.24) has a quite different character in the compound mixed Poisson case. The loss ratio (3.2.23) does not converge towards any unique limit value, but instead it can randomly assume values over a range characterized by the standard deviation σ_q. This means that the law of large numbers does not hold for compound mixed Poisson distributions, as the size of the portfolio increases to infinity. Nor does the aggregate claim amount distribution have the normal distribution as limit distribution, which is a useful fact to remember when methods of approximation are discussed in the next section.

A rigorous treatment of the limit distribution is given in Appendix C where it is shown that the limit d.f. is in fact the mixing d.f. H, that is

$$F_{X/P}(x) = \text{Prob}\{X/P \leqslant x\} \to H(x), \quad \text{as } n \to \infty. \tag{3.2.25}$$

REMARK The considerations above were focused on claims occurring during some fixed time period, for example during one particular year. Hence, the asymptotic behaviour of the compound mixed variable was considered in the case that the size of the portfolio increases towards infinity. However, if the accumulated amount of claims in a compound mixed Poisson aggregate claim amount process is considered over a longer time period, then (supposing the influence of trends and inflation to be eliminated) the mixing process is likely to have different values for different times, so that the law of large numbers may satisfactorily apply to the accumulated claim amount, if the number of years is sufficiently large. A slightly different situation arises when the sum of the aggregate claim amounts of several insurance lines is considered, as will be discussed in section 3.5 in more detail.

(d) On the additivity properties of aggregate claim amount variables. Claims in an insurance portfolio are usually analysed separately for different classes, deriving the claim size d.f. S_j and the d.f. of the aggregate claim amount variable X_j for each class j. Then the aggregate claim amount X of the whole portfolio is simply the sum

$$X = \sum_j X_j \qquad (3.2.26)$$

over all classes j of the aggregate claim amount variables X_j.

One question is whether the sum X of the aggregate claim amount variables X_j has useful additivity properties. For example, under what conditions can the distribution and the basic characteristics, etc., of the sum (3.2.26) be derived from the corresponding class quantities?

It can be shown that the sum of independent compound mixed Poisson variables is generally not a compound variable. Nevertheless, because of independence, the basic characteristics satisfy the general additivity rules in this case. Indeed, whenever the aggregate claim amounts X_j are mutually independent, the main characteristics of their sum X can be expressed in terms of the characteristics of the class variables X_j (1.4.21):

$$\mu_X = \sum_j \mu_{X_j}; \quad \sigma_X^2 = \sum_j \sigma_{X_j}^2; \quad \gamma_X = \frac{\sum_j \sigma_{X_j}^3 \cdot \gamma_{X_j}}{\sigma_X^3}. \qquad (3.2.27)$$

However, under certain conditions the sum X is a compound variable. Cases (1) and (2) below provide examples of such special situations:

(1) *The sum of independent compound Poisson variables.* It is assumed that the X_j are mutually independent compound Poisson (i.e. non-mixed) variables, and it will be shown that the sum X is then also

a compound Poisson variable. For the proof we write the c.g.f. ψ_X of the sum X as the sum of the c.g.f.s ψ_j of the compound Poisson distributed summands (section 1.4(d) and equation (3.2.6));

$$\psi_X(s) = \sum_j \psi_j(s) = \sum_j (n_j \cdot M_j(s) - n_j)$$

$$= n \cdot \left(\sum_j \frac{n_j}{n} \cdot M_j(s) \right) - n = n \cdot M(s) - n, \qquad (3.2.28)$$

where n_j denotes the expected number of claims in the class j, $n = \sum n_j$ and

$$M(s) = \sum_j \frac{n_j}{n} \int_{-\infty}^{\infty} e^{sZ} \, dS_j(Z) = \int_{-\infty}^{\infty} e^{sZ} d \left[\sum_j \frac{n_j}{n} S_j(Z) \right]. \qquad (3.2.29)$$

This proves that M is the m.g.f. of the weighted d.f.

$$S(Z) = \sum_j \frac{n_j}{n} S_j(Z). \qquad (3.2.30)$$

It can be seen from the last expression in (3.2.28) that X is a compound Poisson variable with claim size d.f. given by (3.2.30) and expected number of claims n.

As an immediate consequence of (3.2.30), the moments a_k about zero of the claim size d.f. S can be expressed as

$$a_k = \sum_j \frac{n_j}{n} a_{k,j}. \qquad (3.2.31)$$

The formulae (3.2.27) naturally remain valid in this special case.

(2) *The sum of compound mixed Poisson variables with synchronized mixing, but with otherwise independent summands.* Adding together compound mixed Poisson variables X_j, which all have the same mixing variable q, generates a compound mixed Poisson variable X with the same mixing variable q, provided that the variables X_j are conditionally mutually independent given the value of the mixing variable q. The proof follows by first fixing the value q of the mixing variable and then applying the proof of the previous case (1) with n_j replaced by $q \cdot n_j$; it is observed that the formulae (3.2.30) and (3.2.31) remain unchanged since q is eliminated from the quotient n_j/n.

The difference between the synchronized mixing variable case and the independent mixing variables case is demonstrated in Exercise 3.2.15, which shows that (3.2.27) does not apply in the synchronized mixing variable case, where the aggregate claim amount variables of different classes are correlated through the mixing variable q.

Note that it can be seen from the proof of the synchronized mixing variable case why the sum of independent compound mixed variables is usually not a compound mixed variable. If the mixing variables of different classes j and k have different values q_j and q_k, then the weights of these two classes in (3.2.30) are no longer equal to n_j/n and n_k/n, and therefore (if the special case where every class has exactly the same claim size distribution is excluded) the claim size d.f. S becomes dependent on the values of the mixing variables, which is, by definition, not possible for compound mixed Poisson variables.

Exercise 3.2.1 Show that the c.g.f. of a compound mixed Poisson variable X is given by (3.2.7).

Exercise 3.2.2 Show that the m.g.f. of a compound Pólya variable X is given by (3.2.8).

Exercise 3.2.3 (a) Derive the formulae (3.2.14) for the standard deviation and the skewness of a compound mixed Poisson variable X. (b) Derive the formulae (3.2.15).

Exercise 3.2.4 Give a formula for the cumulant κ_4 and for the kurtosis γ_2 of a compound mixed Poisson variable in terms of the Poisson parameter, the moments a_j of the claim size distribution and the characteristics of the mixing variable.

Exercise 3.2.5 Derive the formula (3.2.18) for the standard deviation σ_X of a compound variable X by using the formulae for conditional expectations presented in section 1.4(g) instead of by using generating functions.

Exercise 3.2.6 Recalculate the values in Table 3.2.1 in the case that the risk index r_2 is equal to 200.

Exercise 3.2.7 Show that, if negative claim sizes are excluded, the risk indices (3.2.16) satisfy $1 \leqslant r_2^2 \leqslant r_3$. (Hint: Use the Schwarz inequality $(E(U \cdot V))^2 \leqslant E(U^2) \cdot E(V^2)$.)

Exercise 3.2.8 Show that a compound Poisson distribution is asymptotically normal as the Poisson parameter n increases to infinity.

Exercise 3.2.9 Let Y be a random variable. Prove that $E(\varphi(Y)) \leqslant \varphi(E(Y))$ for every concave function φ. This is known as the **Jensen inequality**. For simplicity it can be assumed that φ is differentiable, in which case a function is concave if and only if its derivative is a decreasing function.

Exercise 3.2.10 Show that the moments a_2 and a_3 of a non-negative claim size variable Z (or of any non-negative random variable as well) satisfy the inequality $a_2 \leqslant a_3^{2/3}$. (Hint: Use the Jensen inequality $E(\varphi(Y)) \leqslant \varphi(E(Y))$ on the variable $Y = Z^3$.)

Exercise 3.2.11 Show that, if negative claims are excluded, the skewness of a compound Poisson variable always exceeds the skewness of the corresponding Poisson claim number variable.

Exercise 3.2.12 Derive the formula $P = n \cdot m$ for the risk premium and the formulae (3.2.18) for the variance and skewness of a compound variable X by using the formula (3.2.5) for cumulant generating functions.

Exercise 3.2.13 Suppose that X is a compound aggregate claim amount variable, and let k denote the corresponding claim number variable. Show that $\text{Cov}(k, X) = m \cdot \text{Var}(k)$. Write out the corresponding correlation coefficient in the case of mixed Poisson k. In which cases is the correlation large, and when is it small?

Exercise 3.2.14 An insurer's portfolio consists of risks from two different insurance lines $j = 1, 2$. The expected number of claims, n, and the first and second moments, m and a_2, about zero in each insurance line are as shown in the table below.

j	n	m	a_2
1	50	5	30
2	10	10	200

The aggregate claim amounts X_j of each line depend on each other through their mixing variables q_j but are otherwise mutually independent. The standard deviations of the mixing variables are $\sigma_{q_1} = 0.20$ and $\sigma_{q_2} = 0.05$, and the correlation coefficient of the mixing variables is

$$\rho = \frac{\mathrm{Cov}(q_1, q_2)}{\sigma_{q_1} \cdot \sigma_{q_2}}.$$

Calculate the covariance $\mathrm{Cov}(X_1, X_2)$ and the standard deviation of the total aggregate claim amount $X = X_1 + X_2$ as a function of ρ, and evaluate them for $\rho = -1, 0$ and $+1$.

Exercise 3.2.15 Consider a block of $N = 10\,000$ risk units j, each having a distribution of claims of compound mixed Poisson type with the same expected number $n_j = 0.1$ of claims. The standard deviation of the mixing variable q_j of each risk unit j is $\sigma_{q_j} = 0.2$. Furthermore, the joint claim size d.f. is approximated by a discrete d.f., as given in terms of some suitable monetary unit, say £1000, in the following table, where s_i is the probability that the size of a claim is Z_i.

i	1	2	3	4	5
Z_i	1	2	4	8	16
s_i	0.8	0.1	0.05	0.02	0.03

Calculate μ_X and σ_X for the whole block in the following two cases:

(a) assuming that the mixing variable $q_j = q$ is the same for all risk units, but that the risks are conditionally independent given the value of the mixing variable q;
(b) assuming that the mixing variables q_j are mutually independent.

3.3 The claim size distribution

The d.f. S of the size of one claim was introduced in section 3.1(b) and is now considered further. The derivation of claim size distributions from the claim data could be considered to be a separate discipline in its own right, applying the methods of mathematical statistics. It is beyond the scope of this book to provide a comprehensive view of the subject. For further details the reader should refer to an appropriate text-book, such as Hogg and Klugman (1984). In the following presentation we give examples of relevant methods.

Because of the large number of different aspects of the subject, this section is divided into several subsections. These deal, for example, with different methods for finding a d.f. S which fits the observed claim data in a satisfactory manner, and the effect of varying deductibles on the d.f. S. The effects of different reinsurance arrangements on the d.f. S net of reinsurance will be considered in section 3.4.

3.3.1 Constructing the claim size distribution

(a) Claim statistics and other experience. In order to apply the compound aggregate claim amount model one has to know or assume an appropriate claim size d.f. S. The ideal situation is where sufficient statistical data are available from past claims and this data is thought to be a reasonable guide for the intended application. Then the d.f. S can be estimated from the observed data, having regard to special features, such as the elimination of inflation, which will be described later in this chapter.

Insurers normally have data files containing detailed information about policies and claims; these are used to produce many kinds of statistics for accounting, rate-making and other purposes. Claim size distributions and other data needed for risk-theoretical analyses can be obtained, usually after some modifications, as side products from these data-handling procedures.

Unfortunately the claim statistics are often limited. Then the d.f. S has to be based on knowledge of other similar risks. Particular problems arising from large claims will be considered in section 3.3.8. Furthermore, there may also be situations where prior data or experience are not available at all, for example, when a new type of insurance is introduced or when very large special risks are insured. Then a method of (on-site) risk-by-risk evaluation can perhaps be applied, as is discussed briefly in section 3.3.2.

(b) Continuous, discrete and moment-based models. The models used for the d.f. S can be classified into the following three basic types:

(1) The d.f. S is expressed in an **analytic form** which is fitted to the observed data.
(2) The d.f. S is derived directly from statistical data in a **tabular, parameter-free discrete form**.
(3) The d.f. S is not specified explicitly, but the lowest **main characteristics**, in particular the mean, standard deviation and skewness, are derived from the data.

It may sometimes be beneficial to combine the above models, by subdividing the range of the claim size distribution into intervals for which different methods may be employed. For example, the small and medium size claims could be dealt with in a discrete form, using the observed claim size distribution, or just the estimated characteristics, whilst the large claims are treated analytically, for instance by postulating a certain type of analytic distribution and evaluating its parameters. The analytical form is the approach most frequently adopted in the actuarial literature. The problem is to find a suitable analytic expression which fits the observed data well and which is easy to handle. The analytic approach is studied in more detail in sections 3.3.4 to 3.3.7.

The moment-based approach (3) is often appropriate when approximate methods are being used in the computation of the compound d.f. F of the aggregate claim amount. This method requires knowledge only of the key characteristics of the claim size distribution, rather than specification of the whole d.f. S. Constructing S itself might be an unnecessary digression if only the moments are needed.

(c) Elimination of the effect of inflation and of other changes. If the data have been collected over a period during which monetary values have changed, it is important to bring the values onto a common basis by means of a suitably chosen index.

The choice of the index depends on the line of insurance. For example, an index of the cost of construction prices may be suitable for fire and other property insurance, an earnings index for life and accident insurance, and a general price index may be appropriate when a single index is required for several lines or for the whole portfolio.

When analysing claim statistics covering several years, one should bear in mind that, besides the effect of inflation, claim sizes are often subject to other systematic changes. For example, the continuing development of construction techniques may cause a trend in the shape of the distribution of property claims, and in the level of the average claim. Changes in policy conditions, indemnity practices, legal judgments, etc., can have a similar effect. Inflation and the associated problems will be considered in more detail in Chapter 7.

The length of the observation period warrants care. The need to obtain a sufficient number of data points requires the use of a longer rather than a shorter period. However, if the risk structure inside the portfolio is changing, it is desirable to obtain data from as recent a period as possible. Such changes may be difficult to detect. An analysis should be carried out to ascertain whether the data exhibit any trends or temporal changes, and whether possible changes in policy conditions, legal practices or any other factors should be taken into account. If, for example, the relative number of a certain type of risks in the collective has increased significantly over the course of time, and if the claims relating to these risks have a different profile compared with the other risks in the collective, then the change in the structure of the portfolio should be taken into account when S is determined.

Varying deductibles inside the risk collective may lead to an underestimate of the actual number of smaller claims, which are not reported because they fall below the deductible and are therefore missing from the insurer's claim statistics. For some applications such bias should be corrected (section 3.3.10).

3.3.2 Individual evaluation of risks

Individual risk analysis can in some special cases be used in the compilation of the claim size d.f. S, when other methods are not applicable, for example, because of inadequate or missing claim statistics. Then the analysis is usually based on an on-site inspection at the industrial plants, etc. It can be part of the insurer's premium-rating procedure or part of a specific risk management assessment for use by plant managers. The method is simplest in the (rare) cases where no partial claims can occur.

It is useful to present this approach, since it illustrates the connection between the claim size distribution and the structure of the portfolio, thus showing the link between the claim sizes in the classical individual risk approach and the claim size d.f. S of collective risk theory.

The risk units (policies) in the portfolio are numbered by an index i, and the expected number of claims occurring in respect of the risk unit i in one year is denoted by n_i. We will consider first the simple case where no partial claims can occur, which means that only one claim size Z_i is possible for each unit i. The probability that a randomly chosen claim from the portfolio occurred in respect of the ith risk unit is equal to

$$\frac{n_i}{n}, \quad \text{where} \quad n = \sum_i n_i \qquad (3.3.1)$$

provided, for example, that the numbers of claims occurring in respect of different risk units are independent and Poisson-distributed (see

Exercise 3.3.4 or section 3.2(d)). Then the probability $S(Z)$ that the size of a randomly chosen claim is less than or equal to Z is simply . the sum of probabilities (3.3.1) over those risk units which satisfy the condition $Z_i \leqslant Z$, i.e.

$$S(Z) = \frac{1}{n} \sum_{Z_i \leqslant Z} n_i. \qquad (3.3.2a)$$

More generally, if partial claims are also possible, one must specify the claim size d.f. S_i, $S_i(Z) = S(Z \mid$ the claim occurred in respect of the risk unit i) for each risk unit i. Using the probabilities (3.3.1) again, the formula (3.3.2a) generalizes to the form

$$S(Z) = \frac{1}{n} \sum_i n_i \cdot S_i(Z). \qquad (3.3.2b)$$

EXAMPLE As a result of an on-site inspection of three different plants i ($i = 1, 2, 3$) the expected number n_i of claims per year and the d.f. S_i of the claim size for each plant i was obtained, as shown in the table below.

Table *The expected number of claims per year and the d.f. S_i of the chain size for each plant*

Plant number i:	1	2	3
n_i:	0.5	1.5	3.0
EML_i (£1000):	1000	400	200
Claim size profiles:			
Prob $\{Z = 1.00 \cdot \mathrm{EML}_i\}$:	0.05	0.10	0.30
Prob $\{Z = 0.75 \cdot \mathrm{EML}_i\}$:	0.10	0.15	0.10
Prob $\{Z = 0.50 \cdot \mathrm{EML}_i\}$:	0.15	0.25	0.15
Prob $\{Z = 0.25 \cdot \mathrm{EML}_i\}$:	0.70	0.50	0.45

EML_i is the **estimated maximum loss**, i.e. the maximum amount of a claim which can be expected in respect of plant i. For each plant the possible claim sizes were classified into four classes, class limits being chosen to be multiples of a quarter of the EML. (The smallest claims were not taken into account, nor were they included in the expected numbers of claims.) Then we have (Z in £1000):

$$
\begin{aligned}
S_1(Z) &= 0.00, \text{ for } \quad Z < 250 \ (= 0.25 \cdot \mathrm{EML}_1), \\
S_1(Z) &= 0.70, \text{ for } 250 \leqslant Z < 500, \\
S_1(Z) &= 0.85, \text{ for } 500 \leqslant Z < 750, \\
S_1(Z) &= 0.95, \text{ for } 750 \leqslant Z < 1000, \\
S_1(Z) &= 1.00, \text{ for } \quad Z \geqslant 1000.
\end{aligned}
$$

The d.f.s S_2 and S_3 are derived analogously, and thereafter the d.f. S is obtained as the weighted average of the S_i using the weights n_i/n (see Exercise 3.3.3):

$Z =$	50	100	150	200	250	300	400	500	750	1000
$S(Z) =$	0.270	0.510	0.570	0.825	0.895	0.940	0.970	0.985	0.995	1.000

where the value of the d.f. S is given at each step point Z.

In practice the problem with the individual method is that the claim frequencies of claims in different claim size classes are usually not known, and have to be based on more or less subjective evaluations. These evaluations can, however, be supported using the experience derived from other similar or comparable risks. Experienced risk managers and rate-makers may be able to make use of general industry-wide average frequencies and claim size distributions and take into account the effect of local conditions on a case by case basis. The interested reader is referred to the section on individual risk-rating by Tiller in Chapter 3 of *Foundations of Casualty Actuarial Science* (CAS) (1990). It should be noted that, when average data are used, we are, in fact, almost back to the collective approach.

Exercise 3.3.1 An insurer has issued insurance for accidental death, the sums payable being standardized at £100, £250 and £500. The numbers of insured persons in these classes are 5000, 1000 and 2000, respectively. It is known that the rate of death for the two lower amounts can be expected to be equal, but that (owing to antiselection) the rate for the highest amount insured is double that in the lower two classes. Derive the d.f. S.

Exercise 3.3.2 Show that (3.3.2b) reduces to (3.3.2a) if no partial claims can occur, i.e. if only one claim size Z_i is possible for each unit i.

Exercise 3.3.3 Calculate the d.f. S in the Example of section 3.3.2.

Exercise 3.3.4 (a) Let $k = \Sigma k_i$, where k_i are independent Poisson(n_i) claim number variables. Show that $\text{Prob}\{k_i = 1 \,|\, k = 1\} = n_i/n$, where $n = \Sigma n_i$.
(b) Suppose that the claim number variables for different risk units are independent and Poisson-distributed. Show that (3.3.1) holds true.

3.3.3 *Tabular method*

(a) The observed distribution. A natural estimate for the claim size distribution is the observed claim size distribution, i.e. the claim size d.f. S is simply defined to be equal to the sample d.f.

$$S(Z) = \frac{\text{number of claims of size} \leqslant Z}{\text{total number of claims}}. \tag{3.3.3}$$

As a discrete d.f., S can be presented in tabular form, hence the title **tabular method**.

Recall that, if there have been changes in monetary values during the observation period, inflation-corrected data should be used. Similarly, any other systematic changes in the portfolio should be taken into account, as was discussed in section 3.3.1(c).

The tabular method is appropriate only when there is a sufficiently large volume of claim data. This is rarely the case for the tail of the distribution, especially in situations where exceptionally large claims are possible. It is usually advisable, therefore, to divide the range of relevant values of claims, Z, into two parts, treating the claim sizes up to some limit, T, on a discrete basis, while the tail to the right from this limit point T is replaced by an analytic function, for instance by a Pareto curve.

When the number of claims is large, it may be necessary to group the observed claims into classes according to their size (as in Table 3.3.1). In order not to distort the data too much, the lengths of the class intervals should be kept relatively small. The probability mass of each class needs to be positioned in an unbiased way so that the mean claim size remains unchanged. Before grouping the data, therefore, it is important to calculate at least the class averages from the ungrouped data.

(b) An example with real data. In the rest of this section the presentation is supported by an example with actual UK fire claim data. The actual claims of the portfolio in question were obtained in grouped form, as shown in Table 3.3.1. The total number of claims observed during the four year observation period was 16 536, and the mean claim size was equal to £7009.

The claims are classified into groups according to claim size. In this example geometrically increasing class limits (column 2) are used, except for the highest classes. The numbers n_i of observed

Table 3.3.1 *Compilation of claim statistics*

(1) Class i	(2) Upper class limit (£ 1000) Z_i	(3) Class average (£ 1000) $\bar{Z}_i$	(4) Observed number of claims n_i	(5) Expected n_i for the tail $n \cdot \Delta S_i$	(6) Value of the d.f. S $S(Z_i)$
1	0.10	0.041	4319		0.2611877
2	0.14	0.118	795		0.3092646
3	0.20	0.167	910		0.3642961
4	0.28	0.237	962		0.4224722
5	0.40	0.336	1097		0.4888123
6	0.57	0.475	1121		0.5566038
7	0.80	0.673	1046		0.6198597
8	1.13	0.957	969		0.6784591
9	1.60	1.35	843		0.7294388
10	2.26	1.90	805		0.7781205
11	3.20	2.70	694		0.8200895
12	4.53	3.81	602		0.8564949
13	6.40	5.39	480		0.8855225
14	9.05	7.56	382		0.9086236
15	12.80	10.76	329		0.9285196
16	18.10	15.06	273		0.9450290
17	25.60	21.51	214		0.9579705
18	36.20	30.51	172		0.9683720
19	51.20	42.74	136		0.9765965
20	72.41	60.50	108		0.9831277
21	102.40	85.72	88		0.9884495
22	250.00	155.70	117	136	0.9966710
23	500.00	336.90	47	34	0.9987331
24	750.00	635.38	12	9	0.9992801
25	1000.00	845.19	4	4	0.9995179
26	2000.00	1277.01	8	5	0.9998165
27	3000.00	2579.42	3	1.3	0.9998957
28	5000.00		0	0.9	0.9999488
29	10 000.00		0	0.5	0.9999805
30	20 000.00		0	0.2	0.9999926
31	50 000.00		0	0.1	0.9999979
32	100 000.00		0	0.0	0.9999992

claims in each class i are displayed in column 4, and the average size $\bar{Z}_i$ of claims in class i is given in column 3. For example, the number of claims in class 17, between £18 100 and £25 600, is 214.

Grouped claim data are used in this example for convenience. In

practical applications grouping the claims according to size would
not have been necessary; it is obvious that by grouping the data
information is lost (see Sandström, 1991).
The idea is to define the claim size d.f. S by using the observed
discrete distribution for claim sizes below some suitable limit T,
whereas the tail of the d.f. S, starting from the point T, is obtained
by fitting an analytic curve to the observed data. For Z_i-values at
most equal to the chosen limit T, the corresponding $S(Z_i)$ are then
obtained directly from (3.3.3). Note that the probability mass for the
class i should be positioned at the class average $\bar{Z}_i$ rather than at the
class limit Z_i. The fitting of the tail will be described in section 3.3.3(c).

(c) Fitting a tail to the discrete claim size d.f.   In this section the
tail of the claim size distribution corresponding to larger claims is
obtained by fitting a simple analytic curve to the observed data
given in Table 3.3.1. The starting point T of the tail was chosen to be
equal to $Z_{21} = 102.40$ (£1000), since the number of claims in each
class after this limit was considered to be too small, compared to
the relative length of the class interval, to give reliable estimates for

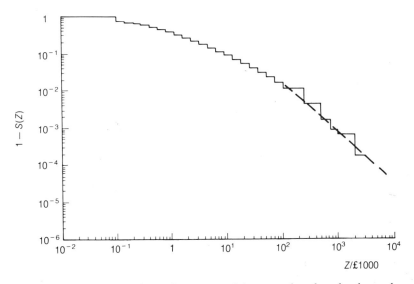

Figure 3.3.1   *The cumulative frequencies of claim sizes based on the observed
crude ratios n_i/n of Table 3.3.1. The tail is fitted with a straight line. Double
logarithmic scale.*

the corresponding claim frequencies (see Figure 3.3.1).

An analytic curve to fit the data can be found as follows. First the graph of the complement $1 - S_{\text{obs}}(Z)$ of the observed sample d.f. $S_{\text{obs}}(Z)$ (for $Z \leqslant T$ we have already defined $S(Z) = S_{\text{obs}}(Z)$) is drawn on a double logarithmic scale. This graph is plotted in Figure 3.3.1. The uncertainty which is characteristic of the upper tail of the claim size distribution can be seen in this figure. Thereafter a straight line is chosen on the double logarithmic scale to fit the observed claim data well, and to go through the point $(T, S(T))$. Transformation back to the linear scale results in our example in $S(Z) = 1 - 7.3208 \cdot Z^{-1.3938}$, for $Z \geqslant T$. This is, in fact, a Pareto d.f. (3.3.14). It was further assumed that the largest possible loss in the portfolio was £100 million, and the Pareto d.f. is, therefore, truncated at this point, i.e. $S(Z)$ is set equal to 1 for $Z \geqslant 100\,000$.

The moments a_j about the origin of the size of a claim (3.2.9) can now be obtained as the sum

$$a_j = \int_{-\infty}^{\infty} Z^j \, dS(Z) \approx \sum_{Z_i \leq T} \bar{Z}_i^j \cdot \Delta S_i + \int_{T}^{\infty} Z^j \, dS(Z), \qquad (3.3.4)$$

in which the class averages $\bar{Z}_i$ and the differences $\Delta S_i = S(Z_i) - S(Z_{i-1}) = n_i/n$ (for $Z_i \leqslant T$) are taken directly from Table 3.3.1, and $T = 102.40$ is the limit point above which the analytic tail distribution is used. The calculation of the value of the integral in the second term is left as an exercise.

Exercise 3.3.5   Calculate the expected value $m = a_1$ according to the formula (3.3.4) for the d.f. S given in Table 3.3.1, with the analytic tail specified in the text (section 3.3.3(c)).

3.3.4   Analytic methods

(a) Merits of the analytic method.   As was outlined in section 3.3.1(b), it is often desirable to try to find an explicit analytic expression for a claim size d.f. S. This is particularly the case if the claim statistics are too sparse to use the tabular method of section 3.3.3. Experience obtained from dealing with other similar risks may suggest some particular form for the claim size d.f.

On the other hand, since an analytic d.f. is often the easiest to handle in computations and programming, one is often also used in those cases where the sample distribution can be expected to provide a satisfactory model for the claim size d.f. Then a suitable analytic d.f. S may be found, for example by curve-fitting, which is very close to the observed discrete d.f., and can therefore be used instead. The characteristic features of the most commonly used analytic d.f.s are readily available from standard text-books. In sections 3.3.5 to 3.3.7 consideration is given to some commonly used analytic distributions; further examples can be found, for example, in Hogg and Klugman (1984).

The techniques of curve-fitting and estimation of the claim size d.f. are well-covered elsewhere in the actuarial literature and are mostly beyond the scope of this book. Guidance can be found in Hogg and Klugman (1984) and Patrik (1980). We will, however, introduce in the following section a useful curve-fitting method, which has proved efficient in applications.

It should be emphasized that many standard estimation methods in statistics are unsuitable for the estimation of the claim size distribution. The main reason for the discrepancy is the strongly skewed nature of typical claim size distributions.

(b) Curve-fitting using the limited expected value function. When one tries to fit an analytic d.f. to the observed data (supplemented, if necessary, by additional knowledge about catastrophic claims), there is always the question of how to measure the goodness of fit. A particular problem in comparing an analytic d.f. with a discrete one is that the latter is a discontinuous step function and, therefore, the two d.f.s will always differ from each other in the vicinity of a step by at least half the size of the step. This problem can be overcome by integrating both d.f.s once, which leads to the so-called limited expected value function. This turns out to be a very useful tool for testing the goodness of fit of an analytic d.f. to the observed claim size d.f.

The limited expected value function L of a claim size variable Z, or of the corresponding d.f. S, is defined by

$$L(M) = E(\min(M, Z)) = \int_{-\infty}^{M} Z \, dS(Z) + M \cdot (1 - S(M)), \qquad (3.3.5)$$

where M is any real number. The value of the function L at a point M is equal to the expectation of the d.f. S truncated at the point M.

In order to fit the limited expected value function L of an analytic d.f. S to the observed data, the limited expected value function L_{obs} (Exercise 3.3.7) of the observed claim size d.f. S_{obs} (3.3.3) is first constructed. Thereafter one tries to find a suitable analytic d.f. S, such that the corresponding limited expected value function L is as close to the observed L_{obs} as possible.

A reason why the limited expected value function is a particularly suitable tool for our purposes is that it represents the claim size distribution in the monetary dimension. For example, we have $L(\infty) = E(Z) = m$. The d.f. S, on the other hand, operates on the probability scale, i.e. takes values between 0 and 1. Therefore, it is usually difficult to see, by looking only at the d.f. S, how sensitive the risk premium is to changes in the values of S, while the limited expected value function L shows immediately how different parts of the claim size d.f. S contribute to the risk premium.

Apart from curve-fitting purposes, the function L will turn out to be a very useful concept in many other connections later in this book, for example in section 3.3.10 dealing with deductibles and in section 3.4.2 when excess of loss reinsurance is considered.

The limited expected value function L (Exercises 3.3.9 to 3.3.12) has the following important properties

(1) The graph of L is concave, continuous and increasing,
(2) $L(M) \to m = E(Z)$, as $M \to \infty$,
(3) $S(M) = 1 - L'(M)$,

where $L'(M)$ is the derivative of the function L at the point M; if S is discontinuous at a point M, then (3) holds true for the right-hand derivative $L'(M+)$.

From the curve-fitting point of view the use of the limited expected value function has the advantage, compared with the use of the d.f.s, that both the analytic L to be fitted, and the corresponding observed function L_{obs} (Exercise 3.3.7), based on the observed discrete d.f. S_{obs}, are continuous and concave, whereas the observed claim size d.f. S_{obs} is a discontinuous step function (see Figure 3.3.2). The property (3) implies that the limited expected value function determines the corresponding d.f. S uniquely. When the limited expected value functions of two distributions are close to each other, not only are the mean values of the distributions close to each other, but the whole distributions are close to each other.

When large claims may occur, then the tail of the claim size distribution needs special attention. If there is information available about

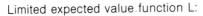

Limited expected value function L:

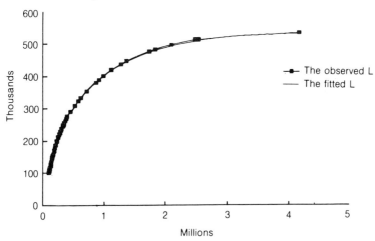

Distribution function S:

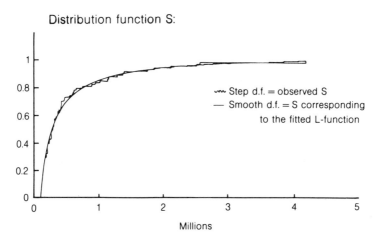

Figure 3.3.2 *The observed and the fitted analytic limited expected value functions* L_{obs} *and* L *(upper graph), and the corresponding d.f.s* S_{obs} *and* S *(lower graph). The black marks in the upper graph denote the 75 observed claims exceeding 100 000 Finnish marks (equivalent to about £12 000 in 1992); claims below this limit were not included, since relatively high deductibles applied in the portfolio, which consisted of certain industrial machinery breakdown policies. There is an excellent fit, because the observed and the fitted L-curves are practically the same. The fitted curve is a censored Pareto curve. Section 3.3.7(c) gives more details on censored Pareto distributions.*

sizes and frequencies of large claims for similar risks in the market, then the observed claim size d.f. S_{obs} should first be modified accordingly. On the other hand, after the fitting has been done, it should always be analysed how the L-function behaves in the range to the right of the point Z_{max} where S_{obs} reaches the value 1, whereafter L_{obs} remains constant. Then, if $L(M)$ still increases significantly for values $M > Z_{max}$, it should be checked that the corresponding probabilities $1 - S(M)$ are credible and in conformity with the observations. The heaviness of the tail of an analytic d.f. S can be altered by using censored distributions, which will be introduced in section 3.3.7(c).

(c) Some remarks on the applicability of different approaches. Even in cases where an analytic distribution seems to fit the observed data well, special attention is required if the volume of claim statistics is so small that a reasonable confidence level cannot be attained. This is often the case in practical applications, especially when it comes to large claims.

The requirements for accuracy of the d.f. S depend on the application in question. For rating a policyholders' deductible, for example, it is important that the distribution should fit the observed data well over the range of smaller claims, and the curve-fitting approach, or alternatively the tabular method, is recommended. On the other hand, when the aggregate claim amount d.f. F is approximated, and when the claim frequency is not too small, then the exact shape of the claim size d.f. S is no longer so critical. It is then more important that the lowest characteristics of S are well estimated and, in the case of a heavy-tailed distribution, that the tail profile is realistic. However, it should be kept in mind that the lowest three moments do not fully define the shape of a d.f. S, and therefore the fit of S to the observed data may be poor.

Goodness of fit of an analytic claim size distribution to the observed data can be seen most easily by comparing the corresponding limited expected value functions L and L_{obs}, as in Figure 3.3.2. In the tail area it is also advisable to compare the analytic and observed d.f.s S and S_{obs}, or rather $1 - S$ and $1 - S_{obs}$, with each other, in order to see the level of uncertainty. For this comparison it is advisable to plot a diagram like in Figure 3.3.1 (observe the double logarithmic scale).

The gamma, log-normal and Pareto distributions, to be dealt with in sections 3.3.5 to 3.3.7 below, are typical candidates to be considered for applications in which it is sufficient for the three lowest moments and the tail profile of the claim size distribution to be in accordance with the observed data. The Pareto and censored Pareto distributions (section 3.3.7) are, in addition, especially suitable for curve-fitting, since the Pareto d.f. is very convenient to deal with in calculations. The main reason for this is that the characteristics of the d.f. S can be calculated in a closed form, and, furthermore, these calculation formulae also apply when any limited range of the d.f. S is considered. If one analytic d.f. fits the observed data only up to a certain claim size value, and another analytic d.f. fits the data for larger claims, then a combined d.f. may be used. The limited expected value function of a piecewisely defined distribution is obtained as is shown in Exercise 3.3.15.

Exercise 3.3.6 Show that, if the d.f. S is a step function, the corresponding limited expected value function L is a piecewise linear function. Draw the graphs of S and L in the case that only the claim sizes 1, 2, 4, 7, 11 and 16 are possible, the respective point probabilities being 0.40, 0.10, 0.25, 0.10, 0.10 and 0.05.

Exercise 3.3.7 Let S be the sample d.f. (3.3.3) for a given sample of n observed claims Z_i, $i = 1, 2, ..., n$. Show that the corresponding limited expected value function L is given by the formula

$$L(M) = \frac{1}{n} \sum_{i=1}^{n} \min(M, Z_i).$$

Exercise 3.3.8 Show that the limited expected value function L_{aZ+b} of a positive (i.e. $a > 0$) linear transformation $a \cdot Z + b$ of Z satisfies the equation

$$L_{aZ+b}(M) = a \cdot L_Z\left(\frac{M-b}{a}\right) + b.$$

Exercise 3.3.9 Prove that the limited expected value function L is an increasing and continuous function.

Exercise 3.3.10 Show that L is a concave function.

Exercise 3.3.11 Show that $L(M)$ approaches the mean claim size m, as M increases to infinity.

Exercise 3.3.12 (a) Show that $S(M) = 1 - L'(M +)$ and $S(M -) = 1 - L'(M -)$, where $L'(M +)$ and $L'(M -)$ denote the right-hand and left-hand derivatives of L.

(b) Show that

$$L(M) = L(0) + \int_0^M 1 - S(Z)\,dZ,$$

where $L(0) = 0$ if negative claims are excluded.

Exercise 3.3.13 (a) Show that $L(M) = M$ if and only if $S(M -) = 0$.

(b) Show that $L(x) - L(y) \leqslant x - y$, if $x < y$.

Exercise 3.3.14 Let L_T be the limited expected value function corresponding to the tail distribution S_T

$$S_T(Z) = \text{Prob}\{Z \leqslant Z | Z > T\} = \frac{S(Z) - S(T)}{1 - S(T)}, \quad Z \geqslant T,$$

obtained by truncating a given d.f. S at the point T. Show that

$$L_T(M) = T + \frac{L(M) - L(T)}{1 - S(T)}, \quad M \geqslant T.$$

Exercise 3.3.15 Let S_i be a d.f., and let L_i be the corresponding limited expected value function, for $i = 1, 2$. Assume that we have $S_1(A) \leqslant S_2(A)$ at a given point A. The d.f. S is defined piecewise by setting $S(Z) = S_1(Z)$ for $Z < A$, and $S(Z) = S_2(Z)$ for $Z \geqslant A$. Show that the limited expected value function of S is given by the formula $L(M) = L_1(\min(M, A)) + L_2(\max(M, A)) - L_2(A)$.

Exercise 3.3.16 Let L be the limited expected value function of the claim size d.f. S of a sum of mutually independent compound Poisson(n_i) variables, and let L_i denote the limited expected value functions of the claim size d.f.s of the summands. Show that $L = \sum (n_i/n) \cdot L_i$.

3.3.5 Three-parameter gamma distribution

(a) The shifted gamma distribution. One candidate for an analytic claim size d.f. S is the shifted gamma distribution, defined as $S(Z) = G(Z - d)$, $Z > d$, where G is the Gamma(r, a) d.f. (section 2.5(a)). The d.f. S has three parameters, a, r and d, and is also referred to as the three-parameter gamma distribution.

Suppose that we are given the three main characteristics, the mean value m, the standard deviation $\sigma = \sigma_Z$, and the skewness $\gamma = \gamma_Z$ (> 0) of the claim size distribution. We want to find the values for the parameters a, r and d of the three-parameter gamma d.f. S so that these three characteristics match. If Y is Gamma(r, a)-distributed, then $Z = d + Y$ is shifted gamma with parameters a, r and d. We have (Appendix B (B.1.3)) the equations

$$m = E(Z) = d + \frac{r}{a}, \quad \sigma_Z^2 = \frac{r}{a^2}, \quad \gamma_Z = \frac{2}{\sqrt{r}}. \qquad (3.3.6)$$

Solving for a, r and d (Exercise 3.3.17) gives

$$r = \frac{4}{\gamma_Z^2}, \quad a = \frac{2}{\sigma_Z \cdot \gamma_Z}, \quad d = m - 2 \cdot \frac{\sigma_Z}{\gamma_Z}. \qquad (3.3.7)$$

The limited expected value function L of the shifted gamma distribution with parameters a, r and d is given by (Hogg and Klugman (1984), p. 226)

$$L(M) = d + \frac{r}{a} \cdot G_{r+1,a}(M - d) - (M - d) \cdot [1 - G_{r,a}(M - d)],$$
$$\text{for } M \geqslant d, \quad (3.3.8)$$

where $G_{r,a}$ denotes the Gamma(r, a) d.f. For numerical calculation of the gamma $G_{r,a}$, see Press et al. (1986) (using their notation, we have $G_{r,a}(x) = P(r, a \cdot x)$).

(b) The shapes of gamma density curves. Examples of gamma densities for different values of the skewness γ are plotted in Figure 3.3.3.

For skewness larger than or equal to 2 the density curve is peaked. In practice the claim size distributions usually have rather high skewness (from 5–10 upwards). The minimum d of a three-parameter gamma distribution is determined by the three basic characteristics of the claim size d.f. It is often the case that the initial point d of the d.f. fits the observed claim data badly. Then, if the range of small claims is important for the application, the three-parameter gamma d.f. has to be rejected. It is, of course, even more fatal if the fit is bad for large claims.

The three-parameter gamma d.f. is sometimes also used as an approximation for the d.f. F of the aggregate claim amount, if the skewness is small (cf. section 4.2.1(b)).

(c) Additivity. The sum of independent Gamma(r_1, a) and

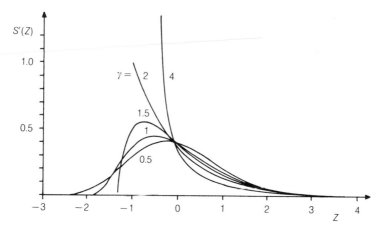

Figure 3.3.3   *Examples of shifted gamma densities having mean = 0, standard deviation = 1 and varying skewness.*

Gamma(r_2, a) distributed random variables is Gamma$(r_1 + r_2, a)$ distributed. This is proved in Appendix B using the moment generating function.

Note that the Gamma$(1, a)$ d.f. S reduces to the exponential d.f. $S(Z) = 1 - e^{-aZ}$, $Z \geqslant 0$. Hence, the Gamma(k, a) d.f. for positive integer values of k is obtained as the kth convolution of the exponential d.f.

Exercise 3.3.17   Prove the formulae (3.3.7).

Exercise 3.3.18   Show that Gamma$(1, a)$ is the exponential distribution.

Exercise 3.3.19   Show that Gamma(r, a) is asymptotically normal as $r \to \infty$.

3.3.6   *Logarithmic-normal distribution*

(a) Transformed normal variable.   A frequently used claim size distribution is the (shifted) **logarithmic-normal**, or briefly **log-normal**. By definition, a claim size variable Z is log-normally distributed if it is of the form

$$Z = d + e^Y, \tag{3.3.9}$$

where $Y = Y_{\mu,s}$ is a normally distributed variable with mean value μ and standard deviation s, and where d is the initial point of the range of the claim size variable Z.

Solving Y from (3.3.9) gives $Y = \ln(Z - d)$. Then $S(Z) = N((\ln(Z - d) - \mu)/s)$, and the density of the variable Z (Exercise 3.3.20) is

$$S'(Z) = \frac{1}{s \cdot (Z - d) \cdot \sqrt{2\pi}} \exp\left[-\frac{1}{2s^2}(\ln(Z - d) - \mu)^2 \right], \quad Z > d,$$

(3.3.10)

where d, μ and s are free parameters. If the mean m, the standard deviation σ_Z, and the skewness γ_Z (>0) of the claim size d.f. S are given, then the corresponding values of the parameters d, μ and s are obtained (Exercise 3.3.21) as follows. First an auxiliary variable η is solved as the real root of the equation

$$\eta^3 + 3\eta - \gamma_Z = 0,$$

(3.3.11)

We then have

$$d = m - \sigma_Z/\eta; \quad s^2 = \ln(1 + \eta^2); \quad \mu = \ln(m - d) - \tfrac{1}{2} \cdot s^2.$$

(3.3.12)

The limited expected value function of the log-normal distribution (see Hogg and Klugman (1984), p. 229) is

$$L(M) = d + (m - d) \cdot N\left(\frac{\ln(M - d) - \mu - s^2}{s}\right)$$

$$+ (M - d) \cdot \left[1 - N\left(\frac{\ln(M - d) - \mu}{s}\right) \right].$$

(3.3.13)

(b) The shapes of the log-normal density curves are plotted in Figure 3.3.4 for different values of the skewness.

The same feature which was noticed concerning the gamma distribution in section 3.3.5(b) is also seen in Figure 3.3.4: when skewness is high, the left-hand tail is packed in a very narrow strip below the mean value as a peak. Therefore, having regard to the rather high skewness of the observed claim size shapes, care is needed in exploring whether these distributions can provide an appropriate model for the claim size d.f. throughout the whole range.

Exercise 3.3.20 Prove that the density of the log-normal distribution is given by (3.3.10).

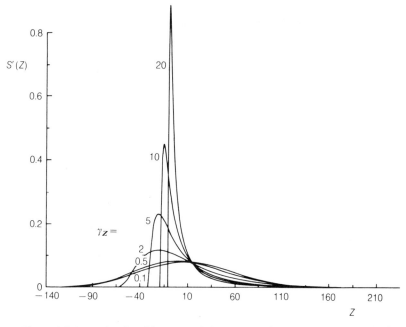

Figure 3.3.4   *A family of log-normal densities with mean* $m = 10$, *standard deviation* $\sigma_Z = 50$, *and varying skewness* γ_Z.

Exercise 3.3.21   (a) The m.g.f. of the normally $N(\mu, s^2)$ distributed variable is

$$M(t) = e^{t\mu} \cdot e^{t^2 s^2/2}.$$

Making use of this, calculate the moments a_k of the log-normally distributed variable Z in the case $d = 0$.

(b) Show that, if Z is log-normally distributed with parameters d, μ, s, then

$$m = e^{\mu} \cdot e^{s^2/2} + d$$
$$\sigma_Z^2 = e^{2\mu} \cdot e^{s^2} \cdot (e^{s^2} - 1)$$
$$\gamma_Z = (e^{s^2} + 2)\sqrt{e^{s^2} - 1}.$$

(c) Prove that, if S is a log-normal d.f. with mean m, standard deviation σ_Z and skewness γ_Z (> 0), then the parameters d, μ and s are given by (3.3.12).

3.3.7 Pareto distribution

(a) **Definition and some basic properties.** One of the most frequently used analytic claim size distributions is the Pareto d.f., or more specifically the Pareto(α, β, D) d.f., which is defined by the formula

$$S(Z) = 1 - \left(\frac{D + \beta}{Z + \beta}\right)^{\alpha} \quad (Z \geqslant D), \qquad (3.3.14)$$

where α, β and D are parameters which satisfy $\alpha > 0$ and $\beta > -D$. The first parameter, α, controls how heavy a tail the distribution has: the smaller the α, the heavier the tail. The second parameter, β, mainly affects the left-hand range of the distribution, and does not essentially change the tail of the distribution in the region where Z is significantly larger than β. The parameter D constrains the range of Z from below (Figure 3.3.5). If only the tail of the claim size distribution is to be fitted, as in section 3.3.3(c), the parameter β can often be omitted.

Experience has shown that the Pareto formula (3.3.14) is often an appropriate model for the claim size distribution, particularly where exceptionally large claims may occur.

One advantage of the Pareto d.f. is that many basic quantities related to the d.f. are elementary and can be derived by straightforward calculations. Examples of such formulae are the expressions for the moments, both in cases where the complete Pareto distribution is involved, and when only the tail of the distribution is employed or when a Pareto-distributed random variable is otherwise confined to an interval (Exercises at the end of section 3.3.7). For example, the main characteristics and the limited expected value function (3.3.5) of the Pareto(α, β, D) d.f. are (Exercises 3.3.24 to 3.3.27)

$$m = \frac{\alpha \cdot D + \beta}{\alpha - 1}, \quad \text{for } \alpha > 1,$$

$$\sigma_Z^2 = \frac{\alpha \cdot (D + \beta)^2}{(\alpha - 1)^2 \cdot (\alpha - 2)}, \quad \text{for } \alpha > 2,$$

$$\gamma_Z = 2 \cdot \frac{\alpha + 1}{\alpha - 3} \cdot \sqrt{\frac{\alpha - 2}{\alpha}}, \quad \text{for } \alpha > 3, \qquad (3.3.15)$$

$$L(M) = \frac{\alpha \cdot D + \beta - (M + \beta) \cdot [1 - S(M)]}{\alpha - 1} \quad (M \geqslant D).$$

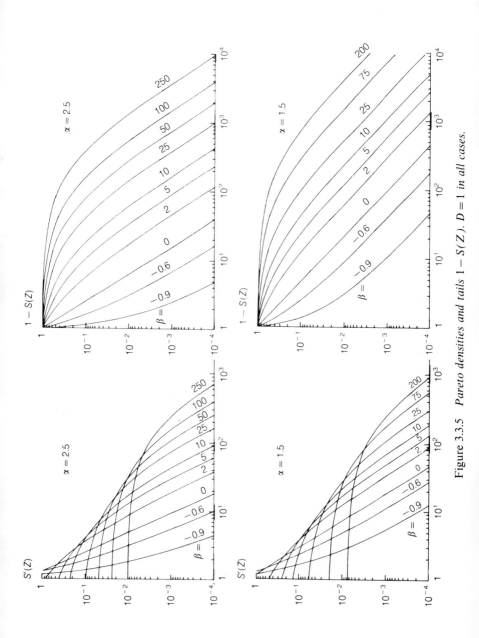

Figure 3.3.5 *Pareto densities and tails* $1 - S(Z)$. $D = 1$ *in all cases.*

In the special case that $\alpha = 1$, $L(M)$ in (3.3.15) is of the form $0/0$, and L must, therefore, be expressed differently (Exercise 3.3.27).

(b) A problem with convergence. For small values of the parameter α the Pareto distribution is heavy-tailed. This is indicated by the feature that the moments a_j about zero of the size of a claim are infinite for those values of j which satisfy $j \geqslant \alpha$ (Exercise 3.3.24). Hence, for $0 < \alpha \leqslant 1$ even the mean value of the Pareto d.f. is infinite, and for the parameter values $2 < \alpha \leqslant 3$, for example, the standard deviation is finite but the skewness and higher order characteristics are infinite. Values of α which are appropriate to fit real data (within the range where claims can be expected to occur) typically vary from less than 1, to larger than 3, and therefore the three basic characteristics in (3.3.15) are in many cases useless.

This problem of infinite moments is, however, manageable, since in practice there is normally some upper bound for possible claim sizes and therefore the top of the tail of the claim size distribution must be cut away. In sections 3.3.7(c) and 3.3.7(d) two different techniques for the purpose are considered. The moments of the resulting d.f.s will be given.

(c) Censored distribution. The censored d.f.

$$S_C(Z) = \text{Prob}\{Z \leqslant Z\} = \frac{S(Z)}{S(C)}, \quad \text{for } Z < C, \qquad (3.3.16)$$

can be a useful modification of (3.3.14), if there is evidence that claim sizes larger than the limit C are excluded, and if the expected number of claims exceeding B ($B < C$) decreases to zero as $B \to C$. Obviously the method of censoring can also be used for other d.f.s S. The moments of a censored d.f. are always finite (Exercise 3.3.29).

In Figure 3.3.2 of section 3.3.4(b) a limited expected value function L of a Pareto$(\alpha, \beta, £12\,000)$ distribution, censored at a point C, was fitted to the observed limited expected value function L_{obs} (Exercise 3.3.7). The fitting was carried out by simple trial-and-error using a computer. The graphs of the observed and the fitted limited expected value functions, L_{obs} and L, were drawn on the computer screen for some guessed starting values of the free parameters α, β and C. Thereafter the values of the parameters α, β and C were varied, until the fit of the two graphs could not be significantly improved. The parameter values $\alpha = 0.9$, $\beta = £7400$ and $C = £0.5$ million resulted. The maximum relative difference between the observed and fitted

L-functions was about 1%. Naturally one could have made use of numerical methods for this fitting, but the trial-and-error method is quite efficient in practice with standard microcomputer software.

(d) Truncated distribution. Truncation is another way to handle a claim size d.f. when claims exceeding some limit Z_{max} are eliminated from the insurer's account, for example, when claim amounts above a certain level are ceded to reinsurers (section 3.4) or when claims above an upper limit are not covered by the indemnity (see also the example of section 1.4(b)). The probability mass $1 - S(Z_{max})$ of the tail $Z \geqslant Z_{max}$ is concentrated into the point Z_{max}, resulting in a d.f.

$$S_{tr}(Z) = \begin{cases} 1, & \text{for } Z \geqslant Z_{max} \\ S(Z), & \text{for } Z < Z_{max}. \end{cases} \qquad (3.3.17)$$

Note that the mean of the truncated d.f. S_{tr} is, by definition, equal to the value of the limited expected value function L of the original d.f. S at the truncation point Z_{max}. The moments of a truncated claim size d.f. are always finite.

The moments $a_k = a_k(Z_{max})$ about zero of a claim size variable $Z_{tr} = \min(Z, Z_{max})$ which is distributed according to the truncated Pareto(α, β, D) distribution S_{tr}, are obtained (Exercise 3.3.28) from the formula

$$a_k(Z_{max}) = \sum_{i=0}^{k} \binom{k}{i} \cdot (-\beta)^{k-i} \cdot \mathrm{E}((Z_{tr} + \beta)^i),$$

with   (3.3.18)

$$\mathrm{E}((Z_{tr} + \beta)^i) = \frac{\alpha \cdot (D + \beta)^i - i \cdot (Z_{max} + \beta)^i \cdot (1 - S(Z_{max}))}{\alpha - i}.$$

If α is an integer, then the case $i = \alpha$ again leads to a singular equation, and the above formula requires some modification (Exercise 3.3.28).

(e) The Burr distribution. There is a need to find distributions which offer greater flexibility than the Pareto d.f. (3.3.14). Such flexibility is provided by the Burr distribution

$$S(Z) = 1 - \left(\frac{\lambda}{(Z - D)^\tau + \lambda} \right)^\alpha \quad (Z \geqslant D) \qquad (3.3.19)$$

where α, τ and λ are positive parameters. This is a generalization of the Pareto d.f. Indeed, with $\tau = 1$ and $\lambda = D + \beta$ we obtain the formula

(3.3.14). The basic characteristics, as well as the limited expected value function, of the Burr distribution can be found in Hogg and Klugman (1984).

Exercise 3.3.22 Show that, if Z is Pareto(α, β, D)-distributed, the shifted variable $Z - c$ is Pareto$(\alpha, \beta + c, D - c)$-distributed.

Exercise 3.3.23 Show that, if Z is Pareto(α, β, D)-distributed, the tail of Z starting at a point $T > D$ (i.e. the conditional variable $Z \mid \{Z > T\}$) is Pareto(α, β, T)-distributed.

Exercise 3.3.24 Let Z be Pareto(α, β, D)-distributed. Show that

$$E((Z + \beta)^k) = \frac{\alpha}{\alpha - k} \cdot (D + \beta)^k, \quad \text{for } \alpha > k,$$

and that the moments for $k \geqslant \alpha$ are infinite.

Exercise 3.3.25 Prove the formulae (3.3.15) for the mean m and the standard deviation σ_Z of the Pareto(α, β, D) distribution.

Exercise 3.3.26 Derive the third cumulant of a Pareto distributed variable Z, and show that the formula in (3.3.15) for the skewness γ_Z holds true.

Exercise 3.3.27 Let L_α denote the limited expected value function of the Pareto(α, β, D) distribution.

(a) Show that $L = L_\alpha$ has the formula (3.3.15), if $\alpha \neq 1$.
(b) Prove that $L_\alpha \to L_1$ as $\alpha \to 1$.
(c) Show that $L_1(M) = D + (D + \beta) \cdot [\ln(M + \beta) - \ln(D + \beta)]$, for $M \geqslant D$.

Exercise 3.3.28 (a) Prove the formulae (3.3.18) for the moments of a truncated Pareto distribution.
(b) Show that for $\alpha = i$ we have $E((Z_{tr} + \beta)^i) = i \cdot (D + \beta)^i \cdot [\ln(Z_{max} + \beta) - \ln(D + \beta)] + (Z_{max} + \beta)^i \cdot (1 - S(Z_{max}))$.

Exercise 3.3.29 Derive an expression for the moments a_k of the censored Pareto d.f. (3.3.16) in terms of the moments $a_k(C)$ of the corresponding truncated Pareto d.f.

Exercise 3.3.30   Let L be the limited expected value function corresponding to a d.f. S, and let L_C be the limited expected value function of the censored d.f. $S_C = S/S(C)$. Show that

$$L_C(M) = \frac{L(M) - M \cdot (1 - S(C))}{S(C)}, \quad \text{for } M \leqslant C.$$

Exercise 3.3.31   Show that the graph of $1 - S(Z)$, where S is the Pareto$(\alpha, 0, D)$-d.f., is a straight line on a double logarithmic scale.

3.3.8   Large claims

(a) The problem of insufficient claim statistics.   As has already been stated above, the lack of sufficient data concerning large claims is a serious problem from both a practical and theoretical point of view. Unless the larger claim size amounts are eliminated by reinsurance, the tail of the d.f. is of critical importance, whenever a heavy-tailed claim size d.f. S is under consideration. Paradoxically, the least-known part of the d.f. S has, in these cases, the greatest effect on the numerical results. It is important, therefore, to make best use of the scattered and often scant information which is available from the insurer's own files and from other sources. Some aspects of this particular problem are discussed below.

(b) Prolonged observation period.   One possibility is to use different observation periods for examining the experience of large and small claims. For small values of Z the experience of a short period may suffice to determine values of $S(Z)$. Since such experience will, in general, include relatively few large claims, a further study of large claims over a longer period may be needed to ensure an adequate body of data.

Thus one year's statistics might suffice for claims less than or equal to £100 000, say, whilst, for claims over £100 000, data for perhaps 20 years or more may be necessary. Of course, in this case the experience of large claims should be adjusted so as to be consistent with the relative amount of business. How far this method can be used depends on how much time elapses before underlying structural changes make aggregation of the experience unreliable. This should be continuously tested, for example following potential alterations in the properties of smaller claims.

(c) Industry-wide experience can support the derivation of assumptions for large claims. For example, aggregate statistics on very large claims may be available for lengthy periods for broadly equivalent conditions. In some countries a number of insurers have established a joint bureau to collect and analyse data for rate-making and reserving. Their files may also be able to produce claim statistics which would be useful for the construction of the basic risk-theoretical functions.

(d) Individual evaluation. Sometimes statistics relating to large claims are unsatisfactory because of relatively rapid changes in the risk structure of the portfolio. In such cases a rough method, such as that which was described as the individual approach in section 3.3.2, may be of use. The largest risk units in the portfolio are considered one by one, and the potential claim sizes and the associated frequencies of occurrence are evaluated. The method is clearly subjective but, in the absence of other methods, it does provide some basis for further calculation.

(e) Shadow claims. If some information is available relating to the potential for large claims, it is sometimes possible to introduce one or more hypothetical shadow claims to the actual claim statistics, which, having regard to the actual portfolio, can be considered realistic although likely to occur very rarely. Instead of the expected numbers of shadow claims it is usually more convenient to think in terms of their frequencies; one could, for example, try to fix the sizes of the largest claims which might occur once every 10, 25, 50 and 100 years in the portfolio.

3.3.9 *Distribution-free evaluations*

Situations may occur in which the magnitudes of the main characteristics of the aggregate claim amount variable X are needed, but the shape of the claim size d.f. S is not known. There are also occasions when a rapid assessment of the approximate size of the moments is required. If it is known that the claims caused by an individual event have an upper limit M, then upper bounds can be found for these basic characteristics. The limit M may be due to the net retention for ceded reinsurance, if the reinsurance cession is arranged on the basis of individual risks (section 3.4.2 deals with excess of loss treaties). A limit M may also arise from special policy conditions or may result from the special character of the portfolio, for example if it consists only of relatively small policies, say home-owners' properties, motor cars, etc.

(a) Approximate formulae for the standard deviation of the aggregate claim amount in the case where the claim sizes have an upper bound, M, will be considered in this section. The approximations will be based on the inequality

$$a_j = \int_0^M Z^j \, dS(Z) \leqslant M \int_0^M Z^{j-1} \, dS(Z) = M \cdot a_{j-1}, \quad (3.3.20)$$

which is valid for any non-negative claim size distribution.

In the compound Poisson case (see (3.2.13)), the above inequality gives

$$\sigma_X = \sqrt{n \cdot a_2} \leqslant \sqrt{n \cdot M \cdot m} = \sqrt{M \cdot P}, \quad (3.3.21)$$

where $P = E(X) = n \cdot m$ is the risk premium. In order to provide an idea of the accuracy of this inequality, an example of the behaviour of the ratio

$$K = \frac{\sigma_X}{\sqrt{M \cdot P}} = \sqrt{\frac{n \cdot a_2}{M \cdot P}} = \sqrt{\frac{a_2}{M \cdot m}}, \quad (3.3.22)$$

of the standard deviation over its upper bound, as a function of M, is shown in Figure 3.3.6.

The ratio K varies, in this example, between the values 0.5 and 0.8 for those values of M which are relevant in most applications. If both negative and positive deviations can be tolerated, then $K = 0.7$ might be accepted as an average value, resulting in an approximate formula

$$\sigma_X \approx K \cdot \sqrt{M \cdot P} \quad (K = 1 \text{ upper bound}; \ K = 0.7 \text{ average}). \quad (3.3.23)$$

It should be appreciated, however, that any value between 0 and 1 for K is theoretically possible.

For the compound mixed Poisson case a similar evaluation is obtained:

$$\sigma_X = \sqrt{n \cdot a_2 + P^2 \cdot \sigma_q^2} \approx \sqrt{K^2 \cdot M \cdot P + P^2 \cdot \sigma_q^2}, \quad (3.3.24)$$

where $K = 0.7$ again gives a rough estimate, while an upper bound is obtained for $K = 1$. Estimates for both M and σ_q are needed to make use of this formula.

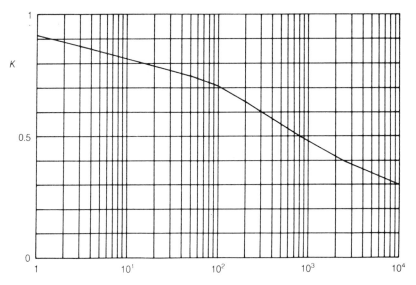

Figure 3.3.6 *The ratio K of (3.3.22) as a function of M. Claim size d.f. as given in Table 3.4.1, M according to an individual risk excess of loss treaty.*

(b) Skewness. In the compound Poisson case an upper bound for the skewness can be derived by using (3.3.20) (see (3.2.13))

$$\gamma_X = \frac{a_3}{a_2^{3/2} \cdot \sqrt{n}} = \frac{a_3}{a_2 \cdot \sqrt{n \cdot a_2}} \leqslant \frac{M}{\sigma_X}. \tag{3.3.25}$$

(c) The most dangerous distribution. The formula (3.3.20) has a nice interpretation, when the equality holds true. Indeed, if negative claims and claim sizes exceeding the limit M are excluded, then it is easily seen that there is equality in (3.3.20) if and only if the whole probability mass is concentrated at the end points 0 and M (Exercise 4.1.6). Among these extreme distributions there is exactly one having mean equal to the given value m, namely the d.f.

$$\mathrm{Prob}\{Z = M\} = m/M = 1 - \mathrm{Prob}\{Z = 0\}. \tag{3.3.26}$$

According to (3.3.21), therefore, in the compound Poisson case the standard deviation σ_X for this particular distribution is larger than the standard deviation of any other distribution restricted to the same limited range $[0, M]$ and having the same mean claim size m. The result is easily generalized to the compound mixed Poisson case

(3.3.24) and also to the general compound case (3.2.18). We conclude that, within the above mentioned limited family of claim size distributions, the d.f. (3.3.26) maximizes the standard deviation of the aggregate claim amount X.

The higher moments a_j are also then maximized by (3.3.20), and, consequently, the distribution (3.3.26) can be considered, in a certain sense, as the most dangerous of all the claim size distributions with mean equal to m, the range of which is restricted to the interval $[0, M]$.

3.3.10 Deductibles

(a) How does the insured risk depend on the deductible? It is common practice in most insurance lines for the coverage to be restricted by a deductible. The idea of a deductible is, firstly, to reduce claim handling costs by excluding coverage for the often numerous small claims and, secondly, to provide some motivation to the insured to prevent claims, through a limited degree of coinsurance or participation in claim costs.

An agreement between the insured and the insurer to have a deductible D means that the insurer pays only the part of the claim which exceeds the amount D; if the size of the claim falls below this amount, then the claim is not covered by the contract and the insured receives no indemnification.

We next consider the claim size d.f. (of claim costs after subtraction of the deductible) as a function of the deductible D. Let S denote the d.f. of the size of a claim in the case of no deductible. If Z is the size of a claim distributed according to the d.f. S, then the insurer pays the amount Z_D in excess of D, or, more precisely, the amount

$$Z_D = Z - \min(D, Z) = (Z - D)^+. \qquad (3.3.27)$$

(Notation: $(x)^+ = \max(0, x)$.) The mean value $E(Z_D)$ of the amount to be paid is then

$$E(Z_D) = L(\infty) - L(D). \qquad (3.3.28)$$

where L denotes the limited expected value function (3.3.5) of the d.f. S, according to which the claim size variable Z is distributed; in particular, $L(\infty) = E(Z) = m$ is the mean of Z. The formula (3.3.28) can be rewritten as

$$E(Z_D) = c(D) \cdot m, \qquad (3.3.29)$$

where the **deductible discount coefficient**

$$c(D) = \frac{L(\infty) - L(D)}{L(\infty)} = 1 - \frac{L(D)}{m} \qquad (3.3.30)$$

is a decreasing and convex function of the deductible D, such that $c(0) = 1$ and $c(\infty) = 0$ (Exercise 3.3.32).

Letting n denote the expected number of claims and $P_D = n \cdot E(Z_D)$, the risk premium for the deductible D, we have

$$P_D = c(D) \cdot P = n \cdot (L(\infty) - L(D)). \qquad (3.3.31)$$

Thus we see that the limited expected value function L is a convenient tool for evaluating the effect of a deductible on the risk premium.

Let S_D denote the conditional d.f. $S_D(Z) = \text{Prob}\{Z \leqslant Z | Z > D\}$ of claim size Z under the condition that the size of a claim exceeds the amount of the deductible D. Then we have

$$S_D(Z) = \frac{S(Z) - S(D)}{1 - S(D)}, \quad Z \geqslant D, \qquad (3.3.32)$$

(cf. Exercise 3.3.14). The excess mean claim size $m_D = E(Z - D | Z > D)$ (after taking into account the effect of the deductible) of a claim larger than the deductible, and the expected number $n_D = n \cdot \text{Prob}\{Z > D\}$ of those events which give rise to a claim larger than D, are given by the equations

$$m_D = \frac{c(D)}{1 - S(D)} \cdot m,$$
$$n_D = (1 - S(D)) \cdot n. \qquad (3.3.33)$$

Note especially the relation $n_D \cdot m_D = n \cdot E(Z_D)$.

(b) The franchise deductible. Sometimes a so-called **franchise deductible** is incorporated in the contract. In this case the insurer pays the whole claim, if the agreed deductible amount is exceeded. If the franchise deductible is D, then the mean m_D in (3.3.33) is replaced by $m_{D(\text{franch.})} = m_D + D$, and we have (Exercise 3.3.33) the formula

$$P_{D(\text{franch.})} = P_D + n_D \cdot D \qquad (3.3.34)$$

for the risk premium, where P_D and n_D are given by (3.3.31) and (3.3.33).

Exercise 3.3.32 Show, by making use of the properties of the limited expected value function L, that the deductible discount coefficient c is a continuous, convex and decreasing function of the deductible.

Exercise 3.3.33   Give a formula for the deductible coefficient $c_{\text{franch.}}(D)$ for the franchise deductible.

Exercise 3.3.34   Consider an insured risk where the probability of one claim in a year is evaluated to be 0.01 and the possibility of more than one claim can be ignored. The size of the claim can be at least $\sqrt[3]{2}$ and at most 10 being distributed in between these limits according to $S(Z) = 1 - 2Z^{-3}$. Calculate the risk premiums in the case where the policyholder's deductible is 2.

3.4   Claims and reinsurance

3.4.1   Modification of the claim size d.f. for reinsurance

(a) Claims net of reinsurance.   In order to keep the variation of the aggregate claim amount X reasonable, an insurer usually takes out reinsurance cover for his insurance portfolio, i.e. he protects himself against losses arising from large, excessively numerous or catastrophic claims by reinsuring large claim amounts or high claim frequency with one or more other insurance or reinsurance companies.

The aggregate claim amount net of reinsurance, $X = X_{\text{ced}}$, is the cedant's part of the total aggregate claim amount X_{tot}, which is retained after the reinsurer's (or reinsurers') share X_{re} of claims is deducted, i.e.

$$X = X_{\text{tot}} - X_{\text{re}}. \tag{3.4.1}$$

Since we are most often dealing with problems concerning the quantities net of reinsurance, our interest is usually focused on the net amount X, which is, therefore, left without subscript in this section.

In this section the main characteristics and distributions of claim variables net of reinsurance, and of the reinsurer's share, are considered. One of the main tasks is the derivation of the d.f. and the risk premium of the net aggregate claim amount X, and of the reinsurer's share X_{re}, from the original total quantities, given the type and parameters of the reinsurance arrangement.

Reinsurance treaties are often arranged on a claim by claim (or event by event) basis. This means that each separate claim is divided between the cedant and the reinsurer. Then, by analogy with (3.4.1), the size $Z = Z_{ced}$ of an individual claim net of reinsurance is

$$Z = Z_{tot} - Z_{re}, \qquad (3.4.2)$$

where Z_{tot} denotes the size of the claim before deducting the reinsurer's share Z_{re}. Excess of loss, quota share and surplus reinsurance, which will be considered in sections 3.4.2 to 3.4.4, are examples of this type of reinsurance arrangement.

A drawback of the claim by claim reinsurance arrangements is that they do not necessarily restrict the impact of exceptionally large values of the aggregate claim amount X_{tot} in all cases; protection may be poor in situations where X_{tot} is large as a result of a high number of claims, rather than because of one or more very large claims. Indeed, it can be shown (see Pesonen, 1984) that, theoretically, in a quite general sense, any optimal reinsurance arrangement is such that it applies directly to the aggregate claim amount X_{tot}, in such a way that X is a function of X_{tot}. Stop loss reinsurance (or aggregate excess of loss reinsurance) is an example of such an aggregate reinsurance arrangement, and will be considered in section 3.4.4. In practice, however, this kind of reinsurance arrangement, giving protection directly against fluctuations of the aggregate claim amount X_{tot}, is problematic, since numerous relevant factors, such as inflation, changes in the business volume or premium levels, changes in the policy terms or in claim settlement practice, etc. may cause unexpected fluctuations in X_{tot}, and the reinsurer often cannot analyse all these risks and factors sufficiently thoroughly to be able to rate and evaluate the risk properly.

In practice, different types of reinsurance are often used for different classes of business and for different types of risk. Sometimes different types are combined.

We first consider each class separately. If the reinsurance for all classes is arranged on a claim by claim basis, it may be possible to construct a weighted claim size distribution net of reinsurance for the whole portfolio, using the additivity of compound variables, provided that the assumptions required for additivity are met (section 3.2(d)). If there are complicated reinsurance covers, possibly including additional aggregate reinsurance covers (like stop loss) for some classes, then simulation may be the only method available to derive the aggregate claim amount d.f. net of reinsurance over all classes of business of an insurer.

3.4.2   Excess of loss reinsurance

(a) Definition.   With an **excess of loss treaty** the reinsurer pays that part of each claim amount which exceeds an agreed limit M, the cedant's retention limit. Excess of loss treaties may be written in a variety of ways. In this section we are concerned with treaties where the cedant's retention is defined for each claim in a certain group of risks, the reinsurer paying the excess over that amount if a claim exceeds the retention level M. This type is called **per claim** (or per risk) excess of loss reinsurance, when it is necessary to distinguish it from event-based variants such as **single event** excess of loss, whereby protection is afforded to a group of risks against catastrophic aggregation of claims arising from a major event such as an earthquake or wind storm.

In excess of loss reinsurance the reinsurer pays the excess $Z_{\mathrm{re},M} = (Z_{\mathrm{tot}} - M)^+$ over an agreed amount M in respect of each claim. The cedant's share $Z_M = Z_{\mathrm{tot}} - Z_{\mathrm{re},M}$ of the claim, net of reinsurance, is then

$$Z_M = \min(M, Z_{\mathrm{tot}}) \tag{3.4.3}$$

and its d.f. S_M, in terms of the d.f. S of the total claim size Z_{tot}, is

$$S_M(Z) = \begin{cases} S(Z) & \text{for } Z < M \\ 1 & \text{for } Z \geqslant M. \end{cases} \tag{3.4.4}$$

The retention limit M is written out as a subscript, since its value is often varied when alternative limits are compared.

From the cedant's point of view, the excess of loss reinsurance does not influence the number of claims. Therefore, if the total aggregate claim amount X_{tot} is a compound variable with claim size d.f. S, and if all risks are reinsured using the same retention limit M, the aggregate claim amount $X = X_M$ net of reinsurance is also a compound variable having the same claim number variable but with claim size d.f. S_M given by (3.4.4). The reinsurer's aggregate claim amount is also a compound variable, but the number of non-zero claims is usually much smaller for the reinsurer.

(b) Basic characteristics.   The mean claim size m_M of the cedant's share (3.4.3) of a claim covered by excess of loss reinsurance is

$$m_M = \mathrm{E}(Z_M) = \int_{-\infty}^{M} Z \, dS(Z) + M \cdot (1 - S(M)). \tag{3.4.5}$$

It follows immediately from (3.4.3), using the definition of the limited expected value function, that

$$m_M = E(\min(M, Z_{tot})) = L(M),$$ (3.4.6)

where L denotes the limited expected value function of the d.f. S of the total claim size Z_{tot}. The properties of the limited expected value function (section 3.3.4(b)) imply that the net mean claim size m_M, and therefore also the net risk premium

$$P_M = n \cdot m_M$$ (3.4.7)

is a concave, continuous, increasing function of the retention limit M.

Moments about the origin (3.2.9) of the cedant's share Z_M of a claim are given by

$$a_k(M) = E(Z_M^k) = \int_{-\infty}^{M} Z^k dS(Z) + M^k \cdot (1 - S(M)).$$ (3.4.8)

Because of the formal analogy between the reinsurer's share $Z_{re} = Z_{re,M}$ of a claim in an excess of loss treaty and the variable (3.3.27) giving the claim size net of a deductible, the results of section 3.3.10(a) concerning the mean claim sizes, risk premiums and claim numbers also apply directly for the reinsurer's share Z_{re}. For example, the expected number of claims exceeding the retention limit M is $n_{re} = n \cdot (1 - S(M))$. The moments of the reinsurer's share Z_{re} can be obtained (Exercise 3.4.4) from the moments (3.4.8) by the formula

$$a_k(Z_{re}) = \sum_{i=1}^{k} \binom{k}{i} \cdot (-M)^{k-i} \cdot [a_i - a_i(M)],$$ (3.4.9)

where $a_i = a_i(\infty) = a_i(Z_{tot})$. Note that the value of Z_{re} is zero whenever the size of a claim is smaller than the retention limit M. The kth moment about the origin of the reinsurer's share of a claim which exceeds the retention limit is

$$a_k(Z_{re} | Z_{re} > 0) = a_k(Z_{re})/(1 - S(M)).$$ (3.4.10)

The denominator $1 - S(M) = \text{Prob}\{Z_{re} > 0\} = \text{Prob}\{Z_{tot} > M\}$ was needed to eliminate the reinsurer's null claims, i.e. those claims which do not exceed the retention limit. Note, however, that it is often more convenient to use the moments (3.4.9) than the corresponding conditional moments (3.4.10), as can be seen, for example, in section 3.4.2(e) below.

Table 3.4.1 Lowest moments and risk indices $r_k = a_k/m^k$ (see (3.2.16)) of the cedant's claim size distribution S_M net of reinsurance, in the presence of an excess of loss treaty with retention limit of M, calculated for the claim size distribution S of Table 3.3.1. Monetary unit is £1000

M	$S(M)$	m_M	$a_2(M)$	$a_3(M)$	$r_2(M)$	$r_3(M)$
0.10	0.2611877	0.085	0.008	0.001	1.09	1.25
0.14	0.3092646	0.113	0.015	0.002	1.15	1.38
0.20	0.3642961	0.153	0.028	0.005	1.20	1.53
0.28	0.4224722	0.201	0.051	0.014	1.27	1.70
0.40	0.4888123	0.266	0.095	0.036	1.34	1.93
0.57	0.5566038	0.347	0.173	0.093	1.44	2.23
0.80	0.6198597	0.441	0.301	0.225	1.55	2.63
1.13	0.6784591	0.556	0.522	0.546	1.69	3.18
1.60	0.7294388	0.694	0.897	1.315	1.86	3.93
2.26	0.7781205	0.855	1.513	3.102	2.07	4.96
3.20	0.8200895	1.043	2.528	7.262	2.32	6.40
4.53	0.8564949	1.256	4.159	16.721	2.64	8.44
6.40	0.8855225	1.495	6.746	37.935	3.02	11.35
9.05	0.9086236	1.764	10.862	85.637	3.49	15.60

12.80	0.9285196	2.066	17.393	192.598	4.07	21.84
18.10	0.9450290	2.395	27.435	425.047	4.78	30.95
25.60	0.9579705	2.754	42.958	933.019	5.66	44.67
36.20	0.9683720	3.140	66.542	2023.655	6.75	65.35
51.20	0.9765965	3.545	101.470	4306.566	8.07	96.66
72.41	0.9831277	3.964	152.490	9017.448	9.71	144.80
102.40	0.9884495	4.381	224.245	18 365.977	11.68	218.42
250.00	0.9966710	5.271	511.096	92 353.348	18.39	630.59
500.00	0.9987331	5.776	869.586	2.910 E + 05	26.07	1510.05
750.00	0.9992801	6.013	1160.741	5.625 E + 05	32.10	2586.83
1000.00	0.9995179	6.160	1415.296	8.957 E + 05	37.30	3831.57
2000.00	0.9998165	6.453	2246.000	2.737 E + 06	53.94	10 186.38
3000.00	0.9998957	6.590	2920.672	5.253 E + 06	67.25	18 354.69
5000.00	0.9999488	6.735	4044.418	1.194 E + 07	89.16	39 083.80
10 000.00	0.9999805	6.890	6248.167	3.636 E + 07	131.61	111 160.67
20 000.00	0.9999926	7.008	9602.810	1.107 E + 08	195.52	321 633.02
50 000.00	0.9999979	7.122	16 865.385	4.823 E + 08	332.49	1 335 137.90
100 000.00	0.9999992	7.185	25 764.811	1.469 E + 09	499.11	3 959 410.62

Notation: xE + n = $x \cdot 10^{n}$.

(c) An example. The lowest three moments $a_k(M)$ for the d.f. S given in Table 3.3.1 of section 3.3.3 are shown in Table 3.4.1 for different values of the retention limit M. Since grouped data were used in Table 3.3.1, and since the class averages $\bar{Z}_i$ and the number of claims in each group were the only information given about the d.f. S, it was assumed that within the range $[0, 102.4]$ the d.f. S is a step function with the steps positioned at the class averages $\bar{Z}_i$. For the tail starting at the point 102.4 the fitted Pareto curve $S(Z) = 1 - 7.3208 \cdot Z^{-1.3938}$, truncated at $Z = 100\,000$, is used (Exercise 3.4.6). For the purpose of checking, it is verified that the double inequality $1 \leqslant r_2^2 \leqslant r_3$ (Exercise 3.2.7) is satisfied for every M in Table 3.4.1.

It can be seen from the figures of Table 3.4.1 that the reinsurance risk premium $P_{re} = n \cdot (m_\infty - m_M)$ will be relatively small compared to the total risk premium $P = n \cdot m_\infty$, if the retention limit M is larger than £1 million. However, the higher moments, and thereby the volatility of the cedant's net risk, increase strongly as the retention limit M is further increased. The claim size d.f. S in Table 3.4.1 is relatively heavy-tailed, the frequency of claims exceeding £20 million being about one in thirty years, and the frequency of claims exceeding £50 million about one in a hundred years. The total number of claims during the four-year observation period was 16 536.

(d) Sensitivity to claim inflation. As was mentioned in section 3.3.1(c), the effect of inflation should always be taken into account when claims are analysed. It should be appreciated that the excess of loss reinsurance risk premium $P_{re,M}$ is particularly sensitive to inflation. In fact, if total claim sizes, Z_{tot}, are multiplied by a factor $r > 1$, then the reinsurance risk premium $P_{re,M}$ increases by a greater factor

$$r^*(M) = \frac{P_{re,M}(r \cdot Z_{tot})}{P_{re,M}(Z_{tot})} = \frac{E((r \cdot Z_{tot} - M)^+)}{E((Z_{tot} - M)^+)} > r, \qquad (3.4.11)$$

where M denotes the cedant's retention limit, which is kept unchanged. To prove this inequality, we first recall that, by (3.4.6), $P_{re,M}(Z_{tot}) = n \cdot [L(\infty) - L(M)]$ and $P_{re,M}(r \cdot Z_{tot}) = n \cdot [L_r(\infty) - L_r(M)]$, where L and L_r, respectively denote the limited expected value functions of the total claim sizes Z_{tot} and $r \cdot Z_{tot}$. We have (Exercise 3.3.8) $L_r(M) = r \cdot L(M/r)$, so that $r^*(M) = r \cdot [L(\infty) - L(M/r)]/[L(\infty) - L(M)] > r$, because L is an increasing function.

To give an idea of the difference between the primary growth factor r and the resulting reinsurance risk premium growth factor r^*, some examples are given in Table 3.4.2, where the original Z_{tot}

Table 3.4.2 *The growth rate $r^* - 1$ for the reinsurance risk premium $P_{re,M}$ in the case of a Pareto($\alpha, 0, 1$) claim size distribution for different values of the Pareto-parameter α. The claim inflation rate $r - 1$ was assumed to be equal to 10%*

Pareto-α:	1.1	1.5	2	3
($r^* - 1$)	11.1%	15.4%	21.0%	33.1%

is assumed to be Pareto($\alpha, 0, 1$)-distributed and where the parameter α is varied. The factor $r^*(M)$ does not depend on the retention limit M in the case of a Pareto distribution with $\beta = 0$; in fact $r^* \equiv r^\alpha$ in this case (Exercise 3.4.7).

It can be concluded that, the lighter the tail of the claim size distribution, the more sensitive the reinsurer's excess of loss risk premium is to inflation. The explanation is that inflation not only increases the claim sizes, but also the expected number $n_{re,M}$ of claims exceeding the retention limit M; the higher the relative amount of smaller claims, the stronger the growth of the expected claim number $n_{re,M}$.

(e) Limited excess of loss cover. In practice excess of loss reinsurance cover is usually limited. This means that the reinsurer pays the possible excess over the retention limit M, but at most up to an agreed amount A per claim. Using standard reinsurance terminology, the treaty covers the **layer** A **xs** M. Then the reinsurer's share of a claim of size Z_{tot} is (Exercise 3.4.8)

$$Z_{re} = \min(A, (Z_{tot} - M)^+) = Z_{M+A} - Z_M, \qquad (3.4.12)$$

where the notation $Z_M = \min(M, Z_{tot})$ of section 3.4.2(a) is used.

In practice, the reinsurance cover may consist of several layers, and risk premiums may have to be calculated for each of them.

By (3.4.6) the reinsurer's risk premium for the layer A xs M is simply

$$P_{re} = n \cdot E(Z_{re}) = n \cdot (L(M + A) - L(M)), \qquad (3.4.13)$$

where n denotes the expected number of (all) claims.

The formulae (3.4.9) and (3.4.10) for the moments of the reinsurer's share Z_{re} generalize (Exercise 3.4.4) to the case of a limited layer A xs M by simply replacing $a_i(M)$ in (3.4.9) by $a_i(M + A)$.

If the total aggregate claim amount X_{tot} is a compound mixed Poisson variable with expected number of claims n, and if there is a limited excess of loss cover for the layer A xs M, then the variance $\text{Var}(X_{re})$ of the reinsurer's share of the aggregate claim amount is (Exercise 3.4.5)

$$\text{Var}(X_{re}) = n \cdot [a_2(M + A) - a_2(M)] - 2 \cdot M \cdot P_{re} + \text{Var}(q) \cdot P_{re}^2, \quad (3.4.14)$$

where q is the mixing variable. The skewness of the reinsurer's share X_{re} can be similarly expressed using the moments $a_k(M)$ and $a_k(M + A)$. However, since the expected number $n_{re} = n \cdot (1 - S(M))$ of claims of a size exceeding the retention limit M is usually rather small, sometimes less than 1, these higher characteristics are not necessarily very useful; it may be more important to know how often the layer is hit by a claim and what the claim size d.f. is.

REMARK   In practice an excess of loss reinsurance cover for a layer A xs M is often also limited with respect to the accumulated claim amount. The treaty might state, for example, that only three reinstatements are permitted during one year, which means that the reinsurer's share of all claims accumulated during one year is limited by the amount $(1 + 3) \cdot A$, i.e. the cover can be reinstated three times. Furthermore, an extra premium may have to be paid for each reinstatement. Sundt (1990) has given some consideration to reinstatements.

(f) Application to the case with a deductible.   As already noted in section 3.4.2(b), there is a formal analogy between the deductible and the excess of loss retention limit. In this section it will be shown how the results of section 3.4.2(e) can be applied to the cedant's claims net of excess of loss reinsurance in the presence of a policyholder's deductible.

Consider a situation where the d.f. S of the total size of a claim is given, and the lowest three moments $a_k(M)$ (3.4.8) of the size Z_M of a claim truncated at M are available for different values of the parameter M. Table 3.4.1 provides an example. Suppose that a deductible D is introduced, and that there is an excess of loss reinsurance with a retention limit M ($> D$). Applying (3.4.9) and (3.4.10) to the truncated d.f. S_M, the three lowest moments $a_k = a_k(D, M)$ of the insurer's share net of reinsurance of each claim exceeding the deductible D are obtained by the formulae

$$a_1 = \frac{m_M - m_D}{1 - S(D)}$$

$$a_2 = \frac{a_2(M) - a_2(D) - 2 \cdot D \cdot (m_M - m_D)}{1 - S(D)},$$

$$a_3 = \frac{a_3(M) - a_3(D) - 3 \cdot D \cdot (a_2(M) - a_2(D)) + 3 \cdot D^2 \cdot (m_M - m_D)}{1 - S(D)}.$$

$$(3.4.15)$$

The denominator $1 - S(D)$ is needed (as in (3.4.10)) to eliminate those claims which are smaller than the deductible D.

If the aggregate claim amount is a compound variable, then the mean, μ_X, the standard deviation, σ_X, and the skewness, γ_X, of the insurer's net aggregate claim amount $X = X_{D,M}$ in the presence of a deductible D and net retention M can easily be obtained by substituting the moments (3.4.15) into the relevant formulae of section 3.2(b), assuming that the claim number distribution is known.

To illustrate the behaviour of the insurer's net claim amount as a function of both the deductible D and the retention limit M, some examples are calculated in Table 3.4.3. The d.f. S of Table 3.3.1 has again been used as the claim size d.f. The expected number of all claims in a year was assumed to be equal to $n = 4134$, the yearly average of the number of observed claims, and, in order to keep calculations simple in this illustrative example, the number of claims was assumed to be Poisson-distributed. Then, according to (3.2.11) and (3.2.13), we have $\mu_X = n_D \cdot a_1$, $\sigma_X = \sqrt{n_D \cdot a_2}$ and $\gamma_X = n_D \cdot a_3 / \sigma_X^3$, where $n_D = n \cdot (1 - S(D))$ denotes the expected number of claims exceeding the deductible D.

Table 3.4.3   *The basic characteristics of the insurer's net aggregate claim amount* $X = X_{D,M}$ *as a function of the deductible D and the excess of loss retention limit M. The numerical data of Table 3.4.1 has been used*

Deductible D ($£$)	Retention M ($£$ million)	n_D	μ_X ($£$ million)	σ_X ($£$ million)	γ_X
0	1	4134	25.5	2.42	0.26
200	1	2628	24.8	2.42	0.26
1 600	1	1119	22.8	2.40	0.26
250 000	1	14	3.7	1.38	0.46
0	5	4134	29.4	8.35	3.43
200	5	2628	28.8	8.35	3.43
1 600	5	1119	26.6	8.34	3.43
250 000	5	14	7.7	7.99	3.82
0	10	4134	29.7	10.32	5.52
200	10	2628	29.1	10.32	5.52
1 600	10	1119	26.8	10.32	5.52
250 000	10	14	7.9	10.02	5.95

Exercise 3.4.1 Assume that the claim size d.f. is exponential $S(Z) = 1 - e^{-cZ}$, and that excess of loss reinsurance is applied with a retention limit M. What is the d.f. of the claim size on the reinsurer's liability for a claim exceeding the retention limit? What is the variance of the reinsurer's share of the aggregate claim amount in the compound mixed Poisson case?

Exercise 3.4.2 Assume that the claim size d.f. is $S(Z) = 0.9 \cdot Z$ for $Z < 1$, and $S(Z) = 1 - 0.1 \cdot Z^{-3}$ for $Z \geqslant 1$, where the monetary unit is £10 000. The layer 3 xs 2 (in units of £10 000) is covered by a limited excess of loss reinsurance treaty. Calculate the reinsurance risk premium P_{re} as a percentage of the total risk premium P_{tot}. Suppose that as a result of inflation the claim sizes are uniformly increased by 10%, but the premium is not changed. What is the reinsurer's expected loss as a result of inflation?

Exercise 3.4.3 Assume that the claim size d.f. S is log-normal with mean $m = 30$, standard deviation $\sigma = 100$ and skewness $\gamma = 76$. The expected number of claims is $n = 100$. Then $P_{tot} = 3000$. Calculate the excess of loss reinsurance risk premium for retention limits $M = 50$, 100, 400 and 2000. Check that these risk premiums follow a convex pattern.

Exercise 3.4.4 (a) Show that the second moment $a_2(Z_{re,M})$ of the reinsurer's share of a claim under an excess of loss treaty is

$$a_2(Z_{re,M}) = a_2 - a_2(M) - 2 \cdot M \cdot (m - m_M).$$

(b) Prove the general formula (3.4.9).

(c) Show that, for the layer A xs M, the corresponding formula is

$$a_k(Z_{re,(A \text{ xs } M)}) = \sum_{i=1}^{k} \binom{k}{i} \cdot (-M)^{k-i} \cdot [a_i(M + A) - a_i(M)].$$

Exercise 3.4.5 Show that the formula (3.4.14) for the variance of the reinsurer's aggregate claim amount X_{re} holds true in the compound mixed Poisson case.

Exercise 3.4.6 Give the calculation formulae for the moments shown in Table 3.4.1. Note that the values of M in Table 3.4.1 are the same as the upper class limits in Table 3.3.1, and recall that the Pareto-tail of S starts at the point 102.4.

Exercise 3.4.7 Show that, if the claim sizes Z_{tot} are Pareto(α, β, D)-

distributed, and $\alpha > 1$, the growth factor for the reinsurance risk premium, defined in (3.4.11), is

$$r^*(M) = r^\alpha \cdot \left(\frac{M + \beta}{M + r \cdot \beta} \right)^{\alpha - 1} \qquad \text{for } M \geqslant r \cdot D.$$

Exercise 3.4.8 Consider the reinsurer's share $Z_{re} = \min(A, (Z_{tot} - M)^+)$ of a claim Z_{tot} in the case of an excess of loss reinsurance treaty which is limited to the layer A xs M. Show that

$$Z_{re} = Z_{M+A} - Z_M = (Z_{tot} - M)^+ - (Z_{tot} - A)^+.$$

Exercise 3.4.9 Prove the formula

$$a_k(M) = a_k(0) + \int_0^M k \cdot Z^{k-1} \cdot (1 - S(Z)) \, dZ,$$

for the moments about zero of the cedant's net claim size Z_M in an excess of loss reinsurance treaty. (Usually $a_k(0) = 0$.)

Exercise 3.4.10 Prove the formula

$$P_{re,(A \text{ xs } M)} = n \cdot \int_M^{M+A} 1 - S(Z) \, dZ,$$

for the reinsurer's excess of loss risk premium for the layer A xs M.

Exercise 3.4.11 Suppose that the layer A xs M is covered by re-insurance, but that the cedant pays the rest of each claim (including the possible excess over the limit $M + A$). Show that the cedant's net risk premium is equal to $P = n \cdot [L(\infty) - L(M + A) + L(M)]$. Illustrate the d.f.s of the total and net claim size in the same picture.

Exercise 3.4.12 Assuming that the claim inflation rate $r - 1$ is positive, consider the reinsurer's share of claims in the case of a limited excess of loss reinsurance treaty which covers a layer A xs M.

(a) Show that the growth rate $(r^* - 1)$ of the reinsurance risk premium is non-negative.
(b) Under what conditions is $r^* = 1$?
(c) Show that $r^*(A \text{ xs } M) = r^\alpha$, if S is Pareto$(\alpha, 0, D)$ and $M \geqslant r \cdot D$.

Exercise 3.4.13 Prove the formula

$$a_k(Z_{re,(A \text{ xs } M)}) = \int_0^A k \cdot Z^{k-1} \cdot (1 - S(M + Z)) \, dZ,$$

for the moments about zero of the reinsurer's claim size Z_{re} for the layer A xs M in a limited excess of loss reinsurance treaty.

3.4.3   Proportional reinsurance

(a) General aspects.   In proportional reinsurance each claim is shared between the cedant and the reinsurer(s) in a proportion which is specified in the treaty. An advantage of this type of reinsurance is that the reinsurance premium rating is easier than in the case of non-proportional reinsurance treaties. Indeed, since the reinsurer pays a certain proportion of every claim, the reinsurance risk premium is the same proportion of the total risk premium.

A drawback of proportional reinsurance is that small claims are shared between the cedant and the reinsurer, as well as the large ones. This means that proportional reinsurance usually reduces the cedant's net business volume much more than, for example, excess of loss reinsurance, which covers only the largest claims. A partial solution to this problem is provided by so-called surplus reinsurance (section 3.4.3(c) below), which assigns different ceded proportions to different risks according to their size. Another weakness of proportional reinsurance is that treaties which deal with claims individually do not provide satisfactory cover against fluctuations in the number of claims.

(b) Quota share reinsurance.   In quota share reinsurance any claim, irrespective of its size, is divided between the cedant and reinsurer in a predetermined ratio. Then the cedant's share $Z = Z_{ced}$ of the total claim amount Z_{tot} is of the form

$$Z = r \cdot Z_{tot}, \tag{3.4.16}$$

where the value of r $(0 < r < 1)$ is fixed, and the reinsurer's share is simply $Z_{re} = (1 - r) \cdot Z_{tot}$.

The claim size d.f. S_r of the cedant's share Z of a claim is

$$S_r(Z) = S(Z/r), \tag{3.4.17}$$

where S denotes the d.f. of Z_{tot}. The claim size d.f. of the reinsurer's share is $S_{re,r}(Z) = S_{1-r}(Z) = S(Z/(1 - r))$.

Exercise 3.4.14   What are the mean, standard deviation and skewness of the cedant's share $Z = r \cdot Z_{tot}$ of a claim in quota share reinsurance, expressed in terms of the corresponding characteristics of the size of the total claim Z_{tot}?

Exercise 3.4.15 An insurer with d.f. S for the total size of claims has a combination of two reinsurance treaties in force:

(1) A quota share treaty, under which the reinsurer pays a proportion r of each claim, and (2) an excess of loss treaty covering the retained business with a maximum net retention M. What is the distribution function of one claim for the insurer's net retention?

(c) Surplus reinsurance treaties are applicable in classes of insurance where for each risk unit there is a defined upper limit Q, such that the size of a claim occurring to that risk unit cannot exceed Q. In property insurance the sum insured is a candidate for Q. However, the so-called **estimated maximum loss** (EML) (example in section 3.3.2) is usually more suitable to be used as the upper limit Q, and in what follows it will be assumed that Q is chosen to be equal to the EML for the risk unit. Note that the EML may well be less than the insured value of the property; in the case of large risk units especially it may be highly improbable that all the insured property could be lost in one accident.

The idea behind surplus reinsurance is to have a proportional reinsurance cover, with the reinsurer's share depending on the risk unit in such a way that, the larger the upper limit Q for the risk unit, the larger the reinsurer's share.

Let M denote the maximum amount the cedant is willing to pay in respect of one claim. In a **surplus reinsurance** treaty with maximum retention M, those risks which are smaller than this limit, i.e. $Q \leqslant M$, are within the cedant's net retention. For larger risk units, where claims larger than M may occur, claims are shared in proportion $r = r_M(Q) = M/Q$, so that the cedant is responsible for an amount $r \cdot Z_{tot}$ of each claim and the reinsurer(s) pay the rest, i.e. $(1-r) \cdot Z_{tot}$. Note the difference as compared to the quota share treaty (3.4.16); in a surplus treaty the ratio r depends on the risk unit through the upper limit Q, whereas in the quota share treaty it is same for all claims.

The larger the upper limit Q of a risk unit, the larger the proportion of each claim occurring to that risk unit which is paid by the reinsurer. In the case of a maximum loss, i.e. when the claim size reaches the upper limit Q of the risk unit, the cedant's share of the claim will be equal to M, assuming that $Q \geqslant M$. On the other hand, the reinsurance cover also extends to partial claims on large risks, because r depends on the pre-determined limit Q (EML) but not on the size of the actual claim.

If we consider a randomly chosen claim occurring in the portfolio, then both the size Z_{tot} of this claim and the respective upper limit Q, depending on the risk unit to which the claim occurred, are random variables. The cedant's share $Z = Z_{ced}$ of the claim net of reinsurance is then given by

$$Z = r_M \cdot Z_{tot}, \qquad (3.4.18)$$

where

$$r_M = r_M(Q) = \min(1, M/Q) \qquad (3.4.19)$$

assumes positive values between 0 and 1, and is a random variable since its value depends on the upper limit Q of the risk.

Suppose that the risk units in the portfolio are divided into classes, j, according to their EML's (upper limits) Q_j, and let $S(Z|Q_j)$ denote the claim size d.f. of class j. By applying the formula (3.4.17) for each class j the d.f. S_M of the net claim size (3.4.18) is obtained from these class distributions as the weighted average

$$S_M(Z) = \sum_j \text{Prob}\{Q = Q_j\} \cdot S(Z/r_M(Q_j)|Q_j). \qquad (3.4.20)$$

If the estimated expected number of claims in each class j is denoted by n_j, then we have the estimate

$$\text{Prob}\{Q = Q_j\} = \frac{n_j}{n} \qquad (n = \sum n_j). \qquad (3.4.21)$$

Using this estimate we obtain the following formula for the cedant's net risk premium:

$$P_M = n \cdot m_M = \sum_j r_M(Q_j) \cdot P(Q_j), \qquad (3.4.22)$$

where $P(Q_j) = n_j \cdot m(Q_j) = n_j \cdot E(Z_{tot}|Q_j)$ is the (total) risk premium in class j. The risk premiums $P(Q_j)$ determine the **EML-profile** of the portfolio, which indicates how the total risk is distributed between different EML-classes.

If the aggregate claim amount is compound mixed Poisson, then (Exercise 3.4.16) the variance of the cedant's share X_M of the aggregate claim amount net of reinsurance is, by (3.2.14) and (3.2.31),

$$\sigma^2_{X_M} = P^2_M \cdot \sigma^2_q + \sum_j (r_M(Q_j))^2 \cdot n_j \cdot a_{2,j}, \qquad (3.4.23)$$

where $a_{2,j}$ is the second moment about the origin of the claim size d.f. $S(Z|Q_j)$ of class j.

Formulae similar to (3.4.22) and (3.4.23) also hold true for the reinsurer's share $X_{re,M}$ of the aggregate claim amount X (Exercise 3.4.17). The conditional claim size distributions $S(Z|Q_j)$ of each class j were needed above because the cedant's and reinsurer's shares of a claim in a surplus reinsurance treaty depend on the EML Q of the risk. The estimation of these class distributions usually requires estimation of the two-dimensional d.f. of the random vector (Z_{tot}, Q). It can be expected that there is a positive correlation between Z_{tot} and Q, and that the d.f. $S(Z|Q)$ changes in a smooth manner as a function of the EML Q. A relatively large volume of claim statistics is needed in order to get a satisfactory estimate for the class distributions.

Since the cedant's share of a claim is limited from above by the maximum retention M, the cedant's claim size d.f. S_M in a surplus reinsurance treaty is easier to estimate than the usually much more skew class distributions $S(Z|Q_j)$. Indeed, if the EML-profile of the portfolio has remained unchanged during the observation period, then the d.f. S_M of the cedant's share of the claim can be estimated directly using the following alternative estimation method, provided that both the size Z_{tot} and the upper limit Q of the corresponding risk unit are recorded for each claim. First fix the retention limit M and then calculate $Z_{ced} = r_M(Q) \cdot Z_{tot}$ for each observed claim. To obtain the desired estimated value of $S_M(Z)$, count the number $N_M(Z)$ of those claims which fulfil the condition $Z_{ced} \leqslant Z$. Letting N denote the total number of claims, we have the estimate

$$S_M(Z) = N_M(Z)/N. \tag{3.4.24}$$

A table similar to Table 3.3.1 can be computed for any fixed retention limit M, allowing Z to vary over a suitable set of discrete values.

Exercise 3.4.16 Prove the formulae (3.4.22) and (3.4.23) for the mean and variance of the cedant's net aggregate claim amount X_M in the case of a surplus reinsurance cover.

Exercise 3.4.17 Give the formulae corresponding to (3.4.22) and (3.4.23) for the reinsurer's share X_{re} of the aggregate claim amount in the case of a surplus reinsurance cover.

Exercise 3.4.18 Consider two alternative reinsurance arrangements, surplus and excess of loss, both with the same retention limit M. Let

$Z_{re,SUR}$ and $Z_{re,XL}$, denote the reinsurer's share of a claim of size Z_{tot} in the surplus and the excess of loss treaty respectively. Show that

(a) If $Z_{re,SUR} = 0$, then $Z_{re,XL} = 0$.
(b) If $Z_{re,SUR} > 0$, then $Z_{re,SUR} \geqslant Z_{reXL}$, where the equality holds only if Z_{tot} is equal to the risk's upper limit Q.

3.4.4   Stop loss reinsurance

(a) Stop loss reinsurance, which is also called aggregate excess of loss cover, is an aggregate type of cover (section 3.4.1(a)) providing protection not only against large individual claims but also against the fluctuation in the number of claims.

In stop loss reinsurance the reinsurer pays the excess $X_{re} = (X_{tot} - M)^+$ of an agreed limit amount M of the cedant's aggregate claim amount X_{tot} accumulated during a certain time period, for example one year. The cedant's share $X = X_{ced} = X_{tot} - X_{re}$ of the claim, net of reinsurance, is then

$$X = \min(M, X_{tot}). \qquad (3.4.25)$$

Note the formal analogy between stop loss and excess of loss reinsurance. In stop loss reinsurance the aggregate claim amount is shared between the cedant and the reinsurer exactly as an individual claim is shared in the case of excess of loss reinsurance. Because of this analogy, the formulae of section 3.4.2 dealing with one claim can be applied to stop loss reinsurance. For example the cedant's risk premium $P = E(X)$ and the reinsurer's risk premium $P_{re} = E(X_{tot}) - E(X)$ can be obtained ((3.4.5), (3.4.6) and Exercise 3.3.12) simply by the formulae

$$P_{ced} = L_F(M) = \int_{-\infty}^{M} X \, dF(X) + M \cdot (1 - F(M))$$

$$P_{re} = L_F(\infty) - L_F(M) = \int_{M}^{\infty} 1 - F(X) \, dX, \qquad (3.4.26)$$

where F denotes the d.f. of the aggregate claim amount X_{tot}, and L_F is the corresponding limited expected value function. More generally, if only a limited layer, A xs M, of the aggregate claim amount is covered by a stop loss treaty, then the reinsurance risk premium, for example, is obtained (cf. (3.4.12)) by the formula $P_{re}(A \text{ xs } M) = L_F(M + A) - L_F(M) = P_{re,M} - P_{re,M+A}$. The higher moments of X and X_{re} can be obtained in a similar manner from the results of section 3.4.2.

(b) An example.   Suppose that the portfolio under consideration is large. If the skewness γ of the aggregate claim amount variable X_{tot} is so small that it satisfies the condition $0 \leqslant \gamma \leqslant 1$, then the so-called NP-approximation, to be presented later in section 4.2, usually provides a relatively good approximation for the tail of the aggregate claim amount distribution. (The NP-approximation is a modified normal approximation, which takes into account the skewness of the distribution to be approximated; for $\gamma = 0$ the NP-approximation reduces to the well-known normal-approximation.)

Without going into any details now we refer to the later section 4.2.4 and Exercise 4.2.1, and just give the result here. We have

$$P_{re} = \sigma \cdot \left(1 + \frac{\gamma}{6} \cdot y_M\right) \cdot \frac{1}{\sqrt{2\pi}} \cdot e^{-(y_M^2/2)} - (M - P_{tot}) \cdot (1 - N(y_M)),$$

(3.4.27)

in the case where $M > P_{tot}$, where σ and γ denote the standard deviation and skewness of the aggregate claim amount X_{tot}, M is the stop loss retention limit, and y_M is the larger root y of the equation

$$\frac{M - P_{tot}}{\sigma} = y + \frac{\gamma}{6} \cdot (y^2 - 1).$$

(3.4.28)

In the special case $\gamma = 0$, the risk premium formula (3.4.27) can be obtained (Exercise 3.4.19) by using the normal approximation.

(c) Hypersensitivity of stop loss reinsurance risk premiums.   Stop loss reinsurance risk premiums turn out to be highly sensitive to changes in the d.f. F of the aggregate claim amount X_{tot}.

Suppose that the aggregate claim amount X_{tot} is multiplied by a factor $r > 1$, because of inflation, for example. If the retention limit M is kept unchanged, then the result of section 3.4.2(d) applied to the aggregate claim amount X_{tot} shows that the reinsurance risk premium $P_{re,M}$ is increased by a greater factor $r^*(M) > r$. As was noted in section 3.4.2(d), the lighter the tail, the larger the difference between $r^*(M)$ and r. Since the skewness of the aggregate claim

Table 3.4.4   *The growth rate for the stop loss reinsurance risk premium P_{re} in the case of 10% inflation*

Skewness:	$\gamma = 0$	$\gamma = 0.2$	$\gamma = 0.5$	$\gamma = 1.0$
Growth of P_{re}:	700%	570%	440%	320%

Assumptions: $E(X_{tot}) = 100$, $\sigma_{X_{tot}} = 10$ and stop loss retention limit $M = 115$. Inflation rate factor $r = 1.1$ (as in Table 3.4.2), growth rate of $P_{re} = r^* - 1$. The approximation formula (3.4.27) was used for calculations.

amount variable is normally much less than the skewness of the size of one claim, it can be expected that the sensitivity ratio r^*/r will be very large for stop loss reinsurance risk premiums. Indeed, Table 3.4.4 shows that the reinsurance risk premium can be dramatically affected by inflation.

Exercise 3.4.19   Derive a general formula for the stop loss risk premium P_{re} for a normally distributed aggregate claim amount. The following steps may be used for the derivation:

(a) Show that the limited expected value function of the standardized normal d.f. N is

$$L_N(M) = M \cdot (1 - N(M)) - \frac{1}{\sqrt{2\pi}} e^{-M^2/2}.$$

(b) Give a formula for the limited expected value function L of the normal distribution $N(\mu, \sigma^2)$ in terms of L_N.
(c) Write the requested formula for the stop loss risk premium in terms of the L-function given above.
(d) Verify that the result is the same as (3.4.27) for $\gamma = 0$.

Exercise 3.4.20   (a) Suppose that X_{tot} is normally distributed having mean value $\mu = 100$ and standard deviation $\sigma = 10$. Calculate the reinsurer's stop loss risk premium $P_{re}(M)$ for $M = 115$.
(b) Next assume that the variable X_{tot} given in (a) is multiplied by the factor 1.1, indicating a 10% increase in claim amounts. Recalculate the stop loss premium $P_{re}(M)$ in this case keeping the stop loss limit $M = 115$ unchanged.

CHAPTER 4

Calculation of a compound claim d.f. F

Numerical calculation of the compound claim amount d.f. F using (3.1.2) is often difficult and may stretch the capabilities of available computing facilities. However, there are some important special cases where the so-called recursion method can make the computations feasible, as will be seen in section 4.1. Otherwise, approximate methods can be used to evaluate the d.f. F; these methods will be studied in section 4.2.

4.1 Recursion formula for F

(a) Assumptions. The most important family of compound aggregate claim amount distributions to which the recursion method is applicable can be specified by the following two conditions:

(1) The claim number probabilities obey the recursion formula

$$p_k = (a + b/k) \cdot p_{k-1} \quad \text{for } k = 1, 2, 3, \ldots \qquad (4.1.1)$$

where a and b are constants specifying the claim number distribution.

(2) The claim size distribution is non-negative, discrete and equidistant (i.e. a non-negative lattice distribution). More explicitly, only claim sizes

$$Z_i = i \cdot C \quad (i = 0, 1, 2, \ldots, r) \qquad (4.1.2)$$

are possible, where C is a positive constant (called the **step length**).

As stated in (2.3.7) and (2.5.9), the condition (1) is satisfied for both the Poisson and the Pólya (i.e. negative binomial) cases. The binomial distribution also satisfies this condition, but no other claim number distribution does (Exercises 4.1.3 and 4.1.8). See Table 4.1.1.

Note that a continuous distribution, and in fact any other type of distribution, can be approximated by an equidistant discrete distribution. Hence the assumption (2) need not be very restrictive. However, a good approximation using an equidistant distribution may require a large value of r, leading to a very time-consuming computation, especially when the expected number of claims is large. We will denote the claim size probabilities by

$$s_i = \text{Prob}\{Z = i \cdot C\}, \qquad (4.1.3)$$

where some s_i, $0 \leqslant i \leqslant r$, are allowed to be equal to 0. For $i < 0$ or $i > r$, $s_i = 0$. It is usually convenient to choose r as the largest i for which $s_i > 0$.

It is important to permit s_0 to be greater than zero, since discretization of a positive claim size distribution often leads to a positive s_0.

The recursion method can be further generalized, to some degree, by relaxing the conditions (1) and (2), as has been shown by Panjer (1981), Jewell and Sundt (1981) and Willmot (1986). Jewell and Sundt, for example, extend the recursion approach to a wider set of claim number distributions which satisfy the recursion condition (4.1.1) for $k \geqslant k'$, where $k' \geqslant 2$. Further details of recursion formulae can be found in Panjer and Willmot (1992).

(b) Recursion formula. With the above conditions the aggregate claim amount distribution is also discrete and equidistant, since only values that are multiples of the step length C are possible. The associated probabilities will be denoted by

$$f_j = \text{Prob}\{X = j \cdot C\} \quad (j = 0, 1, 2, \ldots) \qquad (4.1.4)$$

where X denotes the compound claim amount variable. These probabilities can be calculated recursively from the equations

$$f_j = \frac{1}{1 - a \cdot s_0} \sum_{i=1}^{\min(j,r)} \left(a + \frac{i \cdot b}{j} \right) \cdot s_i \cdot f_{j-i}, \quad j = 1, 2, \ldots, \qquad (4.1.5)$$

with the initial value

$$f_0 = \begin{cases} p_0, & \text{if } s_0 = 0, \\ \sum_{i=0}^{\infty} p_i \cdot s_0^i = M_k(\ln s_0), & \text{if } s_0 > 0, \end{cases} \qquad (4.1.6)$$

or, more explicitly,

$$f_0 = e^{n \cdot s_0 - n} \quad \text{(Poisson case)}$$

$$f_0 = \left(1 + \frac{a \cdot n \cdot (1 - s_0)}{a + b}\right)^{-\frac{a+b}{a}} \quad \text{(Pólya case)} \tag{4.17}$$

$$f_0 = \left(\frac{a - 1}{a \cdot s_0 - 1}\right)^{\frac{a+b}{a}} \quad \text{(binomial case).}$$

The proof is given in Appendix E.

The value of the corresponding compound d.f. F at any given point $X = j \cdot C, j \geqslant 0$, is then obtained from

$$F(X) = F(j \cdot C) = \sum_{i=0}^{j} f_i. \tag{4.1.8}$$

Since F is a step function, formula (4.1.8) also holds true for every X between $j \cdot C$ and $(j + 1) \cdot C$.

REMARK 1   In order to plan computations, it may be useful to know that the number of arithmetic operations needed to calculate the whole distribution (4.1.4) is of the order of $5 \cdot \alpha \cdot r \cdot n \cdot m / C$, where n is the expected number of claims, m is the mean claim size, and α is a multiplier, usually of the order 2...10, indicating up to how many multiples of the expected value $m \cdot n$ of the compound distribution the calculation has to be extended until the remaining probability mass can be regarded as insignificant.

REMARK 2   The remarks in section 2.3(c) concerning computer programming apply also to the recursion (4.1.5).

Table 4.1.1   *Parameters for different claim number distributions satisfying the recursion formula (4.1.1)*

Claim number distribution	a	b
Poisson(n)	0	n
Pólya(n, h)	$\dfrac{n}{n + h}$	$\dfrac{(h - 1) \cdot n}{n + h}$
Binomial(p, N)	$-\dfrac{p}{1 - p}$	$\dfrac{(N + 1)p}{1 - p}$
Geometric(p)	p	0

For the convenience of readers, the values of the parameters a and b for the claim size distributions which satisfy the recursion formula (4.1.1) are given in Table 4.1.1. As is well known, the binomial and geometric probabilities are

$$p_k = \binom{N}{k} \cdot p^k \cdot (1-p)^{N-k} \quad \text{(binomial)}$$

$$p_k = (1-p) \cdot p^k \quad \text{(geometric)} \tag{4.1.9}$$

and the Geometric(p) distribution is, in fact, Pólya($p/(1-p)$, 1). The recursion formula (4.1.5) was introduced by Panjer (1981), although the compound Poisson case was presented earlier by Adelson (1966).

(c) Discretization using the midpoint method. If the claim size distribution S is available in continuous form, or is a discrete distribution based on observed claims, then the first step in the recursion calculation is to discretize it into the form (4.1.3). A straightforward approach is to concentrate the probability mass of each double length interval $(0, 2 \cdot C], (2 \cdot C, 4 \cdot C], \ldots$ into the midpoints of the respective intervals. Then a discrete distribution is obtained as an approximation to the original distribution, with all the probability mass concentrated at the points

$$C, 3 \cdot C, 5 \cdot C, \ldots, r \cdot C \quad (r = 2 \cdot r' - 1), \tag{4.1.10}$$

with probabilities

$$
\begin{aligned}
s_1 &= S(2 \cdot C), \\
s_3 &= S(4 \cdot C) - S(2 \cdot C), \\
s_3 &= S(6 \cdot C) - S(4 \cdot C), \\
&\;\;\vdots \\
s_r &= 1 - S(2 \cdot (r' - 1) \cdot C), \quad (r = 2 \cdot r' - 1),
\end{aligned}
\tag{4.1.11}
$$

and $s_i = 0$ otherwise. The limit r' should be made sufficiently large that the probability mass on the right-hand side of $2 \cdot r'$ is insignificant.

A drawback of the midpoint method, as of all discretization methods, is that rounding inaccuracies arise in $F(X)$, and in its expected value and higher moments. Since the claim density is usually decreasing in the right-hand tail of the claim size d.f. S, there will be, in the tail area, more probability mass in the first half of each interval than in the second half. A systematic bias then typically arises in the right-hand tail, which is the most significant part of the range, when the moments are calculated. A straightforward way to test accuracy is to repeat the calculation using larger and larger r until increasing it further does not affect the results significantly. However, this may lead to such a large value of r that the calculations may be excessively time-consuming. A short cut is to compare the expected value (and higher moments) of the discretized distribution with the characteristics of the original one. A tolerable difference indicates that the approach is acceptable.

Another discretization method, which is introduced below, provides a better solution to the problem of rounding inaccuracies.

(d) Discretization with unchanged expectation. A natural way to approximate a given claim size d.f. S by a discrete equidistant one is to allocate the probability mass relating to the interval between two neighbouring equidistant points to these points in such a way that the expected value of the distribution remains unchanged (the **unbiased expected value** method). Note that when C is relatively large, allocation of the probability mass in the first interval, which starts at zero, may lead to s_0 being significantly different from zero.

Define the approximating lattice distribution with point probabilities s_i and step length C by the formula

$$s_i = r_i + l_i, \qquad (4.1.12)$$

where r_i is the probability mass moved to the point $i \cdot C$ from the interval $(i \cdot C, (i+1) \cdot C]$ on the right, and l_i is the probability mass moved to $i \cdot C$ from the interval $((i-1) \cdot C, i \cdot C]$ on the left.

With the condition that the discretized distribution $\{s_i\}_i$ and the original claim size distribution S should have the same expected value, the following calculation rules can be derived (Appendix E)

$$l_i = d_i - r_{i-1}, \quad r_{i-1} = d_i \cdot i - e_i, \qquad (4.1.13)$$

where

$$d_i = \text{Prob}\{(i-1) \cdot C < Z \leqslant i \cdot C\} = S(i \cdot C) - S((i-1) \cdot C),$$

$$e_i = d_i \cdot \frac{\text{E}\{Z | (i-1) \cdot C < Z \leqslant i \cdot C\}}{C} = \frac{1}{C} \int_{(i-1) \cdot C}^{i \cdot C} Z \, dS(Z). \qquad (4.1.14)$$

REMARK   It is left as an exercise (Exercise 4.1.7) to prove that the variance of the discretized distribution is always at least equal to the variance of the original distribution.

When S is discrete, for example a sample d.f. derived from the claim statistics, then d_i and e_i can be calculated as

$$d_i = \sum_{k(i)} \text{Prob}\{Z = Z_{k(i)}\}, \quad e_i = \frac{1}{C} \sum_{k(i)} \text{Prob}\{Z = Z_{k(i)}\} \cdot Z_{k(i)}, \qquad (4.1.15)$$

where the summations are over all points $Z_{k(i)}$ such that $(i-1)\cdot C < Z_{k(i)} \leqslant i\cdot C$.

EXAMPLE   Let S be the Pareto d.f.

$$S(Z) = 1 - \left(\frac{D+\beta}{Z+\beta}\right)^{\alpha}, \quad (Z \geqslant D),$$

with parameters α, β and D ($\alpha > 0$, $\alpha \neq 1$, $D > 0$, $\beta > -D$). We derive a calculation formula for e_i. By substituting $S(M)$ in (3.3.15) the limited expected value function L of S can be rewritten as

$$L(M) = \mathrm{E}(\min(Z, M)) = \frac{D+\beta}{\alpha-1}\cdot\left(1 - \left(\frac{D+\beta}{M+\beta}\right)^{\alpha-1}\right) + D$$

for $M > D$, and $L(M) = M$ for $M \leqslant D$. Instead of direct calculation of the integral in the second line of (4.1.14), we express e_1 in terms of the functions L and S. Since

$$L(M) = \int_{-\infty}^{M} Z\,\mathrm{d}S(Z) + M\cdot[1 - S(M)]$$

we have, quite generally,

$$\int_{(i-1)\cdot C}^{i\cdot C} Z\,\mathrm{d}S(Z) = L(i\cdot C) - L((i-1)\cdot C)$$

$$+ (i-1)\cdot C\cdot[1 - S((i-1)\cdot C)] - i\cdot C\cdot[1 - S(i\cdot C)]$$

$$= L(i\cdot C) - L((i-1)\cdot C) - C\cdot[1 - S((i-1)\cdot C)] + i\cdot C\cdot d_i.$$

Therefore

$$e_i = i\cdot d_i + \frac{L(i\cdot C) - L((i-1)\cdot C)}{C} - [1 - S((i-1)\cdot C)].$$

Exercise 4.1.1   Calculate the compound Pólya d.f. $F(X)$ for $X \leqslant 6$, when the expected number of claims is $n = 2$, $\sigma_q^2 = 0.1$ and the claim size d.f. S is the two point discrete distribution $s_1 = \mathrm{Prob}\{Z = 1\} = 0.2$, $s_2 = \mathrm{Prob}\{Z = 2\} = 0.8$.

Exercise 4.1.2   Let S be the Pareto$(\alpha, 0, D)$ d.f. with $\alpha = 2$ and $D = 0.5$, truncated at $Z_{\max} = 10$. Discretize S by using (a) the midpoint method, (b) the unbiased expected value method. The step length is $C = 1$. Compare the expected values of the discretized distributions.

Exercise 4.1.3   Show that the binomial distribution satisfies the recursion formula (4.1.1) with $a = -p/(1-p)$ and $b = -a\cdot(N+1)$.

Exercise 4.1.4 Let S be a claim size d.f. Define equidistant d.f.s L (L for left) and R (R for right) with step length C as follows:

$$R(Z) = S((i-1)\cdot C) \quad \text{for } (i-1)\cdot C \leqslant Z < i\cdot C,$$
$$L(Z) = S(i\cdot C) \quad \text{for } (i-1)\cdot C \leqslant Z < i\cdot C.$$

Show that, if Z_L and Z_R are, respectively, L and R distributed random variables then $E(Z_R) - E(Z_L) = C$.

Exercise 4.1.5 Let S, L and R be as in the previous exercise. Let T be the equidistant d.f. with step length $\frac{1}{2}\cdot C$ obtained by using the midpoint method, and let U be the equidistant d.f. (4.1.12) obtained by using the unbiased expected value method. Let F_S, F_L, F_R, F_T and F_U be the corresponding compound d.f.s, each having the same claim number distribution.

(a) In what circumstances does $F_R = F_S$ hold true, and when does $F_T = F_S$?
(b) Show that $F_R \leqslant F_S \leqslant F_L$, $F_R \leqslant F_U \leqslant F_L$, and that, if $S(0) = 0$ and $S(r\cdot C) = 1$, then $F_R \leqslant F_T \leqslant F_L$.

Exercise 4.1.6 Let X and Y be two random variables such that $E(X) = E(Y)$, $m \leqslant X \leqslant M$ and $\text{Prob}\{Y = m \text{ or } Y = M\} = 1$. Prove that $\text{Var}(X) \leqslant \text{Var}(Y)$.

Exercise 4.1.7 Show that the variance of a discretized distribution (4.1.12), obtained by using the unbiased expected value method, is greater than or equal to the variance of the original distribution. (Hint: Use the result of Exercise 4.1.6.)

*Exercise 4.1.8 Show that the Poisson, negative binomial (i.e. Pólya), and binomial distributions are the only claim number distributions which satisfy the recursion formula (4.1.1).

4.2 Approximate formulae for F

4.2.1 Introduction to approximate formulae

(a) Need for approximation. It will be apparent from the foregoing sections that the compound mixed Poisson d.f. F, which gives the distribution of aggregate claim amounts, is complicated to compute.

The recursion method, described in section 4.1, has made a major contribution towards solving this computational problem. Nevertheless, there is still a need for quick and reasonably accurate approximate methods. A shortcoming of the recursion method is that an enormous number of computation steps are required when the risk portfolio is large (as most insurers' portfolios are) or when the claim size distribution has a long tail. This causes serious problems, especially in the simulation of a claim process, where the computation of F is needed repeatedly, often thousands or tens of thousands of times. For this purpose the computation time needs to be of the order of milliseconds rather than seconds.

Approximate methods can also have the merit of suggesting analytic, and sometimes enlightening, relationships between the main variables which control the process.

(b) Symmetrization and moment based methods. A classical approach is to approximate the d.f. F by the normal d.f. with the same mean value and standard deviation. By the central limit theorem of probability calculus the d.f. F is asymptotically normal in the compound (non-mixed) Poisson case (section 2.3(e)). Unfortunately, its area of application is rather narrow, as will be discussed in section 4.2.2, since the d.f. F is usually skew, while the normal d.f. is not.

A family of approximating d.f.s is obtained by applying a suitable symmetrization transformation to the aggregate claim amount variable in order to eliminate the skewness of the d.f. F, and then applying the normal approximation to this transformed distribution. This same idea was, in fact, applied when the Poisson d.f. was approximated by the Anscombe and the Peizer and Pratt approximations (sections 2.3(f) and (g)). When such approximations are applied for the d.f. F, then, in addition to the mean value and standard deviation of F, the skewness of F is also used as a determinant. However, these kinds of approximations are applicable only in cases where the skewness is not large, as will be seen in later sections. The so-called **Normal Power** (briefly NP) method, the **Haldane** method and the **Wilson–Hilferty** formula belong to this group of approximations and will be dealt with in sections 4.2.4 and 4.2.5. Because the parameterization of the approximating functions is eventually based on equating their basic characteristics, i.e. lowest moments, with the corresponding characteristics of the distributions to be approximated, symmetrization is a special variant of this moment method.

Another variant of the moment method is to approximate directly the d.f. *F* by a d.f. chosen from some well-known family of distributions matching the lowest characteristics, usually the mean, standard deviation and skewness. A necessary condition is that the approximating d.f. should tend to the normal d.f. as the skewness approaches zero. This method is also applicable only for relatively small values of skewness. The shifted gamma distribution introduced in section 3.3.5 is an example. It is adapted to this purpose by replacing the characteristics of the claim size variable *Z* in section 3.3.5 by the corresponding characteristics of the aggregate claim variable *X*. The gamma approximation is used by many authors, because it offers the possibility of handling *F* analytically and its properties are well-known and readily available from standard text-books. The suitability of the gamma approximation was investigated by Bohman and Esscher (1964) and was strongly advocated by Seal (1977), although slightly criticized by Pentikäinen (1977). Tests have shown that it is applicable in similar circumstances to the NP-approximation and the Haldane approximation (sections 4.2.4 and 4.2.5).

REMARK Most of the approximate methods are based on the lowest moments (mean, standard deviation and skewness) of the target distribution. Then, self-evidently, the same approximation will be obtained for a whole set of d.f.s with these characteristics equal, but with kurtosis and higher characteristics differing. This provokes the question of how closely the lowest three moments determine the functions belonging to this set. This so-called **moment problem** is well studied in the literature. It is proved that, if no particular restrictive conditions, such as outlined above, are posed, the distributions can be quite far from each other and the standard solutions of the moment problem are not helpful for our purposes (Goevaerts *et al.*, 1984).

(c) Other methods. It is worth mentioning that the d.f. *F* can also be computed using the **Fast Fourier Transformation** (Bohman and Esscher, 1964; Bertram, 1981; and Meyers and Schenker, 1983). This method applies also for more skew distributions.

4.2.2 Normal approximation

(a) Formula. It is common, both in risk theory and elsewhere, to approximate the d.f. *F* of the aggregate claim amount by the normal d.f., as was done in the context of the claim number distribution (section 2.3(e)).

Thus we have

$$F(X) \approx N(v(X)) \qquad (4.2.1)$$

where N is the normal d.f. (see (2.3.9)) and

$$v(X) = (X - \mu_X)/\sigma_X \qquad (4.2.2)$$

standardizes the variable to have zero mean and standard deviation unity. Formulae for the numerical evaluation of N and its inverse are given in Appendix D.

REMARK   Formula (4.2.1) is, for the sake of uniformity, formally shaped in line with the Anscombe and the Peizer and Pratt expressions (2.3.12) and (2.3.13) as well as with the symmetrization formulae of later sections, all of which have N as the outer function and a suitably designed transformation as the inner function.

(b) Discussion about applicability. The normal approximation simplifies the calculations and often makes it possible to analyse problems involving numerous variables and interrelationships in a way which is not otherwise feasible or can be done only with considerable difficulty. For example, the equation

$$\text{Prob}\{|X - \mu_X| \leqslant y_{\varepsilon/2} \cdot \sigma_X\} = \varepsilon \qquad (4.2.3a)$$

gives an estimate of the probability that the standardized aggregate claim amount deviates from its mean by at most $y_{\varepsilon/2}$ times the standard deviation. $y_{\varepsilon/2}$ is the root of the equation $\varepsilon = (1 - N(y_{\varepsilon/2})) + N(-y_{\varepsilon/2}) = 2 \cdot N(-y_{\varepsilon/2})$. More often used in risk theory applications is the one-sided limit equation

$$\text{Prob}\{X \leqslant \mu_X + y_\varepsilon \cdot \sigma_X\} = \varepsilon, \qquad (4.2.3b)$$

which evaluates the probability of excessive claims, y_ε being the root of $N(y_\varepsilon) = 1 - \varepsilon$.

Table 4.2.1   *Approximate solutions for y_ε and $y_{\varepsilon/2}$*

ε	0.05	0.01	0.005	0.001
y_ε	1.64	2.33	2.58	3.09
$y_{\varepsilon/2}$	1.96	2.58	2.81	3.29

Numerical values can found from text-books on probability calculus. For the convenience of readers some common values are given in Table 4.2.1.

Unfortunately, the normal approximation is not usually sufficiently accurate. It is acceptable only if the skewness of X is very small, as will be shown by some examples in section 4.2.6.

4.2.3 Symmetrization

One way to improve the accuracy of the normal approximation is to transform the original variable X into an auxiliary variable y,

$$y = v(X), \tag{4.2.4}$$

by means of a suitably chosen transformation v which makes the distribution approximately symmetrical. In practice this is most often done by imposing in the specification for v the condition that the skewness of the transformed distribution $y = v(X)$ should be equal to zero. Then, further specifying v so as to standardize the mean of y to zero and the standard deviation to unity, it could be expected that the d.f. $\bar{F}$ of the symmetrized variable $y = v(X)$ can be approximated more satisfactorily by the normal distribution:

$$F(X) = \bar{F}(y) \approx N(y). \tag{4.2.5}$$

Different transformation functions v yield a variety of approximations. Two possibilities will be considered in the next two sections. First, in section 4.2.4, the inverse of v is taken as a second degree polynomial, resulting in the so-called NP-formula. Secondly, a general power function, suggested by Haldane, is introduced in section 4.2.5.

A general analysis of the symmetrization approach can be found in Box and Cox (1964).

4.2.4 NP approximation

(a) NP-formula. Following the idea outlined in the preceding section, and choosing the inverse $v^{-1}(y)$ of the transformation function v to be a second degree polynomial, the following formula is obtained

$$x = \frac{X - \mu_X}{\sigma_X} = y + \frac{\gamma_X}{6}(y^2 - 1) \quad (X > \mu_X). \tag{4.2.6}$$

Solving $y = v(X)$ and substituting into (4.2.5) we have

$$F(X) \approx N(v(X)) = N\left(-\frac{3}{\gamma_X} + \sqrt{\frac{9}{\gamma_X^2} + 1 + \frac{6}{\gamma_X} \cdot \frac{X - \mu_X}{\sigma_X}} \right). \qquad (4.2.7)$$

(b) Applicability. The formulae (4.2.6) and (4.2.7) are valid only for the right-hand tail of the distribution and so far as the skewness γ_X does not exceed unity. The former offers a very useful extension of the limits given by (4.2.3) in relation to the normal approximation, for the variation range of the claims. It also takes into account the skewness of the distribution.

The formula (4.2.7) does not have any noticeable advantages compared with the Wilson–Hilferty formula (to be presented in section 4.2.5), particularly since the latter covers the whole range of X, including negative values.

(c) References. The above approximation was found by Kauppi and Ojantakanen (1969) by comparing the outcomes from computing values in parallel by different approximate methods. The formula proved to be a special case of a similar transformation presented by Cornish and Fisher (1937). A good account of the idea is given in Kendall and Stuart (1977), paragraphs 6.25–27.

REMARK   The idea of symmetrization can be extended by introducing further free parameters into the transformation v. For example, it is possible to find a transformed function $\bar{F}$ which has the same kurtosis as the normal d.f. Then the formula can be provided with an additional term controlled by the kurtosis of the original distribution (Beard *et al.*, 1984, p. 117). Tests have shown that this can improve the fit but it works best in cases where the short formula itself gives satisfactory results.

Exercise 4.2.1   Derive the approximate formula (3.4.27) for the stop loss premium.

4.2.5   Wilson–Hilferty formula

(a) Haldane approach. Haldane (1938) constructed the symmetrization referred to in section 4.2.3 by introducing a power expression

$$Y = (X/\mu_X)^h \qquad (4.2.8)$$

where h is an auxiliary parameter to be specified by the condition that the skewness of the transformed variable has to be equated to zero. Further, calculating the mean μ_Y and standard deviation σ_Y for the variable Y and putting

$$y = (Y - \mu_Y)/\sigma_Y = [(X/\mu_X)^h - \mu_Y]/\sigma_Y \qquad (4.2.9)$$

the desired transformation $y = v(X)$ (4.2.4) is obtained.

The derivation of the Haldane formulae is elementary but rather laborious (Pentikäinen, 1987).

(b) Wilson–Hilferty formula. Wilson and Hilferty (1931) employed the same method as Haldane to derive a formula for a numerical approximation of the gamma d.f. It can be obtained as a special case of Haldane's general formula with the parameter h assuming the value $1/3$. Since the gamma d.f. can approximate the d.f. F of the aggregate claim amount, as suggested in section 4.2.1(b), the Wilson–Hilferty (briefly WH) formula suggests itself as an approximation for F.

Examples of both the Haldane and WH approximations will be given in section 4.2.6. Further testing can be found in Pentikäinen (1987). Owing to its greater flexibility, the general Haldane formula is slightly better, but the differences are not significant in the area which is of most practical interest. Having regard to the much simpler structure of the WH formula, therefore, we have adopted it in the remainder of this book as the standard to be employed in applications.

The Wilson–Hilferty formula is given by

$$F(X) \approx N(y) = N(W(x)), \qquad (4.2.10)$$

where x is the standardized variable $x = (X - \mu_X)/\sigma_X$, N is the standard normal d.f. (1.4.22a) and the WH-transformation is

$$W(x) = c_1 + c_2 \cdot (x + c_3)^{1/3}$$

$$c_1 = \frac{1}{3g} - 3g, \quad c_2 = 3g^{2/3}, \quad c_3 = g, \quad g = \frac{2}{\gamma_X}. \qquad (4.2.11)$$

In many problems the value $F(X)$ is given and the corresponding X is required. Solving the above equations, the following calculation steps ensue

(1) Solve y from $N(y) = F(X)$ (see Appendix D)

(2) Set $x = \left(\dfrac{1}{c_2}\right)^3 \cdot (y - c_1)^3 - c_3$ \hfill (4.2.12)

(3) Set $X = \mu_X + x \cdot \sigma_X$.

The derivative of the right-hand side of (4.2.10) is

$$\frac{dN(W(x))}{dx} = \frac{c_2}{\sigma_X \sqrt{18\pi}} \cdot (x + c_3)^{-2/3} \cdot \exp[c_1 + c_2(x + c_3)^{1/3}] \quad (4.2.13)$$

where x is again the standardized variable.

(c) **Applicability.** The comparisons to be considered in section 4.2.6 show that, for slightly skewed distributions, the WH-approximation, as well as both the NP and Haldane formulae, give fairly accurate results. However, when the skewness is large the applicability deteriorates rapidly. If the skewness γ_X exceeds unity or at most the value 1.2, all these methods become unreliable and should not be used.

Another testing method is obtained by calculating numerically the basic characteristics μ, σ, γ and γ_2 from the approximating d.f. $N(v(X))$ and comparing them with corresponding original quantities μ_X, σ_X, γ_X and $\gamma_{2,X}$. It proves that the lowest three characteristics are very close to the originals so long as

$$0 \leqslant \gamma_X \leqslant 1.2, \quad (4.2.14)$$

whereas for more skew distributions even the mean and standard deviations no longer match, indicating that the formulae are not acceptable. Further, the kurtosis γ_2 of the Wilson–Hilferty distribution approximately satisfies the equation

$$\gamma_2 \approx 1.6 \cdot \gamma^2 \quad (4.2.15)$$

when the condition (4.2.14) is satisfied. This can be used to test the applicability of this approximation. If the kurtosis of an aggregate claim amount d.f. F deviates significantly from the kurtosis (4.2.15) of the corresponding Wilson–Hilferty distribution $N(W(x))$, then the fit of the approximation cannot be expected to be good. Typically, if the deviation from (4.2.15) is less than 0.2, the fit is usually very good.

4.2.6   Comparison of approximations

(a) Applicability.   In order to gain insight into the appropriateness of the approximation methods, numerous distributions have been computed using exact methods, mainly by recursion, with the various approximations examined in parallel. A large collection was published by Pentikäinen (1987) consisting of 54 distributions of various sorts. Table 4.2.2 exhibits some typical outcomes.

If an approximation method can be shown consistently to produce acceptable accuracy in numerous relevant areas of application, then its use may be justified for practical calculations. A useful feature has emerged from these tests, as was demonstrated already in section 4.2.5(c): the accuracy is, as a rule, good, or at least satisfactory, in cases where the skewness of the distribution is small. Other provisos are that the number of claims should not be too small and the distribution of the size of one claim not very heterogeneous, i.e. not having a very long tail. The latter condition is usually well satisfied when the peak risks are reinsured.

REMARK   It should be appreciated that the basic data are frequently rather limited, making it spurious to demand excessive accuracy from the calculation technique, particularly if the consequence is a significant increase in computation time.

(b) Conclusions.   Tests so far carried out indicate that there are no very significant differences in accuracy between the NP, WH and Haldane approaches. Hence, in cases where the above outlined criteria of applicability apply, the choice between these formulae should be made on the basis of practical convenience. This points to the WH formula. However, the analytic expression (4.2.6) obtained by the NP method is often very useful, the other approaches having no corresponding feature.

As expected, the normal approximation is rather unsatisfactory, except where the skewness is very small.

Finally, let us note that the approximate and exact methods (and direct simulation, which will be dealt with in section 5.4(a)) often complement each other well. The former are best for large collectives,

Table 4.2.2 Examples of approximations according to Pentikäinen (1987)

Distribution	x	Exact F or 1 − F	Approximations N	NP	WH	HA	Error % ΔN	ΔNP	ΔWH	ΔHA
Example 1	−2.0	0.0205	0.0228	—	0.0205	0.0205	11.0	—	0.1	−0.0
$\gamma x = 0.08$	−1.5	0.0646	0.0668	—	0.0645	0.0646	3.4	—	−0.0	−0.0
	−1.0	0.1586	0.1587	—	0.1586	0.1587	0.0	—	−0.0	0.0
	1.0	0.1586	0.1587	0.1587	0.1586	0.1587	0.0	0.0	−0.0	0.0
	2.0	0.0249	0.0228	0.0249	0.0249	0.0249	−8.6	0.1	0.0	−0.0
	3.0	0.0019	0.0013	0.0019	0.0019	0.0019	−28.9	0.4	0.5	−0.1
	4.0	0.0001	0.0000	0.0001	0.0001	0.0001	−68.3	1.0	1.7	−0.5
Example 2	−2.0	0.0162	0.0228	—	0.0160	0.0160	40.4	—	−1.1	−1.4
$\gamma x = 0.24$	−1.5	0.0595	0.0668	—	0.0594	0.0594	12.3	—	−0.2	−0.2
	−1.0	0.1578	0.1587	—	0.1580	0.1581	0.5	—	0.1	0.2
	1.0	0.1579	0.1587	0.1587	0.1581	0.1582	0.5	0.4	0.1	0.1
	2.0	0.0288	0.0228	0.0289	0.0288	0.0288	−21.0	0.4	−0.1	−0.1
	3.0	0.0031	0.0013	0.0031	0.0031	0.0030	−56.5	−2.1	−1.7	−2.3
	4.0	0.0002	0.0000	0.0002	0.0002	0.0002	−84.2	−8.2	−5.1	−7.2

Example 3 $\gamma_x = 0.59$									
−2.0	0.0078	0.0228	—	0.0059	0.0067	191.7	—	−25.1	−14.4
−1.5	0.0456	0.0668	—	0.0443	0.0447	46.5	—	−2.8	−2.0
−1.0	0.1506	0.1587	—	0.1538	0.1522	5.3	—	2.1	1.1
1.0	0.1529	0.1587	0.1587	0.1554	0.1541	3.8	3.8	1.6	0.8
2.0	0.0362	0.0228	0.0372	0.0361	0.0359	−37.2	2.8	−0.1	−0.6
3.0	0.0068	0.0013	0.0064	0.0064	0.0066	−80.1	−4.9	−5.4	−2.1
4.0	0.0011	0.0000	0.0009	0.0009	0.0010	−97.1	−17.4	−12.2	−0.8
Example 4 $\gamma_x = 1.08$									
−2.0	0.0033	0.0228	—	0.0000	0.0006	589.4	—	−100.0	−83.1
−1.5	0.0322	0.0668	—	0.0145	0.0209	107.5	—	−55.2	−35.1
−1.0	0.1357	0.1587	—	0.1384	0.1307	16.9	—	2.0	−3.7
1.0	0.1376	0.1587	0.1587	0.1491	0.1419	15.3	15.3	8.4	3.1
2.0	0.0376	0.0228	0.0470	0.0431	0.0407	−39.5	25.1	14.7	8.3
3.0	0.0125	0.0013	0.0119	0.0112	0.0116	−89.2	−4.9	−10.4	−7.3
4.0	0.0042	0.0000	0.0027	0.0028	0.0035	−99.2	−35.5	−34.0	−16.5

Notes: x is the standardized variable (4.2.2), F or, for $x > 1$ $(1 - F)$, and is calculated by the recursive formula. The columns N, NP, WH and HA give the approximate values obtained by the Normal, NP, Wilson–Hilferty and Haldane formulae. The claim number distribution was either Poisson or Pólya and the claim size distributions were truncated Pareto or log-normal.

which typically have moderate skewness, whereas the exact methods can be used for small collectives, which are often so skew that approximations are of limited value.

Exercise 4.2.2   Show that the skewness γ and kurtosis γ_2 of a shifted gamma d.f. satisfy (cf. (4.2.15)) the equality

$$\gamma_2 = \tfrac{3}{2}\gamma^2.$$

CHAPTER 5

Simulation

5.1 Introductory remarks

(a) The idea of simulation. The numerical methods presented in Chapter 4 have shown how the evaluation of a claim amount distribution can get quite complicated, even with restrictive assumptions. More serious problems arise when there are correlations and interrelationships between the variables. The problem becomes even more complex, in fact often intractable by conventional methods, when the model is extended to incorporate additional elements such as several years or lines of business, inflation, return on investments, dynamic control, etc. The method of simulation, which will be presented in this chapter, offers a flexible and powerful means of coping with even the most complicated model specifications.

The basic idea of simulation is straightforward; instead of solving the problem analytically or by means of conventional numerical methods, the event of concern is imitated by breaking it down into a chain of primary events, each of which is relatively easy to handle. A random number generator is used to generate each variable required for a particular realization and the whole process is repeated many times.

For example, the distribution of the aggregate claims of a mixed compound Poisson variable can be obtained by first generating a random number to represent the value q for the mixing variable and thereafter a value k for the number of claims, subject to the Poisson parameter nq. Finally, k individual claims are generated and summed to get a value for the aggregate claim amount. The procedure is repeated numerous times to arrive at a sample of simulated aggregate claims, from which, for instance, the compound d.f. F can be estimated using conventional statistical methods, just as if statistical data of a large number of real observed claims were available. In more advanced applications, effects such as those brought about by inflation, business cycles, dynamic control etc. can be incorporated into the process as auxiliary steps without any overwhelming complications, and the process can be continued over several successive periods (accounting years), with the outcome at the end of each period determining the starting point for events in the next period.

A technical tool which is required to produce the many random numbers needed in practical applications is a random number generator, i.e. a computer algorithm which produces sequences of so-called **pseudo-random numbers**. These are numbers which are distributed so that they follow a given distribution to a sufficient degree of accuracy.

Simulation is a frequently used technique in operational research, and an abundant literature is available, including, for example, Rubinstein (1981). We will pick up those items which are appropriate for insurance applications. The presentation has been made self-contained for the convenience of readers, also because the necessary definitions and notation can thus be specified and, in addition, some important modifications made.

The primary building blocks for simulation are given in sections 5.2 to 5.5. Their use in constructing more advanced models is demonstrated in section 5.6, where the merits and demerits of simulation are also discussed. Some details are deferred to Appendix F.

(b) Terminology. The term **simulation** is used in the literature with a variety of meanings. Stochastic elements are often included, but not always. We will employ only stochastic approaches, mostly of the type which are referred to in the general literature as **Monte Carlo methods** (section 5.2). For brevity, we shall speak about simulation, notwithstanding that the term is used more narrowly than might sometimes be the case.

5.2   Random numbers

(a) Sequences of random numbers. The simplest situation where simulation can be employed can be described as follows. Let F denote the d.f. of a given distribution. As a first step, in most applications a sequence

$$X_1, X_2, ..., X_s \qquad (5.2.1)$$

of random numbers is generated, constructed so that they have a common d.f. F and are mutually independent. The process is similar to that involved in a lottery. For example, in the case of a discrete distribution the process is organized so that the probability that a certain value is drawn is the same probability as is assigned to this value according to the given distribution.

(b) Uniformly distributed random numbers are usually a starting point for most simulations. In most applications the range of the admitted values r is limited to the unit interval $(0, 1)$. Applying the above lottery example, the permitted numbers are the decimal numbers of the interval $(0, 1)$ specified to n digits; hence each of them has the probability $1/(10^n - 1)$ of being drawn, if the end points 0 and 1 are excluded, as is often advisable in order to avoid technical problems arising occasionally from these special values. If n is large enough (e.g. $n \geqslant 10$), this discrete uniform distribution can, in practice, be regarded as equal to the corresponding continuous **uniform** distribution having the d.f.

$$R(r) = \begin{cases} 0 & \text{for} \quad r \leqslant 0 \\ r & \text{for} \quad 0 < r < 1 \\ 1 & \text{for} \quad r \geqslant 1. \end{cases} \quad (5.2.2)$$

The distribution is called rectangular, because of the shape of its density function.

Ready-made modules for generating uniformly distributed (pseudo-)random numbers are often included with computers as a standard item of software. A frequently used algorithm is presented in Appendix F, together with some comments.

(c) Random numbers distributed according to a given d.f. F can readily be obtained by first generating a uniformly distributed random number r and then transforming it by the inverse of F

$$X = F^{-1}(r). \quad (5.2.3)$$

Then X is F-distributed, as is illustrated in Figure 5.2.1 (Exercise 5.2.1). If the function F is not one-to-one, then define $F^{-1}(r) = \min\{X : F(X) \geqslant r\}$.

EXAMPLE   Random numbers

$$X = -(1/a)\ln(r) \quad (r \text{ uniform } (0, 1) \text{ distributed}) \quad (5.2.4)$$

have the exponential d.f. (Exercise 5.2.2)

$$F(X) = 1 - e^{-aX} \quad (X \geqslant 0). \quad (5.2.5)$$

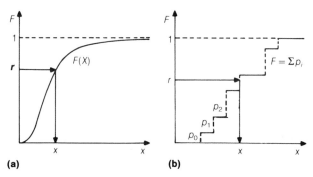

Figure 5.2.1 *Generation of random numbers* X *distributed according to* F; *(a) continuous case, (b) discrete case.*

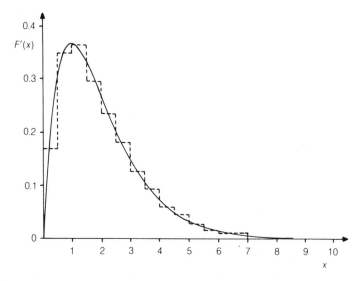

Figure 5.2.2 *Exact and simulated values of the Gamma (2,1) d.f. Sample size 2000.*

The simulation procedure can also be organized so that a numerical estimate is obtained for the density and the d.f. An example is given in Figure 5.2.2. In order to demonstrate the features of the analytical and simulation approaches, they are applied in parallel (notwithstanding that, in as simple a case as is presented here, simulation does not offer any advantage). The simulated X-values are grouped into classes in order to estimate density.

The diagram provides an overall insight into the shape of the target function. On the other hand, the drawback of the simulation approach can also be seen; the outcome is a sample of the true distribution and is, therefore, only an approximation. The accuracy can be evaluated, as will be shown in Appendix F, section F4, and the fit can be improved by increasing the sample size and using shorter class intervals.

(d) Normally distributed random numbers r with zero mean and unit standard deviation, are also important for many applications. Standard programs are available to generate them. The handy **log-and-trig-formula** is set out in Appendix F. It will be used in all of the applications in this book.

Normally distributed random numbers with mean μ_X and standard deviation σ_X are then readily obtained

$$X = \mu_X + \sigma_X r. \tag{5.2.6}$$

Exercise 5.2.1 Prove that X defined by (5.2.3) has F as d.f.

Exercise 5.2.2 Prove (5.2.5).

Exercise 5.2.3 Construct random number generators which produce numbers distributed according to the following distributions: (a) log-normal (3.3.10); and (b) Pareto (3.3.14).

5.3 Simulation of claim numbers

(a) Straightforward simulation. A direct application of section 5.2(c) gives a procedure to generate sequences of claim numbers related either to the Poisson case, the Pólya case or to a general mixed distribution.

The necessary Poisson, Pólya or other relevant parameters n, h, etc. are assumed to be given, so that the d.f. F of the claim number variable is specified ((2.3.8) and (2.5.5)).

The simulation consists of the following two steps:

(1) Generate a uniform $(0, 1)$ random number r.
(2) Find (by calculation or from ready-made tables) the smallest k

which satisfies the inequality

$$F(k) \geqslant r \qquad (5.3.1)$$

and take k as the required random number.

Repeat steps (1) and (2) s times in order to get a sample of size s. This straightforward approach works well so long as a moderate number of claims n is expected. If n is large, then approximate calculation is useful, as has already been considered in the context of claim number distributions in section 2.3. Such an approach is described in the paragraphs which follow.

(b) Simulation of Poisson random numbers. This approach is based on the inverse of the Anscombe transformation (2.3.12), i.e. solving k from

$$F(k) = N(A_n(k)). \qquad (5.3.2)$$

The simulation consists of the steps

(1) Generate an N-distributed random number r using the log-and-trig formula (Appendix F).
(2) Calculate k from

$$k = A_n^{-1}(r) = a \cdot (r + b)^{3/2} - c$$

where $\qquad\qquad\qquad\qquad\qquad\qquad\qquad\qquad (5.3.3)$

$$a = \sqrt{\frac{8}{27}} \cdot \sqrt{n}; \quad b = \frac{3}{2}\sqrt{n} - \frac{1}{24\sqrt{n}} \quad \text{and} \quad c = \frac{5}{8}.$$

Then k is (approximately) Poisson-distributed.

Repeat the steps (1) and (2) s times in order to get a sample of size s.

If a computer program is being constructed to cover collectives of all sizes, including very small ones, it is advisable to combine the approaches of sections 5.3(a, b). A limit n_0 should be stipulated to control which calculation is utilized. If $n < n_0$, then the straightforward approach is used; otherwise the inverse Anscombe. Having regard to Table 2.3.1, the limit can be quite low, say $n_0 = 10$, depending on the accuracy desired.

(c) Simulation of mixed Poisson claim numbers can be achieved by adding to the procedure of section 5.3(b) the generation of the mixing variable q. Hence

(1) Generate q distributed according to the mixing d.f. H (section 2.4(b)).
(2) Generate a Poisson($n \cdot q$) random number as shown in section 5.3(b) above replacing n by $n \cdot q$ in (5.3.3).

Note that, e.g. in the Pólya case, the WH-approach can usually be used to generate q, as will be shown in the fourth paragraph of section 5.4(c).

5.4 Simulation of compound variables

The simulation will now be extended to compound distributions, incorporating claim size into the procedure.

(a) Straightforward simulation. Consider first the simulation of random numbers which represent the aggregate claim amount being distributed according to the general compound distribution (3.1.4).

$$F(X) = \sum_{k=0}^{\infty} p_k S^{k^*}(X). \tag{5.4.1}$$

Assuming that the claim number probabilities p_k and the claim size d.f. S are known, the following steps are required

(1) Generate the number of claims k using the methods of section 5.3.
(2) Generate k numbers $Z_1, Z_2, ..., Z_k$ which obey the claim size d.f. S using the methods described in section 5.2(c).
(3) The sum
$$X = Z_1 + Z_2 + \cdots + Z_k \tag{5.4.2}$$
gives the required compound random number X.

If a sample of size s is needed, repeat steps (1)–(3) s times. If an estimate for the d.f. F aggregate claim amount is also required, it can be achieved by organizing the outputs appropriately. The discrete estimate of Figure 5.2.2 offers one way to proceed. More accurate values for the required characteristics and distributions can be obtained by storing all the X values and computing the relevant quantities therefrom.

The straightforward method is applicable only in the (rare) cases where the number of claims is fairly small and the claim size d.f. is easy to handle. For large collectives, and in the case of long-tailed claim sizes defined in the tabular form of section 3.3.3, the computation time tends to become rather long, suggesting the need for approximate short-cut methods. Such an approach will be given in section

5.4(b) for the compound Poisson distribution, with an extended version for the compound mixed Poisson distribution in section 5.4(c). Unfortunately, these approximate methods are not applicable if the aggregate claim amount distribution is skew enough. If this is the case, then the method introduced in Pesonen (1989) can be applied, which generates compound random numbers so that the processing time is independent of the expected number of claims.

(b) WH-generator for compound Poisson random numbers. Just as the Anscombe formula can be utilized for the simulation of claim numbers (section 5.3), the Wilson–Hilferty formula (section 4.2.5), being of the form

$$F(X) = N(W(x)) \tag{5.4.3}$$

offers a convenient method for generating approximately compound Poisson distributed random numbers. Here N is again the standard normal d.f. (1.4.23), the expression W was given in (4.2.11), and x is the standardized variable (4.2.2).

The primary characteristics of X, the mean μ_X, standard deviation σ_X and skewness γ_X, are needed as initial parameters for the computations. They are supplied by the user of the model, possibly using the formula (3.2.13) (or formulae (3.2.15) and (3.2.16), setting $\sigma_q = 0$).

The simulation consists of the following steps

(1) Generate an N-distributed random number r using the log-and-trig formula (Appendix F).
(2) Solving x from $r = W(x)$, i.e.

$$x = b_1 \cdot (r - b_2)^3 - b_3 \tag{5.4.4}$$

where (see (4.2.11))

$$b_1 = \frac{\gamma_X^2}{108}; \quad b_2 = \frac{\gamma_X}{6} - \frac{6}{\gamma_X}; \quad b_3 = \frac{2}{\gamma_X}. \tag{5.4.5}$$

(3) The required simulated aggregate claim amount is

$$X = \mu_X + x \cdot \sigma_X. \tag{5.4.6}$$

If a sample of size s is required, repeat the steps (1)–(3) s times. If the d.f. F is also required, it can be estimated in the way outlined in the penultimate paragraph of section 5.4(a).

REMARK The WH-generator is a handy tool for simulating other distributions which can be fitted by the gamma d.f., apart from the compound Poisson. The proviso is that the skewness γ_X should not be large, preferably < 0.5 or at most equal to 1 (section 4.2.5(c)). If this condition is not fulfilled, then the straightforward method of section 5.4(a) or other approaches should be used.

It will be shown in the next section how the mixed case can be dealt with. It is convenient to program a module (subprogram) for the WH-generator and to use it for all these various purposes by choosing the input parameters according to the application.

(c) WH-generator for compound mixed Poisson random numbers.
The WH-generator of section 5.4(b) can readily be extended to the mixed case by first generating the mixing variable q. Then the simulation will consist of the following steps

(1) Generate the mixing variable q, as in section 5.2(c).
(2) Calculate the characteristics μ_X, σ_X and γ_X which correspond to the particular value q obtained by replacing n with $n \cdot q$ in (3.2.13).
(3) Apply the WH-generator of section 5.4(b).

In the Pólya case steps (1) and (2) can be replaced by calculating the characteristics directly from (3.2.11) and (3.2.17). Then it is not necessary to simulate q.

The mixing variable q can also be generated (step (1)) using the WH-generator, as was mentioned in the remark of section 5.4(b). For this purpose the input parameters μ_X, σ_X and γ_X in (5.4.5) and (5.4.6) are replaced by the corresponding quantities of q, i.e. $\mu_q = 1$, σ_q and γ_q. The outcome, which was denoted by X in (5.4.6), can then be used as the required q. This approach is particularly suitable in cases where the d.f. is not readily available, but the characteristics can be directly evaluated, for example from actual data (cf. the technique (3) of section 3.3.1(b)).

Note that in long-term simulations q can be simulated by means of time series or derived by other methods to be dealt with in sections 9.2(d, e).

For the convenience of readers, the transformations required for step (2), leading from characteristics for which $q = 1$ to those corresponding to the simulated value of q, are as follows:

$$\mu_X = n \cdot q \cdot m$$
$$\sigma_X = \sqrt{n \cdot q \cdot a_2} \qquad (5.4.7)$$
$$\gamma_X = a_3 / \sqrt{n \cdot q \cdot a_2^3}$$

(d) Conditional WH-generator. In some applications it is necessary to simulate, for each realization, both the claim number k and the related aggregate claim amount X. For example, when the survival of life insurance cohorts is simulated, both the number of deceased cohort members and the associated claim amount are needed (sections 15.2(a, e)). In the straightforward simulation of section 5.4(a), the quantities k and X are both available from the simulation steps. In the case of the WH-generator of section 5.4(b), modifications are necessary as follows

(1) First simulate a value k for the claim number k (section 5.3).
(2) Then simulate a value X for the sum

$$X(k) = Z_1 + Z_2 + \cdots + Z_k \tag{5.4.8}$$

This can be done either by simulating the Z's one by one (as in section 5.4(a)) or, if the skewness of $X(k)$ is not larger than 1, by using the same WH-generator as was presented in section 5.4(b), replacing the input parameters μ_X, σ_X and γ_X by the corresponding characteristics of $X(k)$. These are derived having regard to the fact that the number of terms k in (5.4.8) is now fixed, i.e. not random as it was in the general compound case (3.1.1). Because the Z's are independent and identically distributed, we obtain ((3.2.9) and (1.4.21))

$$\mu_{X(k)} = k \cdot m$$

$$\sigma^2_{X(k)} = k \cdot \sigma^2_Z = k \cdot (a_2 - m^2)$$

$$\gamma_{X(k)} = \frac{k \cdot \kappa_3(Z)}{\sigma^3_{X(k)}} = \frac{\gamma_Z}{\sqrt{k}} \tag{5.4.9}$$

$$= \frac{a_3 - 3a_2 m + 2m^3}{(a_2 - m^2)^{3/2} \cdot \sqrt{k}}$$

REMARK   It is worth noticing that splitting the simulation into two steps can be expected to reduce the inaccuracy involved in using the WH-formula, which, as we have previously noted, increases rapidly as the skewness grows (section 4.2.5(c)). This is due to the fact that the total skewness arises partly from the variation of the claim number (step 1 above) and partly from the variation of the claim size (step 2).

5.5 Outlines for simulation of more complex insurance processes

The simulation methods which have been introduced in the preceding sections are rather simple and many of the target distributions could in fact have been obtained using conventional analytical methods. They are presented as primary building-blocks for the more complex procedures which will be dealt with in subsequent chapters, rather than to be applied as such in isolation. In order to provide an overall insight into advanced simulation techniques, a few examples will be presented in this section, representing various types of application. The examples are constructed very simply in order to highlight the fundamental features. This also offers the opportunity of introducing some useful terms and concepts.

(a) Several years' time horizon. We take the simple underwriting equation (1.3.2), incorporating the yield of interest and allowing for time-dependent changes in P and λ, as an example to demonstrate the use of simulation. For this purpose it is transformed into the algorithm

$$U(t) = (1 + j(t)) \cdot U(t - 1) + (1 + \lambda(t)) \cdot P(t) - X(t) \qquad (5.5.1)$$

where $U(t)$ is the risk reserve at the end of year t (see section 1.3(c)). The profits are accumulated in the risk reserve and the losses are deducted from it. The coefficient $j(t)$ is the rate of interest earned on the reserve, $P(t)$ is the pure risk premium, $\lambda(t)$ a safety loading and $X(t)$ the claims. To perform the simulation the quantities $U(0)$, $j(t)$, $\lambda(t)$, $P(t)$ need to be specified, together with the parameters required for the simulation of $X(t)$, such as n, the characteristics of the claim size distribution and of the mixing variable.

First, for $t = 1$, the claim amount is simulated using one of the generators described in section 5.3 or 5.4. The outcome $X(1)$ is substituted into (5.5.1) and $U(1)$ is obtained. Then the procedure is repeated for $t = 2, 3,..., T$, arriving at the sequence

$$U(0), \ U(1), \ U(2),..., U(T). \qquad (5.5.2)$$

This outcome is a realization of the risk reserve process U. Such a realization is plotted in Figure 5.5.1 by putting it into the ratio form $u = U/P$.

The important advantages of simulation can be seen from this simple example. It makes it possible to split the total process into

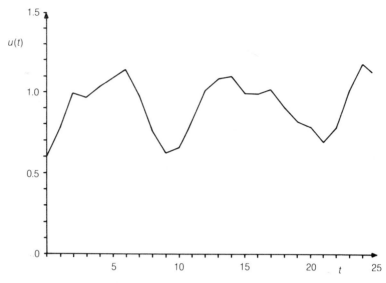

Figure 5.5.1   *A realization of a u-ratio simulation.*

a sequence of steps, in this example relative to the current year t. Each of these steps (the annual underwriting results in the example) can be computed fairly easily. Then, using the algorithm, the whole realization can be derived in a straightforward way, regardless of the number of steps. Conventional analytical methods would be rather laborious and, in fact, often intractable, particularly if external and dynamic feedback effects are incorporated, as will be outlined in subsequent sections.

(b) Monte Carlo method.   Figure 5.5.1, showing a single realization, already gives some information about the process. However, a number of questions remain open – for instance, which of the features exhibited result from the random numbers selected in this particular realization and which are generally characteristic of the process. Furthermore, the range of variation is not clearly indicated. Therefore, it is useful to repeat the simulation, say s times, according to the idea of the **Monte Carlo method** (section 5.1(b)). Then a picture like that in Figure 5.5.2 is obtained, which is more informative. An immediate visual idea is obtained about the character of the process – for example, the range of variation and whether there are any trends. Methods to derive numerical estimates for the relevant characteristics,

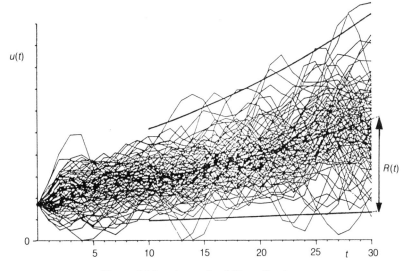

Figure 5.5.2 *A sample of 50 realizations.*

such as the mean and standard deviation of the target variables $u(t)$ and $x(t)$, can also be constructed (section 5.5(e) below), as well as confidence boundaries such as are indicated in Figure 5.5.2.

(c) Outside impulses. The actual insurance business is subject to many kinds of external impulses and influences, such as inflation, booms and recessions of the national economy, competition in the insurance market and movements in the capital market, influencing the return earned on investments. It is a major challenge to incorporate these into insurance models. The simulation technique offers a practical solution, as will be seen in subsequent chapters. We now illustrate the idea by incorporating inflation into the 'mini-model' defined by (5.5.1).

Let $I(t)$ denote the index of monetary value, for example the price index. Furthermore, let $i(t) = I(t)/I(t-1) - 1$ be the incremental rate of claim inflation. We assume here that it is a random variable and model it using our simulation procedure. It should be noted that the successive actual values of this rate of inflation are in fact clearly not mutually independent, as will be dealt with in more detail later in Chapter 7, i.e. there are periods of low and high inflation. For this kind of phenomenon, times series can be developed to generate random number sequences where the successive numbers are **autocorrelated**. The simplest type is the so-called **first order auto-regressive time series** which is constructed as follows

$$i(t) - \bar{i} = a \cdot [i(t-1) - \bar{i}] + \varepsilon(t). \tag{5.5.3}$$

Here a is a coefficient which controls the degree of autoregression (Appendix G), $\bar{i}$ is the average level of the fluctuating $i(t)$ and $\varepsilon(t)$ is a random number which introduces stochastic variation into the scheme. It is often defined to be normally $N(0, \sigma_\varepsilon)$ distributed and is called the noise term. If the behaviour of the phenomenon to be simulated suggests a skew noise, this can be achieved by generating $\varepsilon(t)$ by means of the shifted gamma function (section 3.3.5). It can be defined in terms of the characteristics

$$\mu_\varepsilon = 0, \sigma_\varepsilon \text{ and } \gamma_\varepsilon, \tag{5.5.4}$$

so allowing both the standard deviation and the skewness to be specified. The noise term can be generated by the WH-generator described in section 5.4(c). Then the required index is

$$I(t) = I(0) \cdot \prod_{u=1}^{t} (1 + i(u)). \tag{5.5.5}$$

We assume that the influence of inflation in this mini-model can be represented by transforming the monetary variables of (5.5.1) to reflect values of the index (detailed formulae will be given in section 9.3).

Self-evidently, the variation range of the bundle of realization outcomes is enlarged, because a new source of uncertainty has been incorporated.

This example shows how outside impulses can be introduced. Adding auxiliary sub-modules presents no particular difficulty. The major advantage of the simulation approach is that it is sufficiently flexible to accept an almost unlimited number of operational steps. In that way numerous features can be introduced to enhance its usefulness.

(d) Dynamic control offers further possibilities for extending the insurance model. For example, if there is an adverse development in the financial position, management is likely to intervene with remedial actions such as increasing premium rates, selecting risks more carefully, extending reinsurance cover, etc. On the other hand, if the position becomes very strong, increases in dividends, bonuses and taxes might ensue, premium rates might be reduced or further sales efforts reinforced. In order to make a long-term simulation

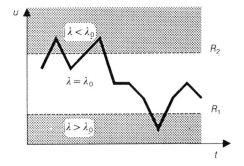

Figure 5.5.3   *Dynamic control.*

model realistic, features such as these should be incorporated in one way or another. An example restricted to a simple premium rate control rule via the safety loading ($\lambda(t)$ in (5.5.1)) will be given here.

The idea is to incorporate measures which will strengthen the financial position if there has been adverse development, and will lead to increased bonuses (dividends) or sales efforts if the position is more satisfactory. As a simple example let us assume that the loading $\lambda(t)$ will be increased if the risk reserve ratio $u(t)$ passes below a certain alarm barrier R_1. On the other hand, if another barrier R_2 is exceeded, $\lambda(t)$ will be reduced (Figure 5.5.3). Accordingly, let

$$\lambda(t) = \lambda_0 + c_1(R_1 - u(t-1))^+ - c_2(u(t-1) - R_2)^+, \quad (5.5.6)$$

and incorporate this as another sub-module into the main model.

Clearly, this kind of dynamic control reduces the variation range of $u(t)$. As will be seen in later chapters, this is often the only way to avoid unlimited spreading of the bundles for long-term simulations and to prevent them from tending to infinity. The various premium rating rules of credibility, experience-rating, etc. are one of the central concerns of the actuarial literature and will be considered briefly in later chapters.

(e) Evaluation of accuracy. Because stochastic simulation produces only a sample of the flows of the target variables, the results are always subject to sample error. A visual investigation of a diagram such as that shown in Figure 5.5.2 can provide a good idea of the accuracy of the sample.

If the outcomes are derived in numerical form, the sample in-accuracy can be estimated directly. For example, if the mean and standard deviation of the risk reserve $u(t)$ are required, they can be calculated from Figure 5.5.2 using standard statistical formulae (see Appendix F, Exercise F.4.1 where a formula for the skewness is also given)

$$\mu_u \approx \frac{1}{s} \sum_{i=1}^{s} u_i(t); \quad \sigma_u^2 \approx \frac{1}{s-1} \sum_{i=1}^{s} (u_i(t) - \mu_u)^2 \qquad (5.5.7)$$

where the $u_i(t)$ are the simulation outcomes of u at t. These characteristics are subject to sampling errors which have standard deviations

$$\sigma_\mu \approx \frac{\sigma_u}{\sqrt{s}}; \quad \sigma_\sigma \approx \frac{\sigma_u}{\sqrt{2(s-1)}}. \qquad (5.5.8)$$

In the case of Figure 5.5.2 $\mu_u \approx 0.5$, $\sigma_u \approx 0.2$, $\sigma_\mu \approx 0.2/\sqrt{100} = 0.02$ and $\sigma_\sigma \approx 0.2/\sqrt{198} = 0.014$.

If it is desired to improve the accuracy by one more decimal place, the sample size s should be increased a hundredfold, i.e. to 5000 instead of 50. This is a general rule applicable to such simulations. Note that besides the sample inaccuracy there are also errors inherent in model identification and parameter estimation (see section 1.3(e)). The problem of evaluating accuracy is dealt with further in Appendix F.

(f) Discussion. The merit of the simulation approach is that it is flexible enough to incorporate models which are subject to numerous conditions and controls, both in time and space, as the foregoing simple examples illustrate. With simulation it is not necessary to be able to write down everything in a condensed global set of equations. Each aspect of interaction can be specified separately, with variability where appropriate. The model proceeds for one period at a time and within that time period can be programmed, for example, to solve simultaneous equations, to take specified actions, to take into account effects such as taxation or accounting policy, and so on. When a large number of simulations are run, the results provide a convolution of a whole series of probability distributions affecting different aspects of the development of business. This enables the outcomes to be expressed in terms of probabilities, whether in relation to continued solvency, continued liquidity, free assets, profit, discounted net worth, etc. Simulation is a powerful tool, which has made it possible to extend risk theory applications far beyond the frontiers which were unavoidably created by the limitations of the conventional, mainly analytical, methods.

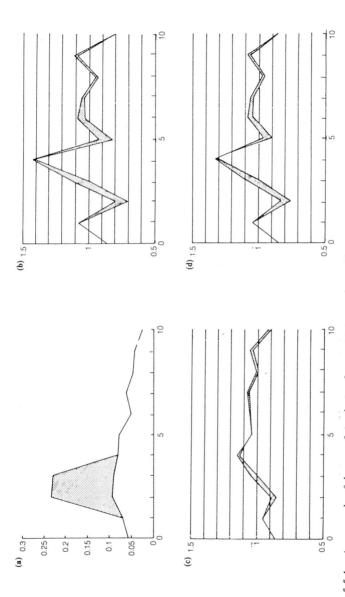

Figure 5.5.4 *An example of the use of simulation for sensitivity analysis. The reserving methods are defined in sections 9.5.2 (a, b, c) (Pentikäinen and Rantala, 1992). (a) The rate of inflation, normal flow and a shock in years 2 and 3; (b) claim ratio X/P with claim reserving by the chain-ladder method; (c) claim ratio X/P assuming the premium-based reserving method; (d) claim ratio X/P assuming the mixed reserving method.*

On the other hand, a drawback of simulation is its inability to give the results other than in the form of samples, with consequent sampling inaccuracy. Moreover, the structures of the model, for example the relationships between the relevant variables, are often not obvious from the numerical outcomes. However, the problem of inaccuracy can be alleviated by error evaluations, which were demonstrated above and are discussed further in Appendix F. The internal relationships can be explored by sensitivity analyses. For example, if it is desired to study the effect of the inflation coefficients in (5.5.3), the simulation can be made for several sets of parameter values in turn and the differences analysed. The impact of the distortion resulting from sampling errors can be greatly reduced by using the same sequence of random numbers for each of the concurrent simulations (i.e. always starting from the same specified seed of the primary random number generator).

Figure 5.5.4 demonstrates a sensitivity analysis. Inflation was first simulated using the standard (5.5.3); then a particular temporary shock for two years was added (Box (a)). After that the claim ratio was simulated for both alternatives. Boxes (b), (c) and (d) show the reaction for three different claim reserving rules. In order to improve comparison between the alternatives the parallel curves were run using the same sequence of the random numbers (Appendix F, last paragraph of section F.1).

Finally, it is worth mentioning that the classical analytical methods and simulation should not be regarded as being in competition. A general rule is that an analytical technique should always be used wherever it is tractable. On the other hand, the temptation should be resisted to manipulate the premises of the model in order to make the analytical calculations possible, if that can only be done at the cost of the applicability of the model to real-world conditions. If that is done, as is often the case in theoretically-orientated risk theory, a warning of the restricted applicability – or non-applicability – should be clearly given. The wide realm of application of simulation methods begins at the frontier where other methods become intractable.

CHAPTER 6

Applications involving short-term claim fluctuation

6.1 Background to the short-term fluctuation problem

In this chapter we consider applications of the techniques which were introduced in previous chapters. At this stage we will restrict consideration to the stochasticity of the claim process and to a short time horizon, in most cases one year. Notwithstanding this limitation, it is still possible to consider some interesting problems, such as evaluating the fluctuation range of claim amounts and assessing the corresponding capital requirements, analysing the effect of reinsurance and the level of net retention, as well as the basic mathematics of rating reinsurance contracts. These topics are also fundamental building-blocks for later chapters, where the time horizon will be extended and vital features such as inflation, cyclical effects, asset risks, and other matters, will be considered, including the mutual interdependence between successive periods.

One should be aware, however, of the weakness of short-term analyses. In practice, adverse fluctuations often occur in consecutive years, giving rise to considerable accumulation of losses. This may not be revealed by an analysis limited to one calendar year.

(a) Risk premium and safety loading. A new quantity, namely **premium income**, will now be incorporated into the model. The aggregate premium income can be decomposed as follows

$$B = P + E + \Lambda, \tag{6.1.1}$$

where

$$P = \mathrm{E}(X) = \mu_X = n \cdot m \tag{6.1.2}$$

is the **risk premium** corresponding to the aggregate claim amount X. E is the amount needed for expenses, which in some applications may include taxes, dividends, etc. (section 1.1(b) and Chapter 11).

The remaining term, Λ, is called the **safety loading**. It is the amount by which the income is expected (*a priori*) to exceed expenditure. It is available to build up the solvency margin or risk reserve (section 1.3(c)) and to finance bonuses and dividends.

The expense term E is fairly stable in practice and can be controlled in the short term by conventional deterministic budget techniques. It will not, therefore, be considered further in this chapter. This restriction can and should be relaxed in longer term analyses (Chapter 11). The impact of investments and other factors will also be deferred to later chapters.

It is convenient to transform the remaining expression, **premium income net of expenses**, into the form

$$P_\lambda = P + \Lambda = \left(1 + \frac{\Lambda}{P}\right) \cdot P = (1 + \lambda) \cdot P \qquad (6.1.3)$$

where λ is the **safety loading coefficient**.

A great variety of methods can be found in the literature for determining the safety loading. The primary concern is to safeguard the solvency (and profitability) of the company, having regard to competitive market pressures and to market cycles. A secondary problem is to divide the loading between the policyholders in an equitable way. These issues will be considered in sections 6.6.6 and 10.2(e, f).

REMARK   The safety loadings are usually assessed separately for different classes and groups within the portfolio. The total safety loading coefficient is obtained from the coefficients λ_i of the groups i in a straightforward way.

$$\lambda = \frac{\Lambda}{P} = \frac{\sum_i \Lambda_i}{P} = \sum_i \frac{P_i}{P} \lambda_i, \qquad (6.1.4)$$

where $P = \sum P_i$.

(b) Basic equation.   A basic process from which to start is the flow of incoming premiums and the outgo in respect of claims. The accumulating balance is denoted by $U(t)$, representing the cumulative underwriting profit (positive or negative), as illustrated in Figure 6.1.1. It can be expressed in terms of the equation

$$(1 + \lambda) \cdot P \cdot t - X(t) = U(t). \qquad (6.1.5)$$

The distribution of U at the end point T of the observation period is of interest and, in particular, the confidence range $[U_{\min}(T), U_{\max}(T)]$

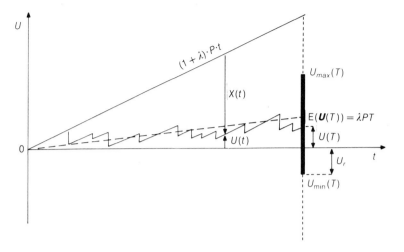

Figure 6.1.1 *An underwriting process.*

inside which $U(T)$ can be expected to fall with some given confidence probability.

In risk theory applications only the lower confidence limit $U_{\min}(T)$ is normally of interest. If the time horizon T is short, the lower limit is usually negative, indicating that there is a possibility, in adverse cases, that the income $\lambda \cdot P \cdot T$ from the safety loading will not be sufficient fully to cover the excess of the aggregate claim amount over the expected value $E(X(T))$.

In the remainder of this chapter the time horizon T will be taken to be one year, i.e. $T = 1$, unless specifically stated otherwise. For brevity, the argument T will be dropped.

A key concept is the **capital at risk**, $U_r = -U_{\min}$. This is defined so that there is a confidence probability of $1 - \varepsilon$ that the insurer's losses will not exceed this amount.

$$\text{Prob}\{U \geqslant -U_r\} = 1 - \varepsilon. \tag{6.1.6}$$

U_r can readily be obtained by solving X_ε from $1 - \varepsilon = F_X(X_\varepsilon)$ and substituting into

$$U_r = X_\varepsilon - (1 + \lambda) \cdot P. \tag{6.1.7}$$

X_ε is the limit under which claim expenditure is expected to fall with the targeted probability $1 - \varepsilon$.

In most applications the business in question is that on the insurer's net retention. However, the same mathematics also applies to the gross business before allowing for reinsurance.

Formula (6.1.7) will be the basic equation for the applications to be considered in this chapter. It gives an expression for the capital at risk or, in other words, the amount of capital which it is necessary to have available in order to meet adverse claim fluctuations. Furthermore, it makes it possible to analyse how the capital requirement depends on various background factors, such as the amount of business, the distribution of claim size and the safety loading λ. The variables involved also depend on the reinsurance arrangements and the level of net retention. Hence, the effect of these different factors can be analysed. Alternatively, it should be possible to determine how the various factors should be controlled in order to manage the business within the framework of the available resources.

REMARK   Comparing the above with section 1.3(c), it can be seen that the model which has been introduced is essentially the same as that which forms the basis of classical risk theory. The model has usually been applied to the whole portfolio of the insurer, ignoring other sources of uncertainty than those which are related to the claims. In these circumstances, U_r is the minimum required capital and ε is the ruin probability.

Our consideration has so far been quite general, so that the analysis may be carried out for any class or subclass of the business, any group of classes, or even for the whole portfolio, if the claim process and its stochastic behaviour is of interest. If several classes are considered together, then there may be interest in amalgamating the distributions, and the characteristics of the combined distribution introduced in section 3.2(d) may find some application.

To apply the basic equation (6.1.7), the values of the aggregate claim d.f. F_X, or its inverse function F_X^{-1}, are required. The techniques presented in previous chapters are available for this purpose. The mutual dependence of the variables involved can often be presented diagrammatically in order to bring out the essential structures of the process. Moreover, there is also a need to have short-cut methods available, notwithstanding that they may be approximate. A useful feature of some of the short-cut methods is that an explicit analytical expression can sometimes be achieved to link together the key variables. A couple of short-cuts will be presented in the following sections.

(c) NP-approximation to the basic equation.   The NP-formula (4.2.6) provides an approximate analytical expression for X_ε in (6.1.7) resulting in the following convenient version of the basic equation

$$U_r = y_\varepsilon \sigma_X - \lambda P + R_\gamma \qquad (6.1.8)$$

where y_ε is the $(1 - \varepsilon)$-fractile of the normal distribution, obtained from $1 - \varepsilon = N(y_\varepsilon)$, and R_γ is the correction term which introduces the effect of the skewness

$$R_\gamma = \tfrac{1}{6}\gamma_X \cdot (y_\varepsilon^2 - 1) \cdot \sigma_X \qquad (6.1.9)$$

In order to relate the results to the primary claim size characteristics, the expressions from (3.2.15) are substituted into (6.1.8) and (6.1.9). The resulting equations can be manipulated into the following form

$$U_r = y_\varepsilon P \sqrt{\frac{r_2}{n} + \sigma_q^2} - \lambda P + R_\gamma$$

where

$$\qquad (6.1.10)$$

$$R_\gamma = P \cdot \frac{(y_\varepsilon^2 - 1) \cdot (r_3/n^2 + 3r_2\sigma_q^2/n + \gamma_q\sigma_q^3)}{6(r_2/n + \sigma_q^2)}.$$

If the correction term R_γ is omitted, the formulae give the normal approximation, which can also be obtained directly from the equations of section 4.2.2.

It is useful to notice that the equations (6.1.10) include, explicitly or implicitly, the following variables

$$n, m, \lambda, \sigma_q, U_r, r_2 \text{ (or } a_2), M \qquad (6.1.11)$$

and further, if the correction term R_γ is included in the calculations,

$$\gamma_q \quad \text{and} \quad r_3 \text{ (or } a_3). \qquad (6.1.11a)$$

When n and m are given, the premium income P is determined by (6.1.2) and will often be one of the primary variables. M is the maximum claim size which can arise from a single event. It is, of course, only applicable if a maximum claim size is defined and known. This limit M can be the maximum net retention with respect to ceded reinsurance if the reinsurance cession is arranged on the basis of individual risks (section 3.4.2 for excess of loss treaties and section 3.4.3 for proportional treaties) or it may arise from special policy conditions.

Most of the problems which will be examined in this chapter are of the type where all but one of the variables (6.1.11) are given and the equations are to be solved for the remaining one, for example U_r or M. A major advantage of the formula (6.1.10) is that it provides many of the relevant relationships in an analytical form, so that the results of the analyses can often be obtained explicitly.

A more sophisticated group of problems arises when two or more of the variables of (6.1.11) are unknown. Then, in order to get a solution, some further conditions are required, for example maximizing the profit subject to the constraint of (6.1.10). This leads to a maximization problem which will be illustrated by considering a group of classes, each with a different net retention (section 6.6.4).

(d) Distribution-free approximation of the basic equation. The distribution-free upper bound which was introduced in section 3.3.9 can be useful in situations where neither the d.f. S nor its moments are known, but it can be assumed that the claim sizes have an upper limit M. Then inequality (3.3.20) immediately implies

$$r_2 \leqslant M/m \quad \text{and} \quad r_3 \leqslant M^2/m^2. \tag{6.1.12}$$

Substituting into (6.1.10) (see (3.3.25)) we obtain, for the compound Poisson case

$$U_r \leqslant y_\varepsilon \sqrt{PM} - \lambda P + R_\gamma \tag{6.1.13}$$

where

$$R_\gamma \leqslant \tfrac{1}{6}(y_\varepsilon^2 - 1) \cdot M. \tag{6.1.13a}$$

Note that M also affects the system through $P = n \cdot m(M)$.

The equality in the above expressions is valid only when all the claims are equal to M. For most of the commonly used distributions the maximum of the formula (6.1.13) gives an overestimate, as was discussed in section 3.3.9(a). By using the average level estimate, introduced by taking the factor $K = 0.7$ as in (3.3.23) and y_ε from Table 4.2.1, the following simplified versions ensue

$$\begin{aligned} U_r &\approx 1.6\sqrt{PM} - \lambda P \quad (\varepsilon = 0.01) \\ &\approx 2.2\sqrt{PM} - \lambda P \quad (\varepsilon = 0.001). \end{aligned} \tag{6.1.14}$$

The correction term R_γ has been omitted. This seems to be justified, because R_γ is significant only for large values of M. The first term in (6.1.13) is an overestimate in so far as it compensates for the omission of R_γ, as Figure 3.3.6 seems to suggest.

In the compound mixed Poisson case the following formula is obtained, substituting (3.3.24) into (6.1.8), and again omitting the correction term

$$U_r \approx y_\varepsilon \sqrt{K^2 MP + P^2 \sigma_q^2} - \lambda P \tag{6.1.15}$$

where K should be taken as 1 for an upper bound evaluation and 0.7 for an estimate of a typical level (3.3.23).

Sometimes it is possible to replace formula (6.1.15) by the simpler (6.1.14). This might be the case, for example, if it is possible to find a conservative estimate q' for the mixing variable q for the period of concern. This may be possible when q arises mainly from cyclical phenomena such as booms or recessions of the national economy, which may lend themselves to a degree of short-term prediction. Then the value of the parameter n should be amended by multiplying it by q' when assessing the premium income P. The safety loading λ should also be adjusted appropriately. In the absence of any better estimate, the amount $q' - 1$ (recall that $E(q) = 1$ according to (2.4.4)) should be subtracted from λ. If this is done, the standard deviation σ_q and the skewness γ_q disappear from (6.1.8) and (6.1.15), the compound mixed Poisson case simplifies to the Poisson case, and formulae (6.1.13) and (6.1.14) apply.

Owing to the rather heuristic derivation of the above inequalities, it is clear that care is required in using them. Table 6.1.1 shows some numerical examples of the outcomes obtained by the various alternative formulae given above.

In Table 6.1.1 U_r^{wh} is obtained from formula (6.1.7), using the Wilson–Hilferty approximation, U_r^{np} is calculated from (6.1.10), using the NP approximation, U_r^n by omitting R_γ (the normal approximation) and U_r^{dfree} by the distribution-free formula (6.1.15), taking K as 0.7. The parameter values (6.1.11) are given in the first few columns, with $\lambda = 0.04$ and $\varepsilon = 0.01$ in all cases. The symbol £m stands for £million. Excess of loss reinsurance is assumed and m, r_2 and r_3 are taken from Table 3.4.1 to correspond to the given M.

The cases with $\sigma_q = \gamma_q = 0$ correspond to the formula (6.1.8) without the correction term R_γ, and the distribution-free U_r^{dfree} can be obtained from the simplified formula (6.1.14).

The differences between the WH-formula and the NP-formula are insignificant. This is to be expected, since these approximations give outcomes which are very close to each other (Pentikäinen, 1987).

As can be concluded from Figure 3.3.6, the distribution-free formula underestimates in these examples as compared to the more exact approximations U_r^{wh} and U_r^{np} for cases of small r_2, i.e. when the maximum net retention M is low, and gives an overestimate when r_2 is high, i.e. when M is high.

Table 6.1.1 *The capital at risk U_r calculated by alternative formulae*

Case no.	M (£m)	n	m (£)	r_2	r_3	σ_q	γ_q	U_r^{wh} (£m)	U_r^{np} (£m)	U_r^{n} (£m)	$U_r^{d\,free}$ (£m)
1	0.1	100	4381	11.7	218	0.04	0.25	0.39	0.39	0.33	0.33
2	0.1	1000	4381	11.7	218	0.04	0.25	1.07	1.07	1.00	0.98
3	0.1	10 000	4381	11.7	218	0.04	0.25	3.89	3.89	3.61	3.56
4	0.1	100 000	4381	11.7	218	0.04	0.25	27.81	27.83	24.72	24.65
5	0.1	100	4381	11.7	218	0.00	0.00	0.39	0.39	0.33	0.32
6	0.1	10 000	4381	11.7	218	0.00	0.00	1.79	1.79	1.73	1.66
7	1.0	100	6160	37.3	3832	0.04	0.25	1.27	1.32	0.85	1.25
8	1.0	1000	6160	37.3	3832	0.04	0.25	3.04	3.05	2.58	3.84
9	1.0	10 000	6160	37.3	3832	0.04	0.25	8.61	8.61	8.00	11.55
10	1.0	100 000	6160	37.3	3832	0.04	0.25	43.18	43.20	39.02	45.51
11	1.0	100	6160	37.3	3832	0.00	0.00	1.27	1.32	0.85	1.25
12	1.0	10 000	6160	37.3	3832	0.00	0.00	6.75	6.76	6.29	10.32
13	5.0	100	6735	89.2	39 100	0.04	0.25	2.63	3.62	1.45	2.96
14	5.0	1000	6735	89.2	39 100	0.04	0.25	6.45	6.61	4.45	9.20
15	5.0	10 000	6735	89.2	39 100	0.04	0.25	15.47	15.50	13.38	27.85
16	5.0	100	6735	89.2	39 100	0.00	0.00	2.62	3.62	1.45	2.96
17	5.0	10 000	6735	89.2	39 100	0.00	0.00	14.25	14.28	12.11	27.19

Note: In order to demonstrate the outcomes of the variants, the rules were applied irrespective of whether or not the approximation formulae, by which they were derived, are permissible, owing to the high skewness values.

6.2   Evaluating the capital at risk

We will now illustrate how the formulae which were derived in the previous section can be used, with some numerical examples of the evaluation of the variation range of the aggregate claim amount and the capital required to cover such variation. The constraints of section 6.1 still apply, i.e. we are concerned only with the stochasticity of the claim process.

(a) Reference insurer. In these examples, a model or reference insurer is chosen first, and the results calculated. Then the control parameters, for example the size of the portfolio, the safety loading coefficient, the level of reinsurance cession, etc. are varied in order to demonstrate the structural features of the business and the sensitivity to the different parameters.

According to (6.1.11), the following data are sufficient to define the reference insurer for the application of the formulae of section 6.1, assuming the claim size distribution to be that of Table 3.4.1

$$n = 10\,000, \; M = \pounds 1 \text{ million}, \; \lambda = 0.04, \; \sigma_q = 0.04, \; \gamma_q = 0.25, \; \varepsilon = 0.01.$$
$$(6.2.1)$$

Then we have (see Tables 3.4.1 and 6.1.1, line 9)

$$P = \pounds 61.6 \text{ million}, m = \pounds 6160, r_2 = 37.3, r_3 = 3832, U_r = \pounds 8.61 \text{ million}.$$

(b) Dependence on the business volume and on the level of reinsurance. Figure 6.2.1 shows the dependence of the capital at risk U_r on the volume of business written, measured in terms of the net premium income P (see (6.1.2)), and the net retention M (Table (3.4.1)). This is for the reference insurer with excess of loss reinsurance.

This kind of diagram can be used to evaluate the required amount of capital, when constructed on the basis of actual data. Recall that $P = n \cdot m(M)$ also depends on M. This feature will be dealt with in section 6.3(c).

Another example is given in Figure 6.2.2 to demonstrate the effect of the safety loading coefficient λ. The relative amount of capital, i.e. the ratio U_r/P, is plotted on the vertical axis.

Two very important features of general validity can be deduced from the above diagrams. Firstly, the need for capital is, as a rule, an increasing function of the size of the portfolio. However, the requirement is not linearly proportional to the volume, the latter being measured by P in Figure 6.2.1; it is in fact a convex curve close to the shape of a parabola.

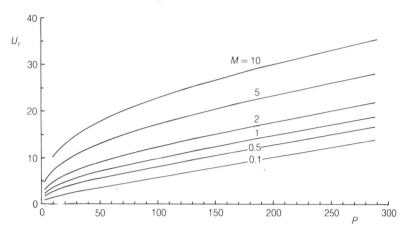

Figure 6.2.1 U_r as a function of the premium income P and the net retention M according to (6.1.8) and using the WH-formula. Monetary unit £ million, reference data (6.2.1) and Table 3.4.1.

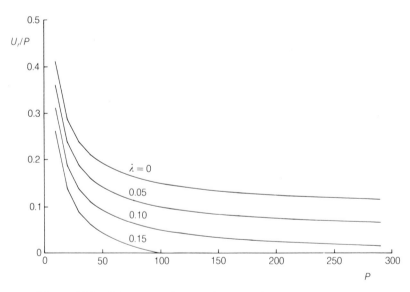

Figure 6.2.2 The ratio U_r/P as a function of P and λ. Data (6.2.1).

Secondly, the ratio U_r/P is a decreasing curve for increasing P, with the horizontal line

$$U_r/P = y_\varepsilon \cdot \sigma_q - \lambda \tag{6.2.2}$$

as an asymptote, as follows from (6.1.10) with $n \to \infty$. If $\lambda > y_\varepsilon \sigma_q$, there is a critical limit for P, above which no initial capital is needed, i.e. the accruing safety loading is sufficient to cover the claim expenditure, even in adverse cases, at the confidence level $1 - \varepsilon$.

(c) Rules for statutory minimum solvency margins for general insurance which are laid down in some countries, in particular in the directives of the European Community (EC), have a shape which is close to the parabolic form seen in Figure 6.2.1. The parabola is approximated by a broken line of the type

$$U = U_0 + aB - b(B - B_0)^+ \tag{6.2.3}$$

as shown in Figure 6.2.3. The premium income is now denoted by B, because the formula is applied to the gross premiums, including the loading for expenses, as distinct from the net premiums used in (6.1.15) and in other formulae in section 6.1. For the EC case the coefficients have the values $a = 0.18$, $b = 0.02$ and B_0 is 10 million monetary units (écus). There is an alternative formula of the same structure based on claim expenditure, but this will not be discussed here.

It should be appreciated that (6.2.3) is a considerable simplification and does not stand up well to deeper analysis. For example, the solvency margin requirements are independent of the company's

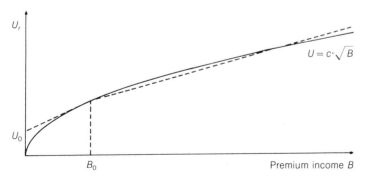

Figure 6.2.3   *A model for the statutory solvency margin.*

risk distribution, asset risks are ignored, as is the effect of business cycles. It is obvious that no simple rule can be found which would be satisfactory in all cases. Insurance portfolios and markets differ so much that the main emphasis in solvency regulation should be on an expert analysis performed for each insurer in order to have adequate regard to the individual features of each case (Pentikäinen (1952), Daykin *et al.* (1984, 1987), Pentikäinen *et al.* (1989) section 5.3, and Daykin and Hey (1990), further discussion in section 14.6).

(d) Profile figure.   One way to obtain an overall view of the multi-variable structure is to construct a profile diagram like that in Figure 6.2.4. The idea is to calculate U_r for a specified standard set of values, which might correspond, for instance, to average circumstances in the environment of concern, and then to change the relevant variables one by one in turn in order to illustrate the sensitivity of the system to various background factors.

Figure 6.2.4 shows that the capital requirement depends on several factors, none of which can be distinguished as clearly predominant in these examples. This is in line with the remarks in the last paragraph of section 6.2(c) above. However, if the size of the company

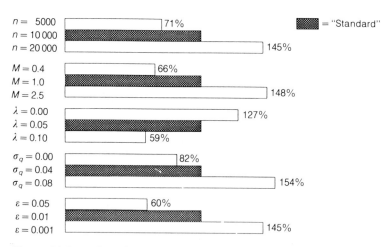

Figure 6.2.4   *A dependence profile of U_r with regard to some of the control variables. The shaded line corresponds to the reference values (6.2.1). Each of the parameters is then changed as shown in the diagram and the resulting U_r is given as a percentage of the value in the standard case. The standard U_r is £8.6 million.*

increases significantly, with the parameter n exceeding, say, $1\,000\,000$, then the dominant factor in the determination of U_r becomes the size of the company, as measured by the premium income. Therefore, when for administrative reasons simple rules have to be devised for the minimum solvency margin requirement, the size of the company is clearly likely to be chosen as the relevant factor. This is the case, for example, in the rule for general insurance laid down in the directives of the European Community, as stated in section 6.2(c) above. This does not, however, eliminate the need for the individual expert analysis referred to in the same section.

(e) Merger of portfolios. How the need for reserve funds is dependent on the size of the company can be illustrated by means of the following example. Let C_i, $i = 1$ or 2, be two insurance companies, and let $U_{r,i}$ denote the minimum reserve fund for company C_i, computed by means of the approximate formula (6.1.8), i.e.

$$U_{r,i} = \left(y_\varepsilon + \frac{\gamma_i}{6} \cdot (y_\varepsilon^2 - 1) \right) \cdot \sigma_i - \lambda_i \cdot P_i \quad (i = 1, 2), \qquad (6.2.4)$$

where λ_i is the safety loading coefficient, $P_i = E(X_i)$ is the risk premium income, σ_i the standard deviation, and γ_i the skewness of the distribution of the aggregate claim amount X_i of the company C_i. The question can now be asked: How is the need for reserve funds affected if these two companies are merged into a single company C?

Assuming that the portfolios are independent, the following expression is obtained for the minimum reserve U_r of the merged company C by means of (6.1.8), provided that the security level $1 - \varepsilon$ is kept unchanged:

$$U_r = \left(y_\varepsilon + \frac{\gamma}{6} \cdot (y_\varepsilon^2 - 1) \right) \cdot \sigma - \lambda \cdot P, \qquad (6.2.5)$$

where we have $\lambda \cdot P = \lambda_1 \cdot P_1 + \lambda_2 \cdot P_2$ for the safety loading, with $P = P_1 + P_2$, and, because of independence and (1.4.21),

$$\left. \begin{array}{l} \sigma = \sqrt{\sigma_1^2 + \sigma_2^2} < \sigma_1 + \sigma_2 \\[2mm] \gamma \cdot \sigma = \dfrac{\kappa_3}{\sigma^2} = \dfrac{\kappa_{3,1}}{\sigma^2} + \dfrac{\kappa_{3,2}}{\sigma^2} < \dfrac{\kappa_{3,1}}{\sigma_1^2} + \dfrac{\kappa_{3,2}}{\sigma_2^2} = \gamma_1 \cdot \sigma_1 + \gamma_2 \cdot \sigma_2 \end{array} \right\} \qquad (6.2.6)$$

Consequently (assuming $\varepsilon < 0.15$ to guarantee the inequality

$y_\varepsilon^2 - 1 > 0)$,

$$U_{r,1} + U_{r,2} - U_r = y_\varepsilon \cdot (\sigma_1 + \sigma_2 - \sigma) - (\lambda_1 \cdot P_1 + \lambda_2 \cdot P_2 - \lambda \cdot P)$$
$$+ \tfrac{1}{6} \cdot (y_\varepsilon^2 - 1) \cdot (\gamma_1 \cdot \sigma_1 + \gamma_2 \cdot \sigma_2 - \gamma \cdot \sigma) > 0, \quad (6.2.7)$$

giving

$$U_r < U_{r,1} + U_{r,2}. \tag{6.2.8}$$

The conclusion is that the reserve needed by the merged company C is less than the sum of reserves needed by the separate companies C_1 and C_2, assuming that the security level $1 - \varepsilon$ is to remain unchanged. The result is of general interest. The same fact can also be seen from (6.1.10), which shows that the relative minimum required capital U_r/P is a decreasing function of the volume variable $n = P/m$.

The result generalizes immediately to the case of several companies. It is also possible to generalize the result to cases where there is dependence, for example through the mixing variables, between the merging companies (Exercises 6.2.5 and 6.2.6).

One can say that a merger enables the existing reserves to be used in a more efficient way or, alternatively, that a merger releases idle reserves if the security level is left unchanged. In practice it may not be necessary for the companies actually to merge since the same advantages can be obtained by exchanging reinsurance on a reciprocal basis. This problem will be considered further in section 6.6.3.

Exercise 6.2.1 A compound mixed Poisson model is used for the aggregate claim amount of an insurer. The estimated key characteristics are $n = 10\,000$, $m = £10\,000$, $r_2 = a_2/m^2 = 30$, $\lambda = 0.05$. The capital at risk U_r is equal to £20 million. Let ε denote the corresponding 'ruin' probability defined by (6.1.8) without the skewness term R_y. How much smaller could U_r be, provided that the security is maintained at the original level $1 - \varepsilon$, if the risk properties of all the risks in the portfolio were improved so that the expected number of claims decreases by 10% (while the premium rates are kept unchanged)?

Exercise 6.2.2 A friendly society grants funeral expense benefits and each member of the society can choose a benefit of either £100 or £200. It is assumed that $\lambda = 0.5$, $n = 20$, $\sigma_q = 0$ and $\varepsilon = 0.01$. How large should the minimum required capital U_r be according to (6.1.8) without the skewness term R_y if it is not known in advance how many members will choose the option 100 and how many the option 200 and, consequently, what mix of these options is to be assumed so as to maximize the risk?

Exercise 6.2.3 Assume that the distribution of one claim is exponential $S'(Z) = e^{-Z}$ $(Z \geqslant 0)$ (taking the average size of claims as the monetary unit). The expected number of claims n is 1000 and $\sigma_q = 0.04$. How large should the safety loading λ be according to (6.1.8) without the skewness term R_y if there is no reinsurance and no capital U_r, and if ε is fixed at 0.01?

Exercise 6.2.4 The following characteristics are computed from the statistics of an insurance company C_1: $m_1 = \text{£}1000$, $r_{2,1} = 40$, $r_{3,1} = 400$, $\sigma_{q,1} = 0.05$, $\gamma_{q,1} = 0.1$ and $n_1 = 1000$. The company has capital at risk $U_{r,1} = \text{£}500\,000$ and safety loading $\lambda_1 = 0.1$.

Another insurance company C_2 with the following characteristics: $m_2 = \text{£}500$, $r_{2,2} = 50$, $r_{3,2} = 500$, $\sigma_{q,2} = 0.1$, $\gamma_{q,2} = 0.5$, $n_2 = 200$ and $\gamma_2 = 0.05$ is merged with company C_1. The portfolios are mutually independent.

If the ruin probability ε according to (6.1.10) of the former company is not allowed to increase following the merger, how large should the reserve fund U_r be for the merged company?

Exercise 6.2.5 Suppose that the variables X_1, X_2 and their sum X are normally distributed. The variables X_i need not be independent of each other. Show that the inequality $U_r \leqslant U_{r,1} + U_{r,2}$ holds true if the security level $1 - \varepsilon$ is kept fixed. In what circumstances is there equality?

Exercise 6.2.6 Suppose that insurance companies C_1 and C_2 each have several lines of business. Suppose also that, except for one line, called A, which is assumed to be independent of other business lines for both companies, the portfolios (or more precisely, the corresponding aggregate claim amounts) are independent. In addition, let the aggregate claim amount X_{Ai} of line A of each company C_i be a compound mixed Poisson variable, each with the same mixing variable q_A but conditionally independent of each other under the condition $q_A = q$ for every possible value q. Prove that the inequality

$U_r < U_{r,1} + U_{r,2}$ holds true. It can be assumed that the NP-approximation is applicable for the conditional aggregate claim amount distribution under the condition $q_A = q$ for both of the original companies as well as for the merged company.

Exercise 6.2.7 Let X_1 and X_2 be independent, suppose that $\text{Prob}\{X_i = 0\} = 0.9$ and $\text{Prob}\{X_i = 1\} = 0.1$ for both companies C_i, and let $\varepsilon = 0.11$. Show that in this case the inequality $U_r \leqslant U_{r,1} + U_{r,2}$ does **not** hold true.

6.3 Rules for maximum retention

The main purpose of reinsurance is to reduce the impact of adverse fluctuations in the cedant's claim expenditure to an acceptable level in the context of the cedant's financial resources and business philosophy. Evaluating the effectiveness of the various forms of reinsurance and establishing a suitable level of net retention are among the most important applications of risk theory. The problem of finding a proper level for the net retention is dealt with in this section and the rating of some types of reinsurance treaty in subsequent sections.

(a) Specification of the problem. The basic equation (6.1.7) will be utilized. The net retention M is incorporated into it through the claim size distribution $S = S_M$, as was shown in section 3.4. Allowing M to vary, the moments of the claim distribution also vary and consequently the premium income P (6.1.2) and in particular the capital at risk U_r, which was introduced in section 6.1(b). In most common situations U_r is given, i.e. chosen by the cedant, and a reinsurance arrangement has to be found which limits the capital at risk to this amount in an optimal way.

For the sake of simplicity we will consider only the case of excess of loss reinsurance. Other forms can in principle be handled in a similar way. As will be stated later, an excess of loss treaty can be expected to give a more conservative outcome than, for example, a surplus treaty (sections 3.4(c) and 6.3(h)). The retention limits which are obtained for the former can, therefore, serve as upper bound estimates for the latter.

(b) The capital at risk U_r depends on the resources of the cedant and his readiness to accept risks. U_r may include the company's so-called hidden reserves, i.e. margins in technical reserves, in asset valuations and other balancing items, in addition to specific reserves such as equalization reserves. A conservative approach is to include in U_r, if possible, only the hidden reserves, whilst treating the equalization reserve, if separately identifiable, as part of the technical reserves.

Furthermore, we recall the weakness of short-term considerations which ignore the risk of consecutive adverse years. Therefore, U_r should be determined sufficiently cautiously that the total resources of the insurer will not be exhausted in any one year, leaving it vulnerable to forthcoming risks. This problem can be treated more appropriately when the time horizon is extended from one year to longer periods, as will be done in section 13.3. However, it is often useful to know the range of annual fluctuation. Therefore, considerations limited to one year are also of interest and will be examined in this chapter. In this way the results can be derived more easily and a better qualitative general view obtained of the structures.

(c) Capital at risk as a function of the net retention. A straightforward solution can be obtained by plotting U_r as a function of M using the basic equation (6.1.7), as shown in Figure 6.3.1.

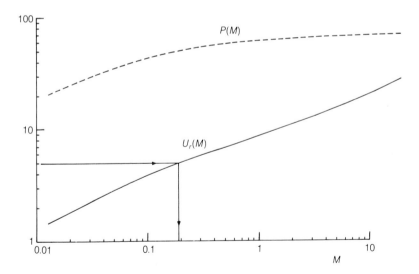

Figure 6.3.1   *The capital at risk U_r as a function of the net retention M. Data according to (6.2.1), excess of loss treaty, d.f. S_M from Table 3.4.1, monetary unit £million, logarithmic scale.*

If the value of U_r is given, the corresponding value of M can be read from the graph. For example, if the insurer wishes to protect his underwriting business so that it is 99% certain that the loss of capital will be no more than $U_r = £5$ million, then $M = 0.19$.

For more advanced applications it is also useful to determine the dependence of the premium income $P = n \cdot m(M)$ on M, as is shown by the dotted line in Figure 6.3.1. If the retention M is decreased in order to reduce the capital at risk U_r, P should also be reduced and, depending on the conditions of the treaty, the corresponding safety loading and loading for expenses is lost by the cedant. This should be taken into account when the price of reinsurance protection is evaluated. A general rule is that it is profitable for the cedant to fix the retention limit as high as resources permit. It should be appreciated, however, that the real life situation may be more complicated than that assumed above. For example, the cedant may have an opportunity to obtain satisfactory reciprocity against the ceded business and in this way balance profitability. Risk exchange will be discussed in section 6.6.3.

The problem of the optimal level of reinsurance has been examined, for example, by Pentikäinen and Rantala (1982, Vol. 2, Chapter 6) and is exemplified in Exercise 6.3.5. Balancing the contradictory business objectives of profitability and solvency will be considered in more detail in Chapter 14.

(d) M as a function of the business volume, the capital at risk and the safety loading.   The effect of different background factors can be examined by letting one or more of the relevant variables of (6.2.1) vary, with the remaining ones being kept constant. Figure 6.2.1 showed the dependence of the capital at risk on the premium income P and on the level of net retention M. Note that P also depends on M, as is shown in Figure 6.3.1.

In Figure 6.3.2 M is shown as a function of U_r and λ. A considerable degree of dependence on the background factors can be seen in all of these graphs, for example with regard to λ in Figure 6.3.2. This confirms the observations from the profile shown in Figure 6.2.4.

A noticeable feature of Figure 6.3.2 is the very steep gradient of the curves as U_r decreases and the fact that, for small values of U_r, there is no reasonable solution at all for the net retention. The reason can be seen from equations (6.1.10) and (6.1.15). It arises from the mixing variation introduced into our model by the standard deviation σ_q and from the well-known weakness of forms of reinsurance which

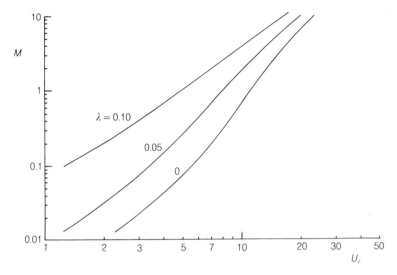

Figure 6.3.2   *The net retention M as a function of U_r and λ. Reference data (6.2.1) and logarithmic scales.*

operate on a claim-by-claim basis. These types of treaty do not provide adequate protection against the excessive claim expenditure which results from a substantial increase in the number of claims. The mixing phenomenon generates such increases from time to time and little protection is provided against them by per claim excess of loss reinsurance.

(e) The case of multiple lines.   For brevity, the problem of finding the maximum net retention was considered above in relation to a single line of business, or several lines together for which the same retention limit M was assumed to apply. The problem will be generalized in section 6.6.4 with differing limits M_j for each line j. The main result, subject to some theoretical provisos, suggests that the M_j should be proportional to the safety loadings λ_j of the respective lines.

(f) Distribution-free approximations.   The interrelationships between the relevant variables and parameters can be seen more clearly if the distribution-free approximate equation (6.1.15) is solved for M

$$M = \frac{(\lambda^2 - y_\varepsilon^2 \sigma_q^2)P^2 + 2\lambda UP + U^2}{K^2 y_\varepsilon^2 P}. \tag{6.3.1}$$

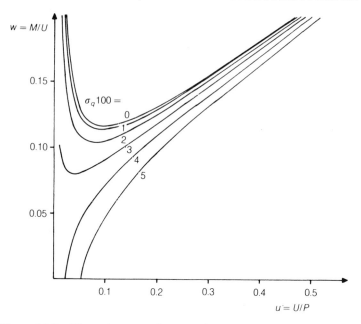

Figure 6.3.3 *The ratio w as a function of σ_q and u, according to (6.3.3).*

Note that, according to (6.1.2), P depends on M. This equation is useful, therefore, only if this dependence can be expected to be so weak that it can be ignored in this evaluation, which is in any case rather approximate (Figure 6.3.1).

The relevant structures are illustrated in Figure 6.3.3. In order to bring out the relationships more clearly, the main variables in equation (6.3.1) are transformed into the ratios

$$w = M/U_r \quad \text{and} \quad u = U_r/P. \tag{6.3.2}$$

Then (6.3.1) can be written in the form

$$w = \frac{1}{K^2 y_\varepsilon^2}\left(\frac{\beta}{u} + u + 2\lambda\right), \tag{6.3.3}$$

where

$$\beta = \lambda^2 - y_\varepsilon^2 \sigma_q^2. \tag{6.3.4}$$

The ratio w is a hyperbolic function of the solvency ratio u. If β is positive, the curve has a minimum at the point given by

$$u = \sqrt{\beta}; \quad w = \frac{2(\lambda + \sqrt{\beta})}{K^2 y_\varepsilon^2}. \tag{6.3.5}$$

When $w = M/U_r$ is at its minimum, the inverse has a maximum. Hence, an upper bound approximation for U_r can be found. The rule of thumb in the next section is based on this feature.

Furthermore, (6.3.5) applies only if β is positive. Otherwise the curves have no turning point, as can be seen from Figure 6.3.3. A positive value of β is guaranteed in the Poisson case where $\sigma_q = 0$, provided that the safety loading λ is positive. The fact that no positive M may be found was stated already in section 6.3(d).

(g) Rules of thumb. Earlier, when effective techniques for the examination of the real effect of reinsurance arrangements had not been developed, or at least were not generally known, the approximate retention level M was often determined by means of various simple rules. Some of them suggested that M should be a small proportion of the premium income, i.e. of the type $M = c \cdot P$; others that it should be proportional to the capital at risk, $M = d \cdot U_r$. It is interesting to find that the dependence of M on the relevant variables leads to a simple approximate rule of the latter type.

A conservative lower limit for the maximum net retention M can be obtained from the minimum point of the $w = M/U$ variable in Figure 6.3.3. According to (6.3.2) and (6.3.5) we have

$$M \geqslant \frac{2(\lambda + \sqrt{\beta})}{K^2 y_\varepsilon^2} \cdot U_r. \tag{6.3.6}$$

This can be further simplified by omitting the influence of the mixing variation (putting $\sigma_q = 0$), perhaps incorporating its effect in a conservative estimate of the safety loading λ (thereby glossing over the inconvenience of the condition $\beta > 0$ for the existence of a minimum point in Figure 6.3.3). Then we have

$$M \geqslant \frac{4}{K^2 y_\varepsilon^2} \cdot \lambda U_r. \tag{6.3.6a}$$

For $K = 0.7$ and $\varepsilon = 0.01$ this gives

$$M \geqslant 1.5\lambda \cdot U_r. \tag{6.3.6b}$$

Recall that this inequality was derived by excluding the mixing variation (shown in the last paragraph of section 6.3(d) and the restrictions relating to (6.3.5)).

These estimates may be useful when the order of magnitude of the net retention has to be obtained quickly or when information regarding the claim size distribution and the relevant parameters is not available. However, these rules are based on such a weak premise that a more detailed analysis, using the methods introduced in this chapter, should be used whenever possible.

REMARK   A general feature of most of the outcomes for M from the basic equation (6.1.7) is that they exceed, sometimes to a considerable extent, the level which might be applied in practice by many insurers, at least in some countries. Rules of thumb have perhaps been used along the lines referred to above. These rules might derive from tradition rather than from any rational analysis of their actual efficiency. For example, the level obtained by the formulae of this chapter would suggest that life insurance does not need risk reinsurance at all, except for exceptionally high risks, since for the large bulk of ordinary policies the distribution of claim sizes is very homogeneous. Hence the level of M resulting from risk theory considerations exceeds by far most of the gross sums at risk in the portfolio.

(h) Robustness of the retention results.   Most of the numerical applications and examples given in this chapter have been based on Table 3.4.1. The results are, therefore, related to the excess of loss reinsurance used in the table. Other types of reinsurance can clearly be handled in the same way, for example the surplus treaty (section 3.4.3). In order to test to what extent the features indicated may depend on the type of reinsurance, Heiskanen (1982) calculated examples using both excess of loss and surplus reinsurance in parallel (Pentikäinen and Rantala, 1982, Vol. I, section 4.2.3). This proved that the results were fairly robust for the choice of type of reinsurance as long as M is not too large. The excess of loss case can be expected to provide conservative estimates, i.e. an upper bound for U_r and a minimum for the retention limit M.

This observation justifies the use of excess of loss results as an approximate guide to the impact of surplus treaties. This is useful in situations where the specially tailored d.f. S for the latter type of reinsurance is not readily available.

Exercise 6.3.1   The aggregate claim amount X of an insurance company is assumed to be a compound Poisson variable. The claim size d.f. S is continuous, $S(0) = 0$, and the skewness of X is so small that the d.f. F_X can be approximated by the normal d.f. The portfolio is protected by an excess of loss reinsurance treaty with retention limit M. Prove that the derivative of the minimum required capital

$U_r = U_r(M)$, defined by (6.1.10) with $R_\gamma = 0$, is

$$U_r'(M) = \left[\frac{y_\varepsilon \cdot M \cdot \sqrt{n}}{\sqrt{a_2(M)}} - \lambda \cdot n \right] \cdot (1 - S(M)).$$

Exercise 6.3.2 Find conditions according to which the derivative in Exercise 6.3.1 has a negative value at $M = 0$.

Exercise 6.3.3 Examine the general shape of the function $U_r(M)$ defined in Exercise 6.3.1. When does this function have a minimum for a finite M? Note that the type of the solution depends on the sign of the derivative at the origin (Exercise 6.3.2).

Exercise 6.3.4 Consider the insurer depicted in Figure 6.3.1. Evaluate from the graph the maximum net retention M corresponding to capital at risk $U_r = 10$ (in units of £million) and the premium income $P(M)$. Try also formulae (6.3.1) and (6.3.6), first applying the standard value 0.7 for the coefficient K, then a value which fits the order of magnitude of M according to Figure 3.3.6.

Exercise 6.3.5 The reference insurer considered in section 6.2(a) allocates from its freely disposable profit a small increment ΔU_r to its reserve U_r and increases the net retention M by an amount ΔM according to the basic equation (6.1.10) taking $\sigma_q = 0$ and omitting the correction term R_γ. The premium income net of reinsurance P then increases by ΔP, including the safety loading for the reinsurer's profit, solvency and expenses. What is the rate of return i, i.e. the expected increment of the cedant's profit margin divided by ΔU_r? The formula derived for Exercise 6.3.1 may be used.

Exercise 6.3.6 Verify that the return rate i introduced in Exercise 6.3.5 is a decreasing function of the retention M when the moment $a_2(M)$ is taken from Table 3.4.1. Find the largest M for which the rate $i \geqslant 0.05$.

Exercise 6.3.7 A new policy is added to a life insurance portfolio. The annual rate of death for it is q and the insurer will accept a sum assured M for net retention. What should M be if the ruin probability ε calculated according to (6.1.10), without the skewness term R_γ, were not changed? The new policy has the same safety loading λ as for the average of the portfolio. Derive a simple rule for M if q is small, the expected number of claims n large and $\sigma_q = 0$ (M is to be expressed as a function of the variables λ, n, U_r, P and the moments about zero of the claim size d.f. S).

Exercise 6.3.8 Let the claim size d.f. be Pareto$(\alpha, 0, 1)$. Excess of loss reinsurance is arranged with maximum net retention M. Calculate M to one decimal place from (6.1.10) when $\alpha = 1.5$, $U_r = 30$, $n = 100$, $\lambda = 0.03$, $\sigma_q = 0$ and $y_\varepsilon = 2.33$. Hint: Derive an expression for U_r as a function of M and then find (by trial and error) the requested numerical value of M.

6.4   An application to rate-making

(a) The problem.   It is not necessary to limit the application of the theory of risk to the claim process or to underwriting for the whole portfolio or of some specific line. On the contrary, the theory is applicable to any insurance collective. An example is in the conventional treatment of insurance statistics for rate-making. The risk units are arranged into homogeneous groups and the problem arises as to how large such a group must be in order to give an adequate statistical basis for setting premium rates. The standard methods of mathematical statistics are, of course, available, but the formulae derived in section 6.1 can be useful for making a quick calculation. One benefit is that they make use of the special properties of the claim process, which can improve the efficiency of the approach, in particular having regard to the usual skewness of claim distributions.

As an example, consider a group of similar fire or other property insurance policies which have been observed during a certain period. Suppose that the total amount of claims has been $X = 8.00$, using £million as the monetary unit. With the same monetary unit, the total sum insured for the group of policies is $S = 7000$. The so-called **burning cost** can be estimated as $f = X/S = 1.14$ per thousand. The problem is to assess the accuracy of this estimate of the underlying claim experience.

(b) Confidence limits.   Assuming that the compound mixed Poisson distribution applies, the confidence limits are obtained as a direct application of (6.1.7). The aggregate amount of claims X which is observed during the period of concern can be expected to fall within the interval

$$X_1 \leqslant X \leqslant X_2 \tag{6.4.1}$$

with probability $1 - 2\varepsilon$, where the limits are defined by the equations

$$F(X_1) = \varepsilon \quad \text{and} \quad F(X_2) = 1 - \varepsilon. \tag{6.4.2}$$

Hence, the burning cost is, with the same probability, within the interval

$$X_1/S \leqslant f \leqslant X_2/S. \tag{6.4.3}$$

Assume in the above example that the following numerical parameter values are applicable ((6.1.10), (6.1.11) and (6.1.11a))

$$n = 5000, \; r_2 = 40, \; r_3 = 4000, \; \sigma_q = 0, 1, \; \gamma_q = 0.5, \; \varepsilon = 0.025. \tag{6.4.4}$$

The observed mean claim size is $X/n = 0.00160$, which can be used as an estimate for the mean m. Furthermore, the standard deviation σ_X and the skewness γ_X can be calculated from (3.2.15), obtaining the values 1.07 and 0.37 respectively. Then, applying the inverse WH-formula of section 4.2.5(b), the following numerical result is obtained for (6.4.3)

$$6.09/7000 = 0.87‰ \leqslant f \leqslant 10.28/7000 = 1.47‰. \tag{6.4.5}$$

Using the normal approximation, i.e. taking

$$X_{1,2} = \mu_X \pm y_\varepsilon \sigma_X, \tag{6.4.6}$$

the limits would be 0.84 and 1.44 per thousand. The difference from the above estimate is due to the fact that the normal approximation ignores the effect of skewness.

If the skewness is slight, as it was in the above example, then the normal approximation can conveniently be used. If the skewness is large, say 0.5 to 1, the WH-formula is recommended. If the skewness exceeds 1 then more accurate methods are necessary, for example resorting to the recursive approach of section 4.1.

Exercise 6.4.1 It is known that for certain fire risks $f \approx 0.1\%$ and the number of claims per annum is about 1000. How many years of statistics are needed to estimate f with an accuracy of 20% on 10% confidence level? The risk index r_2 is estimated to be 100 and $\sigma_q = 0$. The normal approximation can be used.

6.5 Experience-rating

6.5.1 Introductory comments

(a) The concept and scope. A universal problem in rate-making is the fact that the statistical basis is often scanty, uncertain and sometimes nearly non-existent. It is common practice to combine similar risks into tariff classes and to assess average rates for each class. However, considerable differences in risk propensity can occur within each such class. Therefore, some policies may be overcharged or undercharged to the benefit or the detriment of other policies in the class. To alleviate this problem methods have been developed to provide individual *a posteriori* corrections to the initial rates on the basis of actual claim experience, in the form of continual amendments to the rates or through profit returns, discounts and bonuses or maluses. In other words, the class tariffs are replaced by **individual rating** in cases where each insured risk unit has a claim history which is sufficient for conclusions to be drawn about the risk level.

We shall use **experience-rating** as an umbrella term for all of these methods. It can be subdivided into several subdisciplines, such as profit return (section 6.5.2), exponential smoothing (section 6.5.3) and credibility (6.5.4).

Experience-rating is suitable for relatively large policies where small claims occur continually. An example might be personal accidents in respect of the staff of an industrial or commercial firm, or the fleet of cars of such a firm, etc. Traditionally workers' compensation insurance has been considered as amenable to this approach.

Experience-rating can also be applied to a continual (semi-automatic) adjustment of general tariffs by controlling the rates separately in different tariff groups, for example brick houses in some specified area or particular classes of motor cars.

(b) Merits and demerits. From the point of view of policyholders, experience-rating helps to promote fair and equitable rating. To the extent that it is technically feasible, each risk unit should pay its own cost in the long run. It is also in the interests of the insurer for the rating to be implemented in as reliable a way as possible.

Antiselection can be a problem in cases where general level tariffs are applied for risk units for which the policyholder (or competitors) might be able to assess whether what they are being charged is fair. Those policyholders being overcharged will tend to move their policies to companies with a more flexible rating system (or might establish a captive company of their own). This will lead to a situation where a growing proportion of policies are undercharged, leading to poor results, even though the class tariffs might be correct as average rates.

If the premiums have to be approved by regulatory authorities, or if they are subject to frequent negotiation between the insurer and the customer, then an advance agreement about the rules to be followed will strengthen the consistency of the process and enhance the trust of the customers in the credibility and fairness of the rating. Experience-rating methods might be appropriate for this purpose.

The major difficulty with experience-rating is to achieve sufficiently reliable individual rating, having regard to the normally limited statistical basis of the claim history. In particular, the risk of potential large claims is often not evident from the individual past experience. This problem can be alleviated by supporting the experience-rating with an excess of loss treaty. Large claims become the responsibility of the reinsurer, whereas small and medium size claims are dealt with by experience-rating. This extends the scope of application of the method.

(c) References. Experience-rating has a long tradition in the USA, where it is normally called **credibility**. A presentation of the state-of-the-art is given by Venter in Chapter 7 of *Foundations of Casualty Actuarial Science* (1990), with particular regard to practical applications. A presentation of the theory and a list of references are given by Goovaerts and Hoogstad (1987). All in all, experience-rating has grown as a wide-ranging independent topic which is no longer usually viewed as a branch of risk theory. It is beyond the scope of this book to provide any detailed treatment. However, having regard to the fact that some of the basic functions which were introduced in the previous sections have found application in experience-rating, a brief review of the subject will be given in this chapter.

6.5.2 Profit return

(a) Profit return is a type of bonus which is sometimes found in reinsurance treaties and also in some direct insurance contracts. The uncertainty involved in determining the premium rate $P = E(X)$ is compensated for by fixing the safety loaded premium $P_\lambda = (1 + \lambda)P$ at a conservative level, with the proviso that a certain ratio k of any profit which emerges under the contract will be returned to the insured.

As an example, consider the profit formula

$$G = k \cdot (P_\lambda - X)^+, \tag{6.5.1}$$

where the superscript '+' indicates that, if the expression in the brackets is negative, it should be replaced by zero. The problem is to find a relationship between the return coefficient k and the safety loading λ, subject to the condition that the expected profit to be retained by the insurer should be at least a specified amount $\lambda_0 P$.

The underwriting profit for the insurer is then defined by

$$U = P_\lambda - X - G. \tag{6.5.2}$$

Its expected value is

$$E(U) = P_\lambda - E(X) - k \cdot \int_0^{P_\lambda} (P_\lambda - X)\,dF(X). \tag{6.5.3}$$

It is helpful to bring out the relationship between this expression and the expression for a stop loss reinsurance premium. The integral expression in (6.5.3) can be elaborated as follows.

$$\int_0^{P_\lambda} (X - P_\lambda)\,dF(X) = \int_0^\infty (X - P_\lambda)\,dF(X) - \int_{P_\lambda}^\infty (X - P_\lambda)\,dF(X)$$

$$= E(X) - P_\lambda + P_{re} \tag{6.5.4}$$

where $P_{re} = P_{re}(P_\lambda)$ is the stop loss premium (3.4.26) with stop loss limit $M = P_\lambda$. Replacing $E(X)$ with P and P_λ with $(1 + \lambda)P$ and substituting (6.5.4) into (6.5.3) we obtain

$$E(U) = \lambda P - k(P_{re} + \lambda P) = P \cdot [\lambda \cdot (1 - k) - k \cdot P_{re}/P]. \tag{6.5.5}$$

From the condition that the expected profit to be retained by the insurer should be at least the target amount $\lambda_0 P$, the following inequality ensues

$$k \leqslant \frac{\lambda - \lambda_0}{\lambda + P_{re}/P}. \tag{6.5.6}$$

(b) Discussion. The above derivation assumed that the relevant distributions and parameters were known and available. In fact, return of profit arrangements are usually applied in an environment where this is not the case. Often only an approximate average risk premium P for a group of similar insured risk units is known. However, the risk propensity within the group may vary a great deal. The insurer, therefore, has to use the average rate P, and the average profit requirement $\lambda_0 P$, for all the individual policies in the group. Formula (6.5.6) shows that there is an upper limit for the return coefficient k. If this limit is exceeded, then the insurer can expect to make an underwriting profit of less than the target amount $\lambda_0 P$. The extreme position is illustrated by the hypothetical choice of $k = 1$, i.e. always returning the whole of any profit. Then $E(U)$ is $-P_{re}$, i.e. the expected profit is negative, and the amount of the loss is equal to the stop loss premium. This shows that, if the group of insured units, with joint premium P_λ and return coefficient k, is very heterogeneous, i.e. the unknown $E(X)$ varies considerably between the units, it is difficult to find any λ and k which will avoid some of the more risky units giving an expectation of an underwriting loss to the insurer. However, the situation might still be better for the insurer than if there were no profit return arrangement and a much lower premium had been charged.

Other experience-rating models can give rather more flexibility than the return of profit method in cases where the cover is usually renewed for a number of years.

6.5.3 Exponential smoothing

An alternative approach to return of profit is to arrange for the premiums which are rated for subsequent years to be adjusted according to the accumulating claim experience obtained from the policy's own statistics. The variant which we now introduce is called **limited fluctuation credibility** by Venter (1990, Chapter 7). This name refers to the condition which aims to constrain the changes in tariff rates within limits which are not too inconvenient for the policyholder.

Consider a risk or a group of risks which is insured by a policy which is to be experience-rated. An initial premium P_0 is agreed and it is further agreed that the rates $P_1, P_2, \ldots$ for subsequent years will be calculated according to the algorithm

$$P_t = Z \cdot X_{t-1} + (1 - Z) \cdot P_{t-1}, \qquad (t = 1, 2, \ldots) \qquad (6.5.7)$$

where X_t is the total claim amount in year t arising from this risk collective. The coefficient Z is called the **credibility weight** or, briefly, **credibility**. It is chosen from the interval

$$0 < Z \leqslant 1,$$

and will be fixed sufficiently small to eliminate excessively large random fluctuations. More precisely, it is chosen subject to the condition that pure random fluctuations in the total claim amount will not, with probability $1 - \varepsilon$, result in a change in the premium in excess of $100p\%$ of the expected value of claims. Expressed in symbols, this is the case if the constant Z satisfies the condition

$$Z \cdot \Delta X \leqslant p\mathrm{E}(X), \tag{6.5.8}$$

where ΔX is obtained from

$$F(\mathrm{E}(X) + \Delta X) - F(\mathrm{E}(X) - \Delta X) = 1 - \varepsilon, \tag{6.5.9}$$

where the d.f. F of X is known or presumed. Then the absolute value of the deviation ΔX of the actual claim amount X from the expected $\mathrm{E}(X)$ can be larger than the ΔX determined by (6.5.9) only with probability $1 - \varepsilon$.

If it can be assumed that the NP approximation gives a satisfactory approximation for F, then, confining the analysis to upwards jumps of X, we obtain (see (6.1.8) and (6.1.9))

$$Z x_\varepsilon \sqrt{r_2/n + \sigma_q^2} = p, \tag{6.5.10}$$

where, according to (4.2.6),

$$x_\varepsilon = y_\varepsilon + \tfrac{1}{6}\gamma_x(y_\varepsilon^2 - 1), \tag{6.5.11}$$

and y_ε is the root of $1 - \varepsilon = N(y_\varepsilon)$ and r_2 is again the risk index (3.2.16). Hence we have

$$Z = \frac{p}{x_\varepsilon \sqrt{r_2/n + \sigma_q^2}}. \tag{6.5.12}$$

REMARK   The above considerations are strictly correct only if $P_{t-1} = \mathrm{E}(X_{t-1})$. In practical applications this condition is not fully satisfied, P_{t-1} varying in the vicinity of $\mathrm{E}(X_{t-1})$. Hence, the resulting formulae are only approximate, the more so because the limits of the permissibility of the NP-formula (section 4.2(b)) are often not honoured.

If the mixing variation is insignificant (see Table 3.2.2), the formula can be simplified by putting $\sigma_q = 0$. Then

$$Z = \frac{p}{x_\varepsilon} \sqrt{\frac{n}{r_2}}. \tag{6.5.13}$$

The expected number of claims n which makes $Z = 1$, that is

$$n_0 = \frac{r_2 x_\varepsilon^2}{p^2 - x_\varepsilon^2 \sigma_q^2}, \tag{6.5.14}$$

is of special interest. Following the terminology of American credibility theory, it is said that there is full credibility if $Z = 1$.

Table 6.5.1 *Values of n_0 for full credibility (constant claim size and no mixing variation)*

p	ε		
	10%	5%	1%
0.01	27 057	38 416	66 347
0.05	1082	1537	2654
0.1	271	384	663
0.2	68	96	166

In the special case where $\sigma_q = 0$ and the risk sums are all equal or, if only the number of claims is recorded for calculating the frequency of the claims, then $r_2 = 1$ and the values of n_0 which are large enough for full credibility can immediately be obtained from a table of the normal distribution, without use of the NP formula, as in Table 6.5.1.

In most practical cases the risk sums are not equal and hence r_2 is not 1. The variation in the value of this quantity depends significantly on the degree of heterogeneity of the risk sums and consequently the limit of full credibility can be considerably larger than is given in Table 6.5.1. The values of r_2 may often be of the order of 5 to 10, but in cases where large risk sums can occur the values can be much larger (Table 3.4.1).

If the expected number of claims n is smaller than the value obtained from (6.5.14) then the constant Z has values smaller than 1 and the term **partial credibility** is used. If $\sigma_q = 0$, then one of the well-known formulae of credibility theory can immediately be obtained from (6.5.13) and (6.5.14) by eliminating the coefficient of $\sqrt{n}$ in (6.5.13).

$$Z = \sqrt{n/n_0}. \qquad (6.5.15)$$

In the foregoing it was assumed that the NP approximation would be used. However, owing to the small size of the risk collective which often arises in cases subject to experience-rating or to credibility theory, the NP formula can be of doubtful applicability, even if very small values of ε are not needed, for which the accuracy of the formula is most unsatisfactory. The uncertainty can of course be avoided by calculating the quantity x_ε by some other method of computation, such as the recursion approach of section 4.1.

The experience of American actuaries might suggest that even the normal approximation gives values which are applicable in practical work notwithstanding that the formula is often used far beyond the theoretical limits of acceptable accuracy (section 4.2(b)).

(c) Limit value. Applying (6.5.7) for a sequence of t years, and developing the algorithm into a series it follows that

$$P_t = ZX_{t-1} + (1-Z)P_{t-1}$$
$$= Z \sum_{i=1}^{t} (1-Z)^{i-1} X_{t-i} + (1-Z)^t P_0. \tag{6.5.16}$$

If the expected value $\mu = E(X_i)$ is assumed to be equal for all i values, then

$$E(P_t) = Z \sum_{i=1}^{t} (1-Z)^{i-1} \mu + (1-Z)^t P_0$$
$$= [1 - (1-Z)^t]\mu + (1-Z)^t P_0, \tag{6.5.17}$$

which tends to μ as $t \to \infty$. Thus, in the long run the expected value of P_t tends to the theoretically correct unknown mean value μ. Hence the formula fulfils the requirement of fairness. The coefficient Z regulates the fluctuation of the sliding premium rate.

Owing to the fact that P_t depends on the claim amounts X_{t-i} of the preceding years by means of weights having the elapsed time i in exponents, the algorithm (6.5.16) and the formula (6.5.7) are sometimes called **exponential**.

Bonsdorff (1990) considered the bias arising when the coefficient Z is not constant.

6.5.4 Credibility theory

The problem dealt with in section 6.5.3 can be approached from a slightly different angle.

(a) The Bayesian approach. It is often the case that of seemingly similar risk units one may be more risky than another, although the insurer may not be able to see this difference when rating the risk. For example, heterogeneity of a motor-car insurance portfolio was dealt with in section 2.6.

If no claim history or other relevant information is available for a new risk unit, the insurer may simply assume that its risk propensity is at the average level. On the other hand, the quality of a risk unit which has been in the portfolio for a longer period of time may be able to be estimated by using its own claim history.

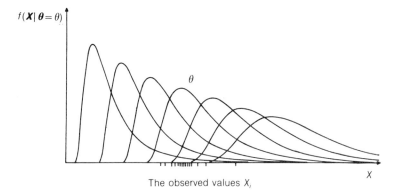

Figure 6.5.1 *Alternative densities f = F' curves of a postulated distribution and a cluster of observed yearly aggregate claim amounts of a risk unit.*

Consider a collective of apparently similar risk units. For each risk unit we assign a fixed parameter θ which indicates the 'quality' of the risk, for example, the expectation of the actual yearly aggregate claim amount for the risk unit. It is assumed that the d.f. $F(X|\theta)$ of the yearly aggregate claim amount for a risk unit is known if the value θ of the quality parameter is known.

The idea of the derivation of an estimate for θ is illustrated by Figure 6.5.1.

The density f is plotted for a set of different values of the parameter θ which determine the position of the curve on the X-axis and its shape. The observed values X_i of the risk unit are marked on the X-axis. Intuitively it is natural to choose an estimate for θ such that the curve is centred over the cluster of observed values.

This leads to a type of statistical inference which is known as **Bayesian**. The shape of the required distribution is postulated but its relevant parameters are unknown. The problem is to derive estimates for these parameters.

(b) The credibility formula. Suppose now that we have a randomly chosen risk unit in the portfolio. The quality parameter $\boldsymbol{\theta}$ of the risk can then be thought as a random variable, and suppose that the distribution $\boldsymbol{\theta}$ is known. The risk unit's identically distributed aggregate claim amounts X_i for different years i depend on each other through the parameter $\boldsymbol{\theta}$ but are assumed to be otherwise independent, i.e. the X_i are conditionally independent, given $\boldsymbol{\theta}$.

The problem now is that, if we are given a risk unit together with its

t years claim history

$$X_1, X_2, \ldots, X_t, \qquad (6.5.18)$$

what is the best estimate for the unknown risk premium $\mu(\boldsymbol{\theta}) = \mathrm{E}(X_i \,|\, \boldsymbol{\theta})$ of the risk unit? Theoretically the conditional expectation

$$\mathrm{E}(\mu(\boldsymbol{\theta}) \,|\, X_1, \ldots, X_2, \ldots, X_t) \qquad (6.5.19)$$

is the best (minimum variance) estimate for $\mu(\boldsymbol{\theta})$. Unfortunately, it is not easy in practice to solve (6.5.19) explicitly, since the joint d.f. of the $(t + 1)$-dimensional random vector $(\boldsymbol{\theta}, X_1, X_2, \ldots, X_t)$ is needed for the solution. However, if we restrict ourselves to those estimates of $\mu(\boldsymbol{\theta})$ which are of the linear form

$$a + b_1 \cdot X_1 + b_2 \cdot X_2 + \cdots + b_t \cdot X_t, \qquad (6.5.20)$$

then we can find the best estimate P for $\mu(\boldsymbol{\theta})$ such that the mean square deviation $\mathrm{E}[(P - \mu(\boldsymbol{\theta}))^2]$ is minimized. By symmetry, the best linear estimate must be such that b_i is the same for every i. Consequently we may restrict ourselves to solutions of the form

$$P = a + b \cdot \bar{X}, \qquad (6.5.21)$$

where $\bar{X} = \bar{X}_t$ denotes the mean value $(X_1 + X_2 + \cdots + X_t)/t$. The problem is now reduced to a form which is equivalent to finding the linear function of $\bar{X}$ which minimizes the mean square deviation from $\mu(\boldsymbol{\theta})$. It is well known that the solution is given by

$$P = \mu + Z_t \cdot (\bar{X} - \mu) = (1 - Z_t) \cdot \mu + Z_t \cdot \bar{X}, \qquad (6.5.22)$$

where $\mu = \mathrm{E}(\mu(\boldsymbol{\theta})) = \mathrm{E}(X_i) = \mathrm{E}(\bar{X})$ is the average $\mu(\boldsymbol{\theta})$, and

$$Z_t = \frac{\mathrm{Cov}(\bar{X}, \mu(\boldsymbol{\theta}))}{\mathrm{Var}(\bar{X})}. \qquad (6.5.23)$$

The value of the credibility coefficient Z_t is always between 0 and 1. The closer it is to 1, the better estimate $\bar{X}$ is for $\mu(\boldsymbol{\theta})$.

We shall now prove that, under the above assumptions, the **credibility coefficient** Z_t satisfies the formula

$$Z_t = \frac{\mathrm{Var}(\mu(\boldsymbol{\theta}))}{\mathrm{Var}(\mu(\boldsymbol{\theta})) + \dfrac{\mathrm{E}(\sigma^2(\boldsymbol{\theta}))}{t}}, \qquad (6.5.24)$$

where $\sigma^2(\boldsymbol{\theta}) = \mathrm{Var}(X_i \,|\, \boldsymbol{\theta})$. For the proof it is first noted that $\mathrm{E}(\bar{X} \cdot \mu(\boldsymbol{\theta})) = \mathrm{E}(X_i \cdot \mu(\boldsymbol{\theta})) = \mathrm{E}(\mu(\boldsymbol{\theta})^2)$. Therefore,

$$\begin{aligned} \mathrm{Cov}(\bar{X}, \mu(\boldsymbol{\theta})) &= \mathrm{E}(\bar{X} \cdot \mu(\boldsymbol{\theta})) - \mathrm{E}(\bar{X}) \cdot \mathrm{E}(\mu(\boldsymbol{\theta})) \\ &= \mathrm{E}(\mu(\boldsymbol{\theta})^2) - \mu^2 = \mathrm{Var}(\mu(\boldsymbol{\theta})). \end{aligned} \qquad (6.5.25)$$

On the other hand, the conditional variances formula (1.4.38) gives

$$\mathrm{Var}(\bar{X}) = \mathrm{E}(\mathrm{Var}(\bar{X}|\boldsymbol{\theta})) + \mathrm{Var}(\mathrm{E}(\bar{X}|\boldsymbol{\theta}))$$
$$= \mathrm{E}[(t/t^2)\cdot\mathrm{Var}(X_i|\boldsymbol{\theta})] + \mathrm{Var}[(t/t)\cdot\mu(\boldsymbol{\theta})]$$
$$= \mathrm{E}(\sigma^2(\boldsymbol{\theta}))/t + \mathrm{Var}(\mu(\boldsymbol{\theta})). \qquad (6.5.26)$$

Substituting these in (6.5.23) gives (6.5.24).

EXAMPLE Suppose that for each risk unit the aggregate claim amount is a compound Poisson($n\cdot q$), where q is fixed (but unknown) for each risk unit, but varies between different risk units according to a structure d.f. H (section 2.6). If the claim size d.f. S does not depend on the risk unit, then we may choose $\boldsymbol{\theta} = q$, where q denotes the structure variable defined in section 2.6.

(c) Discussion. When the number of observation years $t \to \infty$, then $Z_t \to 1$, as it should by the law of large numbers.

The variance $\mathrm{Var}(\mu(\boldsymbol{\theta}))$ can be thought to measure the heterogeneity between different risk units, whereas $\mathrm{E}(\sigma^2(\boldsymbol{\theta}))$ gives the average variation of the yearly aggregate claim amount of one risk unit. Therefore we may conclude from (6.5.24) that, for a fixed number of observation years t, the credibility coefficient Z_t is larger, the larger the heterogeneity of the portfolio is compared with the average variation within one risk unit.

6.6 Optimal risk sharing

6.6.1 Measures of risk

(a) Variance as a measure of risk. In the preceding sections the variation range of claims or of other relevant variables was a key concept when the capital at risk, ruin probability, etc. were deduced. When alternative policy strategies are compared, their optimality can often be assessed on the basis of which of the alternatives is the least volatile, i.e. leads to the smallest range of variation. The standard deviation σ_X, or for computational reasons, its square, i.e. the variance

$$\sigma_X^2 = \int_{-\infty}^{\infty} (X - \mathrm{E}(X))^2 \, \mathrm{d}F(X) \qquad (6.6.1)$$

is commonly used as the measure of variation of the relevant variable X (the aggregate claim amount, for example); the smaller the variance the smaller the risk. This will be illustrated in sections 6.6.2 and 6.6.3

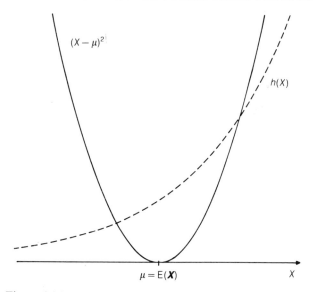

Figure 6.6.1   *Mean square deviation and a harm function* $h(X)$.

by comparing different forms of reinsurance and searching for the one with the smallest variance.

(b) More general measures of risk.   The usefulness of the variance as a measure depends on how the relevant random variable is spread around its mean. Positive and negative deviations have the same weight and both tails of the distribution together determine the size of the variance. In many applications, however, only the positive deviations (large claim amounts) are harmful, whereas small values, i.e. negative deviations, may even be beneficial. In this case variance is not a proper measure of risk, because it does not differentiate between positive and negative deviations. A more satisfactory measure of the 'harm' caused by a loss X is, therefore, obtained if the mean square deviation $(X - \mu)^2$ is replaced by an increasing, convex function $h(X)$ of X. The dotted line in Figure 6.6.1 illustrates such a non-symmetric harm function h. Instead of the variance (6.6.1) the 'harmfulness' of a loss X is now measured by the expected harm

$$H_X = \mathrm{E}(h(X)) = \int_{-\infty}^{\infty} h(X)\,\mathrm{d}F(X) \qquad (6.6.2)$$

of the loss variable X. The smaller the expected harm the better the risk.

The convexity of a harm function h means that the derivative h' is increasing, which in turn is equivalent to the condition

$$h'' \geqslant 0. \qquad (6.6.3)$$

The convex shape of the harm function is motivated by the general observation that the larger the value of X the more harmful an additional loss will be perceived.

The degree of convexity of the chosen harm function depends on how much risk the risk carrier is willing to take. The limit case $h'' \equiv 0$ means that h is no longer convex, but a linear function, in which case the risk measure H is equivalent to the mean value, i.e. $H_X \leqslant H_Y$ if and only if $E(X) \leqslant E(Y)$. This extreme risk measure does not have any regard to stochastic variations.

A drawback of the variance as a measure of risk is that it measures the deviations from the mean value but completely overlooks the expected value, whilst the measure H takes both into account.

Note that in conceptual terms we have come into the realm of so-called **utility functions**, which will be briefly considered in section 6.6.6. Turning a one-dimensional concave utility function u upside-down gives us the corresponding convex harm function h.

6.6.2 Optimality of stop loss reinsurance

(a) **The problem.** Consider an insurer who wants to find a re-insurance policy which gives the smallest variance σ^2, or more generally the smallest expected harm H (section 6.6.1(b)), for fixed reinsurance risk premium P (without any safety or expense loadings). Such a policy can be considered optimal from the cedant's point of view because it provides the best protection against the volatility for a fixed net risk premium.

Let X_{tot} be the total aggregate amount of claims during a year and $X = X_{ced}$ the cedant's share. The conditions

$$0 \leqslant X \leqslant X_{tot} \quad \text{and} \quad E(X) = P, \qquad (6.6.4)$$

are assumed to apply, where P is fixed. The type of reinsurance is not otherwise restricted. The problem is to find the reinsurance arrangement which has the smallest variance σ_X^2 or the smallest expected harm H_X.

Before solving this specific minimization problem, a general inequality for convex functions is first derived in section 6.6(b) below.

(b) A generalization of the Jensen inequality. The graph of a convex function $h(X)$ is always above its tangent line $T(X) = h(M) + h'(M) \cdot (X - M)$ drawn at a point M (see Figure 6.6.2). Therefore, the inequality $h(X) \geqslant h(M) + h'(M) \cdot (X - M)$ holds for any random variable X. If we choose $M = E(X)$ and take expected values of the both sides, then the last term on the right-hand side vanishes, since $E(X - M) = 0$, and we obtain the **Jensen inequality** $E(h(X)) \geqslant h(E(X))$ (Exercise 3.2.9).

More generally, if we have two random variables X and X^* and a constant M such that

(1) $E(X) = E(X^*)$ and
(2) X^* is always between X and M, i.e. for all outcomes either $X \geqslant X^* \geqslant M$ or $X \leqslant X^* \leqslant M$,

which means that X^* is more concentrated around M than X is, then the following generalized Jensen inequality holds:

$$E(h(X)) \geqslant E(h(X^*)). \qquad (6.6.5)$$

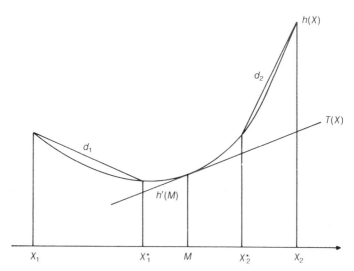

Figure 6.6.2   *The derivative $h'(M)$ of a convex function h compared with the difference quotients d_1 and d_2 of the two segments of the graph of h on both sides of the point M.*

In the special case $X^* \equiv E(X)$ the ordinary Jensen inequality results. From Figure 6.6.2 it can be seen that

$$d_1 = \frac{h(X_1^*) - h(X_1)}{X_1^* - X_1} < h'(M)$$

$$d_2 = \frac{h(X_2) - h(X_2^*)}{X_2 - X_2^*} > h'(M).$$

To prove (6.6.5) we note, by condition (2), that X and X^* are always either both to the left or both to the right of M. Checking both cases separately, it is seen from Figure 6.6.2 that

$$h(X) - h(X^*) \geqslant h'(M) \cdot (X - X^*). \tag{6.6.6}$$

Taking the expected values of both sides of (6.6.6), and recalling (1), we get $E(h(X)) - E(h(X^*)) \geqslant h'(M) \cdot E(X - X^*) = 0$, which gives (6.6.5).

(c) Optimality of stop loss.   It turns out that under the assumptions (6.6.4) the stop loss reinsurance treaty (section 3.4.4)

$$X^* = X_{\text{ced}} = \min(X_{\text{tot}}, M) \tag{6.6.7}$$

is the required optimal arrangement, where the retention limit M is chosen so that the condition $E(X^*) = P$ is satisfied. This condition uniquely determines the limit M, since $E(\min(X_{\text{tot}}, M))$ increases strictly and continuously from 0 to $E(X_{\text{tot}})$ as a function of M.

For the proof we use the fact that the stop loss net retention X^* is more concentrated than that of any other reinsurance treaty X satisfying the conditions (6.6.4). More precisely, we will show that the conditions (1) and (2) of the generalized Jensen inequality (6.6.5) are satisfied. Since $X^* = M$ in the event that $X_{\text{tot}} \geqslant M$, only the case that $X_{\text{tot}} < M$ remains to be checked. But then we have, by (6.6.4), $X \leqslant X_{\text{tot}} = \min(X_{\text{tot}}, M) = X^* < M$, and again X^* is closer to M. Therefore, if h is a convex harm function, we have $H_{X^*} = E(h(X^*)) \leqslant E(h(X)) = H_X$, by (6.6.5). Similarly, the choice of the convex function $h(X) = (X - \mu)^2$ gives $\sigma_{X^*} \leqslant \sigma_X$. Hence the stop loss treaty is the desired optimal solution.

(d) Stop loss treaties are problematic in practice.   Even though the optimality of stop loss reinsurance under the given assumptions is theoretically interesting, it has little practical application, because it is based only on the risk premiums. The fact that strong variability of the reinsurer's share leads to considerable rating uncertainty was completely ignored. Owing to this uncertainty and the high sensitivity to inflation and to other changes in risk exposure (section 3.4.4(c)), stop loss premiums are usually heavily loaded and unlimited stop loss covers are often not available in practice.

6.6.3   Optimal risk exchange

(a) The problem.   Consider two insurers C_1 and C_2 which would like to exchange reinsurance on a reciprocal basis. Let X_1 and X_2, respectively, denote the aggregate claim amounts of the companies.

The companies want to find an optimal reciprocal risk exchange (Y_1, Y_2) (condition (1) below) such that the claim amounts Y_1 and Y_2 after the risk exchange have variances as small as possible. Thus we have to find variables Y_1 and Y_2 satisfying the following optimality conditions:

(1) $Y_1 + Y_2 = X_1 + X_2$
(2) $\sigma_{Y_i} \leqslant \sigma_{X_i}$ for both i,
(3)   There is no other risk exchange giving smaller variance for both companies C_i.

The problem can be reformulated in an obvious way to give the case where both companies C_i use harm functions h_i (section 6.6.1(b)) to measure their losses instead of the variance.

(b) Minimizing the variances.   Let us denote the total aggregate claim amount by $X = X_1 + X_2$. Suppose that we are given a risk exchange (Y_1, Y_2) satisfying the condition (1).

We will first consider only those risk exchanges (Y_1, Y_2), which satisfy the additional condition that $\sigma_{Y_1} = \sigma$, where σ, $0 \leqslant \sigma \leqslant \sigma_X$ is a fixed constant. Since $Y_2 = X - Y_1$, we then have

$$\sigma_{Y_2}^2 = \sigma_X^2 + \sigma^2 - 2 \cdot \text{Cov}(X, Y_1) = \sigma_X^2 + \sigma^2 - \rho(X, Y_1) \cdot \sigma_X \cdot \sigma, \quad (6.6.8)$$

where $\rho(X, Y_1) = \text{Cov}(X, Y_1)/(\sigma_X \cdot \sigma)$ is the correlation coefficient. As is well known, $\rho(X, Y_1)$ attains its maximum value, 1, if and only if Y_1 is an increasing linear function of X. Hence, the proportional risk exchange

$$Y_1^* = c \cdot X, \quad Y_2^* = (1 - c) \cdot X, \quad (6.6.9)$$

where $c = \sigma/\sigma_X$ (in order to have $\sigma_{Y_1^*} = \sigma$), maximizes the correlation coefficient and, therefore, minimizes the variance of company C_2 among all those risk exchanges (Y_1, Y_2), which satisfy the condition $\sigma_{Y_1} = \sigma$.

(c) The Pareto optimal. Note that the standard deviations of the proportional shares Y_1^* and Y_2^* in the risk exchange (6.6.9) are $c \cdot \sigma_X$ and $(1 - c) \cdot \sigma_X$, respectively, where c $(0 \leqslant c \leqslant 1)$ is a parameter. It can be seen that if the value of c (i.e. σ) is changed, then the standard deviations change by equal amounts but in opposite directions. Therefore all risk exchanges (6.6.9) satisfy condition (3) of section 6.6.3, and we can conclude that these are the optimal risk exchanges.

A set of optimal solutions such that any other solution is worse for at least one of the parties is called **Pareto optimal**.

The remaining condition (2) restricts the values of c to the interval

$$1 - \frac{\sigma_{X_2}}{\sigma_X} \leqslant c \leqslant \frac{\sigma_{X_1}}{\sigma_X}. \tag{6.6.10}$$

as a straightforward calculation shows. Note that since, quite generally, $\sigma_{X_1 + X_2} \leqslant \sigma_{X_1} + \sigma_{X_2}$, where the equality holds true only if X_1 and X_2 are already positively linear functions of each other, the equations (6.6.10) really define a proper interval of values of the parameter c.

If the variance is used as a measure, then every Pareto optimal risk exchange $(c \cdot X, (1 - c) \cdot X)$ satisfying (6.6.10) is better for both companies than the original situation. The final choice of the parameter c to be used is to be negotiated by the parties. Some simple rules can be suggested in order to reach a unique solution. In the present case, for example, we could decide to choose the optimal solution corresponding to the midpoint of the interval (6.6.10). Such problems of choice belong to the realm of so-called game theory (section 6.6.6).

REMARK 1   In practice a risk exchange of the aggregate claim amounts between two insurance companies means a very close relation between the companies, because each party accepts all risks underwritten by the other. Partial risk exchanges (i.e. pools), which only deal with larger claims for certain specified types of risks are more common in practice.

REMARK 2   The variance is usually a suitable measure of risk only in cases where the skewness of the total aggregate claim amount X is small. Therefore, additional reinsurance covers for exceptionally large or catastrophic claims are usually needed in addition to a risk exchange arrangement.

All the results of this, section 6.6.3 also holds true in the case of more than two companies C_i.

(d) Generalizations. Suppose, more generally, that both companies

C_i want to find a risk exchange (Y_1, Y_2) that minimizes the expected harm H_{i, Y_i} (6.6.1(b)) of their shares. Since the harm functions h_i are convex, the theorem of Borch (Gerber, 1979) gives the result that there is again a one-parameter family of Pareto optimal risk exchanges (Y_1^*, Y_2^*) such that the shares of both companies are increasing functions of the total aggregate claim amount $X = X_1 + X_2$. However, these functions are not usually linear functions of X as they were when variance was used as a measure of risk. On the contrary, it can be shown that, given any risk exchange (Y_1, Y_2) such that the shares of both parties are increasing functions of X, there exist convex functions h_1 and h_2 such that this particular risk exchange (Y_1, Y_2) is optimal if these convex functions are used as harm functions (Pesonen, 1984). The conclusion, therefore, is that no more can generally be said about optimal risk exchanges than that the shares of both parties in an optimal risk exchange are increasing functions of the total amount X.

(e) Optimal reinsurance.   Note that all the results of this section can be applied to the case of reinsurance by simply substituting $X_2 \equiv 0$.

If the variance is used as a measure, condition (2) of section 6.6.3(a) can be removed, since the variance of the reinsurer's share Y_2 in a risk exchange obviously increases as compared with the case that the reinsurer carries no risk at all (from this cedant). But the reinsurer compensates for the increased risk by loading the reinsurance premium more, the larger the variance of Y_2. It can be seen from the result of section 6.6.3(b) that the quota share reinsurance is still optimal in this case.

It is important to note that in most cases the variance of the retained aggregate claim amount is not a satisfactory measure of risk when an insurer is comparing different reinsurance covers. The reasons for this are those already mentioned in Remark 2 at the end of section 6.6.3(c). Indeed some individual risks in an insurance portfolio may be very large, or there may be a danger of accumulation of claims caused by a storm, for example. Then a claim or a catastrophe may occur which gives rise to loss amounts which exceed by many times the maximum loss the insurer can carry without major financial difficulties. This would mean that if the reinsurance were arranged through a quota share treaty, the retained quota should be very small or, in other words, most of the premium income should go to the reinsurers.

If the expected harms H_i of both companies are used as a measure of risk, then, theoretically, a parametrized family of optimal reinsurance treaties is obtained. For practical applications, however, the problem referred to in Remark 2 of section 6.6.3(c) still remains.

Exercise 6.6.1 Suppose N insurance companies C_i have mutually independent aggregate claim amounts X_i such that the standard deviation $\sigma_{X_i} = \sigma$ is the same for every i. The companies want to minimize their variances by a reciprocal risk exchange arrangement. The companies have agreed to choose the optimal risk exchange, which reduces the standard deviations of all parties by the same ratio r. Show that $r = 1/\sqrt{N}$.

6.6.4 Retention limits in a multiline portfolio

(a) Problem. As a further example of how risk measures can be used in solving optimization problems, we consider a multiline port-folio with a view to find a rational method of setting the net retention limits for each line, having regard to the relevant characteristics of the line, so that an optimal profit for the whole business can be obtained. For simplicity the variance is used as a measure of risk.

Let the portfolio be subdivided into lines $j = 1, 2, ..., k$. The expected number of claims n_j, safety loading λ_j, claim size d.f. S_j and the standard deviation σ_{q_j} of the mixing distribution of the compound aggregate claim amount distribution are assumed to be known for each line j. It is assumed that each line j has an excess of loss re-insurance cover, the respective retention limits being denoted by M_j.

The retention limits M_j are to be determined so that the expected amount of return (profit)

$$r(M_1, ..., M_k) = \sum_j \lambda_j \cdot n_j \cdot m_j(M_j) \qquad (6.6.11)$$

is maximized subject to the condition that the ruin probability does not exceed a fixed level ε.

If, for simplicity, the retained aggregate claim amount d.f. of the whole portfolio is approximated by the normal distribution, then the skewness term R_y in the basic equation (6.1.8) vanishes. By (6.1.8) the condition that the ruin probability is equal to ε is then equivalent to the condition

$$Q(M_1, ..., M_k) = U - y_\varepsilon \cdot \sigma_X(M_1, ..., M_k)$$
$$+ \sum_j \lambda_j \cdot n_j \cdot m_j(M_j) = 0, \qquad (6.6.12)$$

where $m_j(M_j) = a_{1,j}(M_j)$ (see (3.4.5)) and

$$\sigma_X(M_1,...,M_k) = \sqrt{\sum_j (n_j \cdot a_{2j}(M_j) + n_j^2 \cdot m_j(M_j)^2 \cdot \sigma_{q_j}^2)}. \qquad (6.6.13)$$

The second moments a_{2j} are obtained from (3.4.8).

(b) Solution. We have to find the point $(M_1,...,M_k)$ on the $(k-1)$-dimensional surface defined by the equation $Q = 0$ where the return function r reaches its maximum. This maximization problem can be solved by using the Lagrangian method, introducing a function

$$F = r - \rho \cdot Q \qquad (6.6.14)$$

where ρ is an auxiliary variable. The partial derivatives of F relative to the variables M_j are formed and equated to zero. First note that

$$\frac{da_{ij}(M_j)}{dM_j} = \frac{d}{dM_j} \left[\int_{-\infty}^{M_j} Z^i \, dS_j(Z) + M_j^i \cdot (1 - S_j(M_j)) \right]$$

$$= i \cdot M_j^{i-1} \cdot (1 - S_j(M_j)) \qquad (6.6.15)$$

(an alternative proof is given in Exercise 3.4.9). The existence of the necessary derivatives is assumed, as well as the independence of the safety loadings λ_j relative to the retention limits M_j. Then

$$\frac{\partial F}{\partial M_j} = n_j \cdot (1 - S_j(M_j)) \cdot \left[(1 - \rho) \cdot \lambda_j + \frac{\rho \cdot y_\varepsilon}{\sigma_X} M_j + \frac{\rho \cdot y_\varepsilon}{\sigma_X} n_j \cdot m_j(M_j) \cdot \sigma_{q_j}^2 \right]$$

$$= 0 \qquad (6.6.16)$$

for each j. The required solution, if it exists for the given values of the parameters, can be found among the points where these derivatives and Q are zero. Because the target variables M_j are contained within the expressions for the moments, this may lead to technical problems, which are not dealt with here. However, the above expression permits some conclusions to be drawn about the character of the solution.

Suppose that the expression in brackets in (6.6.16) is equal to zero. Solving for M_j therefrom we have

$$M_j = \frac{(\rho - 1) \cdot \sigma_X(M_1,...,M_k)}{y_\varepsilon \cdot \rho} \lambda_j - n_j \cdot m_j(M_j) \cdot \sigma_{q_j}^2. \qquad (6.6.17)$$

Substituting these into (6.6.12), the value of ρ can be determined.

Note that the multiplier for λ_j in (6.6.17) is the same for all lines j. Hence, this term contributes to the amount of M_j in proportion to the safety loading, i.e. the better the expected profitability of the line the higher the suggested retention.

In the particular case where the mixing is negligible, i.e. $\sigma_{q_j} \approx 0$ for every j, the following theorem ensues: The retention limits M_j should be chosen in proportion to the corresponding safety loadings λ_j.

The second term of the right-hand side of (6.6.17) reduces the required retention in proportion to the variance of the mixing variable attributable to this line. It may even make M_j negative if the mixing variation is large. This is not an unexpected result, since the assumed excess of loss treaty does not give protection against variation in the number of claims.

(c) Discussion. It should be recalled that the above conclusions are dependent on the assumptions, particularly that the reinsurance premium does not depend on the profitability of the ceded business. In practice this might often not be the case, which makes the real-life situation much more complicated. However, the result might be generally relevant in that it is in the interest of the cedant to keep on net retention as much as possible of the profitable parts of his business. This contradicts any doctrine that the same retention limit should be applied consistently over the whole portfolio.

6.6.5 The concept of utility

It was noted in section 6.6.1(b) that the variance is not ideal as a general optimality measure. Instead it is useful to construct a measure that compares more satisfactorily the desirability of the various alternatives between which a choice has to be made in a decision-making situation. For this purpose the concept of utility is introduced and mathematically formulated. It will be introduced by means of a simple example which follows the idea presented by Jewell in his lecture at the Oberwolfach seminar in 1980.

(a) Policyholder's decision-making. Assume that a potential policy-holder is deliberating whether or not to take out insurance cover for a particular risk object of value X. The probability of an accidental loss of the object is p and the policyholder's initial capital is U_0. The policyholder has to choose between the alternatives

(I) Take out insurance and pay a premium B.
(II) Save the premium B and take the risk of loss of the amount X.

Table 6.6.1   *Decision options of a potential policyholder*

	The policyholder's wealth U if	
	(I) *Insurance* *taken*	*(II)* *Insurance* **not** *taken*
In case of		
(i) no incidence of loss	$U_0 - B$	U_0
(ii) incidence of loss	$U_0 - B$	$U_0 - X$

The utility function $G(U)$ is now constructed to describe the decision-making situation (Table 6.6.1) and to give the problem a mathematical formulation. It attempts to weigh up how desirable the different outcomes are for the decision-maker. Clearly it is better for the policyholder the larger his or her wealth U. It is, however, reasonable to assume that desirability is not directly proportional to U. If the decision-maker loses a substantial part of the wealth, it may cause considerable inconvenience or even hardship. A consequence, for example, may be the loss of the home or loss of the ability to afford education for the children. Hence special weight should be given to the outcomes where the wealth falls to a very low level (not to say negative, for instance if a loan was originally taken to finance the property; in the case of fire the property is lost but the loan remains!). The desirability of such events is best described by low values of the utility function. On the other hand, growing wealth is generally accepted as a positive event, but the wealthier someone becomes, the less weight can be assumed for the desire to increase the wealth still more by a certain fixed amount ΔU. These personal aspects, conveniences and inconveniences will be described by constructing a function $G(U)$ to evaluate numerically the desirability or, as it is called, the **utility** of different outcomes.

The utility function $G(U)$ can be assumed to be an increasing function of wealth U, and the aspects discussed above suggest that it should be concave, as illustrated in Figure 6.6.3, i.e.

$$G'(U) > 0 \quad \text{and} \quad G''(U) < 0. \tag{6.6.18}$$

The alternatives available to the potential policyholder in the above simple example are depicted in Figure 6.6.3, relating them to a hypothetical utility function. It seems natural to characterize the two decision alternatives, i.e. to insure or not, by taking into account

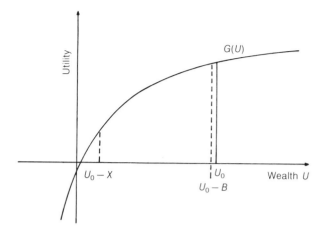

Figure 6.6.3 *Schematic presentation of the options available to a policy-holder.*

on the one hand the utility function G and on the other hand the probability that the particular outcome may occur. This suggests the concept of **expected utility**, which in option I is

$$G_I = (1 - p) \cdot G(U_0 - B) + p \cdot G(U_0 - B) = G(U_0 - B) \qquad (6.6.19)$$

and in option II, i.e. not to insure, is

$$G_{II} = (1 - p) \cdot G(U_0) + p \cdot G(U_0 - X). \qquad (6.6.20)$$

Insurance should be taken if $G_I > G_{II}$, otherwise not. This depends on the selection of the utility function, as well as on the initial capital and the values of p, X and B.

Note that in the particular case in which G is the linear function $G(U) = U$ (note that then G is no longer strictly concave, since $G'' = 0$), the expected utility is, in fact, the expected value and we have

$$G_I - G_{II} = -B + p \cdot X. \qquad (6.6.21)$$

Because $p \cdot X$ is the risk premium $E(X)$ and hence, if the rating is correct, less than the gross premium B, this expression is always negative. This means that on the average the policyholder is losing out. Insurance is, however, meaningful if the possible loss, X, is large compared with the wealth U_0 and the consequences are harmful, maybe extremely so, if not covered by insurance. A linear utility function is not, therefore, suitable and the concave shape is likely to describe the actual situation more accurately.

The above example was simplified in that only two outcomes were assumed. A more general situation is where several, perhaps continuously varying, outcomes are possible. Then the probabilities p and $1 - p$ of different outcomes U_0 and $U_0 - X$ are replaced by a d.f. F of a stochastic outcome U which results from the decision made, and the expected utility can be expressed as

$$G_U = \text{E}(G(U)) = \int_{-\infty}^{+\infty} G(U)\,\mathrm{d}F(U). \qquad (6.6.22)$$

It is called the utility of the outcome U.

The simple Figure 6.6.3 is now replaced by Figure 6.6.4. Option I is conservative, the distribution not extending to very small values, whereas option II implies both very large positive and very small, even negative, outcomes. The weighted utility for option II is, however, larger.

As was mentioned earlier, the expected harm introduced in section 6.6.1 is conceptually very close to the concept of utility.

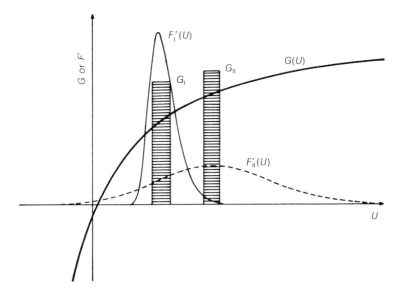

Figure 6.6.4   *The idea of utility applied in a decision situation where two strategy options are available, conservative (I), and risk-taking (II).*

(b) Utility applied to the insurer. The utility concept is useful to describe rational behaviour in simple situations such as in the case of the policyholder, notwithstanding the fact that constructing the policyholder's utility function may be difficult.

On the other hand, many management situations in an insurance company, not to speak of the overall control of an insurance company, are so complicated and continually changing under the pressure of external and internal forces and impulses that it has not been possible in practice to find a utility function to describe the total control of a company. This topic will be discussed further in section 14.5 in the context of business planning.

(c) References. The idea of a utility function dates back as far as 1738, when a famous article by D. Bernoulli (1738) was published in St. Petersburg.

The utility concept is widely used in the economics literature in attempting to describe the behaviour of individuals, institutions and whole markets. Borch (1961) and (1974) was a pioneer in applying the utility approach in the insurance environment.

In this section only the utility of monetary outcomes has been considered. The modern axiomatic utility theory, introduced by von Neumann and Morgenstern (1947, reprinted 1964), deals, however, with a very general concept of utility.

Exercise 6.6.2 A potential policyholder has a risk of loss of amount 75 (in suitable monetary units) with evaluated probability $p = 0.01$. His initial wealth U_0 is 100 and utility function $1 - e^{-aU}$, coefficient a being 0.01. How much at most can the premium be so that, in terms of utility, it is preferable for him to insure?

Exercise 6.6.3 Prove that a utility function G and its linear transform $\bar{G} = a \cdot G + b$, with $a > 0$, both lead to the same order of preferences, i.e. $G_U > G_V$ if and only if $\bar{G}_U > \bar{G}_V$.

6.6.6 Allocation of the safety loading – a link to the theory of multiplayer games

(a) Safety loading problem. An insurance company needs a solvency margin U as protection against risk fluctuations. Formulae of type (6.1.8) were used as an approximate measure for the purpose in previous sections and the problem will be considered in greater detail in section 13.2. Omitting the skewness term in (6.1.8), the required minimum capital is assumed to be

$$U = y \cdot \sigma_X = y \cdot P \cdot \sqrt{r_2/n + \sigma_q^2} \qquad (6.6.23)$$

where y is a proportionality factor, say 3. In the case of a proprietary company it is expected that the portfolio will provide a return on this capital. In the case of a mutual company, U will be created by self-financing and maintained from the safety loading λ, which gives a total yield λP proportional to U, or, as is approximately the same according to (6.6.23), to the standard deviation σ_X of the aggregate claims.

Another way to arrive at the same result is to suppose a hypothetical situation where a group of policyholders is going to establish an insurance company and the minimum initial capital U is to be raised by means of a levy from each member of the group.

(b) Multiplayer approach.   Note that the above formulae and reasoning suggest only that the total safety loading income should be proportional to the required minimum capital U or to the standard deviation σ_X, i.e.

$$\lambda P = kU \qquad (6.6.24)$$

where k is a proportionality factor. Because P is the total premium income associated with the whole portfolio, this does not determine how the total loading is to be distributed among individual policies or policy groups. A natural approach is to load all the policies equally in proportion to the standard deviation or, possibly, to some other measure of risk. A rationale for this is to let each policy pay for the contribution which its volatility makes to the total capital requirement of the portfolio. In fact, this is a standard procedure often proposed in the actuarial literature. We will now demonstrate some of the consequences, using a simple example.

Let us assume that the portfolio is composed of three groups of insureds. Group 1 comprises small risks, such as motor cars, family property, etc., group 2 contains large risks such as industrial plants, ships, aircraft, etc. and group 3 is characterized by exceptionally great short-term variation such as is typical for forest insurance or other coverage against natural events. The basic characteristics are given in Table 6.6.2.

The quantity U_j is the required capital (6.6.23) if each group j constitutes an insurance collective. Without loss of generality,

Table 6.6.2 *Portfolio composed of three groups $j = 1, 2, 3$. Combined groups are denoted by ih. The expected number of claims n, risk indices r_2 (3.2.16), standard deviations of the mixing variable σ_q, mean claim size m and the premium income $P = nm$ are given for the groups. Monetary unit $£10^6$*

1	2	3	4	5	6	7	8	9	10	11	12	13
j	n_j	r_{2j}	σ_{q_j}	m_j	P_j	U_j	ih	U_{ih}	G_{ih}	G_j	G_j^σ	G_j^V
1	10 000	5	0.05	0.0028	28.0	4.6	12	26.8	4.2	2.0	1.2	0.3
2	5000	150	0.10	0.0088	44.0	26.4	13	8.1	3.2	5.1	7.1	9.2
3	1000	5	0.80	0.0028	2.8	6.7	23	27.2	5.9	3.0	1.8	0.6
Σ	16 000	–	–	–	74.8	37.7				10.1	10.1	10.1

the proportionality factor k is taken to be 1. U_{ih} is the capital requirement if two of the groups constitute a collective as a pair (section 6.2(e)) and finally $U_{123} = 27.6$ is the capital for the whole combined portfolio.

It is important to note that the combined capital requirement is considerably less than the sum of the separate components in column 7. This is the same feature which was shown in section 6.2(e). The difference

$$G_{123} = U_1 + U_2 + U_3 - U_{123} = 10.1 \qquad (6.6.25)$$

is called the **gain** achieved by cooperation among the groups. The problem is how to divide this gain amongst the groups. A natural condition is to require that the share G_j of each group should be positive, subject to the condition

$$\sum_{j=1}^{3} G_j = G_{123} \qquad (6.6.26)$$

so that it pays to join the combined portfolio rather than to establish a separate one. Another condition is that it should not pay for any two of the groups to establish their own company, excluding the third, i.e.

$$G_i + G_j > G_{ij} \quad (i \neq j; \ i, j = 1, 2, 3) \qquad (6.6.27)$$

where G_{ij} is the gain achieved if groups i and j are combined, as compared to the capital required if they are separate. For example, $G_{12} = U_1 + U_2 - U_{12} = 4.2$, as shown in Table 6.6.2.

The above conditions are not sufficient to determine the shares G_j uniquely. As a first proposal let these shares be assessed proportional to the standard deviation σ_X or, equivalently, to the separate capital requirements U_j. The results are shown in column 12 of Table 6.6.2. It can be seen that the condition (6.6.27) is not fulfilled for the combination 1, 3. The groups 1 and 3 together would gain less ($= 1.2 + 1.8$) than if they had established their own company without group 2 ($G_{13} = 3.2$). This suggests that the standard deviation rule overloads the groups of small risks in favour of the big risks (group 2).

The situation is even more inequitable if the variance is used as a measure of risk, i.e. the loading is defined in proportion to the variance. This is shown in column 13 of Table 6.6.2.

There are, however, combinations (G_j) which satisfy all the conditions postulated. Column 11 illustrates one. In the theory of games such a set of possible solutions is called the **core** of the game. To arrive at a unique solution further conditions are required. The situation is similar to that of the Pareto optimal in section 6.6.3(c). It is, however, beyond the scope of this book to consider the different philosophies for sharing the core between the participants. Reference can be made to the literature on game theory, such as the items referred to in section 6.6.6(e) below.

Extending the problem to the case of more than three companies is straightforward.

(c) Aspects of equity. The above example introduced to the problem of loading premiums a feature which is important from the point of view of both theory and practice. The loading rules, whatever they might be, should avoid loading any of the types of insurance or groups of risks excessively so that the group or combination of groups could achieve cheaper insurance separately, for example, by establishing their own company (captive or other). In practice, however, inequitable premium loading may not give rise to the birth of a new company. Instead it can be expected that some competing insurers will establish a more equitable tariff structure. Then those policyholders who are being overcharged will probably move their policies to that company. A **selection** phenomenon may arise, in that those policies which are undercharged will remain and those overcharged will go elsewhere. If this trend grows, it could lead to severely impaired profitability.

The above considerations in relation to safety loadings are equally applicable for expense loadings, with similar caveats. This will be dealt with in more detail in Chapter 11.

(d) Company size. Another interesting observation which can be drawn from (6.6.24) is that the safety loading

$$\lambda = kU/P \qquad (6.6.28)$$

is smaller the larger the portfolio. This means that in theory a large company can get by on a lower safety loading than a small insurer, other things being equal.

(e) References. The fundamental work on game theory was presented by von Neumann and Morgenstern (1947). Owen (1969) and Driessen (1988) can be referred to as study books.

The idea of applying game theory to the insurance environment was suggested by Borch (1960, 1961).

Part Two

Stochastic Analysis of Insurance Business

In Part Two consideration is extended to
the analysis of variability of
assets and liabilities
with a short- and long-term horizon
and to
practical aspects of modelling the insurance business

CHAPTER 7

Inflation

7.1 Introductory remarks

(a) Changes of value. Both when analysing the behaviour of an insurance undertaking and when developing appropriate models it is important to allow adequately for changes of monetary value. Inflation affects many different aspects of an insurer's financial situation. Expenses of operation depend to a large extent on the salaries of employees, which are affected by general earnings inflation. Sums insured, and hence premiums receivable, can be expected to increase as the value of the insured property, or required indemnity, increases. In some cases, for example with employers' liability business, coverage and premiums may depend directly on the salary roll. In many lines of business the amount which has to be paid to settle claims often increases the longer settlement is delayed.

(b) Consumer price inflation is in many countries the best documented time series of changes in monetary value. An index of consumer price inflation is often calculated and published monthly, on the basis of a weighted average of the costs to the consumer of purchasing a specified basket of items, which usually includes services as well as goods. Figure 7.1.1 shows the growth in the consumer price index over the period from 1960 to 1991 in six economies. Over this period the index of consumer prices in Finland increased more than tenfold, while that in Germany increased about fourfold, reflecting very different general levels of inflation. All countries have, however, experienced considerable variation from year to year in the rate of inflation, as can be seen in Figure 7.1.2, which illustrates the annual increases in the consumer price indices in each of the six countries over the 30-year period. Even this picture conceals a great deal of the true month-to-month variability. Figure 7.1.3 shows the monthly changes in the UK retail price index from 1960 to 1991.

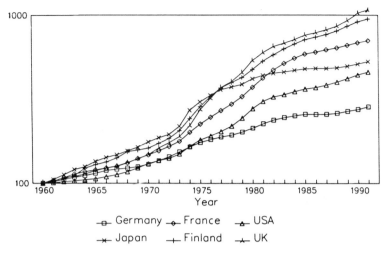

Figure 7.1.1   *Consumer price indices for six countries from 1960 to 1991.*

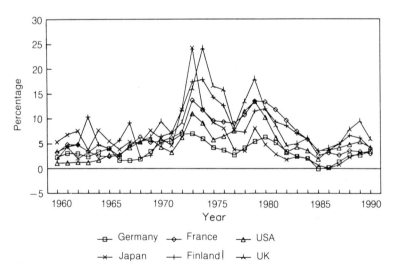

Figure 7.1.2   *Annual increases in consumer price indices for six countries from 1960 to 1991.*

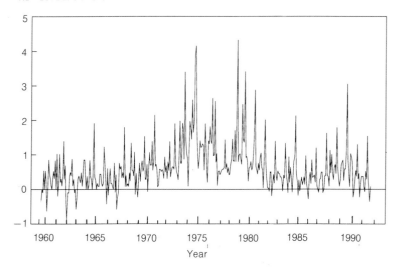

Figure 7.1.3 *Percentage change in the UK monthly consumer price index from 1960 to 1991.*

(c) Assumptions about the future rate of inflation are often made by actuaries, for example when assessing funding rates for pension schemes or profit-testing life insurance products. In many cases the most important assumption is the real rate of return net of inflation. The experience in some developed economies has tended to suggest that the real rate of return is a less variable quantity than either nominal rates of return or the rate of inflation. Although for some applications it may be adequate to assume a constant future level of inflation or real rate of return, actual outcomes can be expected to show a high degree of variability. For some purposes, therefore, it is essential to develop a stochastic model for inflation. A stochastic approach may be of particular value in testing resilience to the range of possible outcomes, rather than for making specific forecasts or projections.

(d) Inflation and the economy. Reference should be made to economic text-books for discussion of the factors which give rise to inflation. Some have argued that there is a close link between inflation and economic cycles; others that the rate of inflation is strongly influenced by the rate of expansion of the money supply. Inflation is a symptom of instability in an economic system. It may reflect imbalance between supply and demand, or between wage increases and productivity improvements, exchange rate pressures or management of the domestic economy. Inflation may be caused, or exacerbated, by exogenous effects, such as sudden movements in the prices of major commodities, for example oil. Historically, negative inflation has occurred as much as positive inflation. However, few industrialized countries have experienced negative inflation since the Second World War.

7.2 Inflation and insurance

(a) Expense inflation.   The largest single component of the expenses of an insurer relates to the salaries of staff. In an economy where there is real growth, earnings increases will exceed increases in consumer prices over most periods, reflecting real increases in the standard of living of the economically active population. Figure 7.2.1 shows

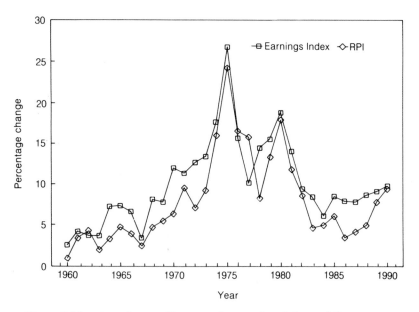

Figure 7.2.1   *Annual rates of increase of an earnings index and the consumer price index in the UK from 1960 to 1990.*

the annual increases in an earnings index for the UK over the period 1960 to 1990, with the corresponding annual increases in the consumer price index shown for comparison.

Neither the earnings index nor the consumer price index will necessarily give a good indication of changes in an insurer's total costs. Although these will be dominated by salaries, they will also reflect items which may not appear with appropriate weightings in the consumer price index, such as property costs, heating and lighting, computers and other office equipment, stationery and postage. The development may, on the other hand, be more favourable than might be indicated by movements in an index, because of gains from efficiency and productivity. All these factors would need to be taken into account in modelling the future expense development of an insurer (Chapter 11).

(b) Claim inflation. In many classes of general insurance business the insurer provides an indemnity to the policyholder against economic loss. The cost of rebuilding a house that has been damaged by fire, for example, or of repairing a vehicle that has been in an accident, can be expected to increase year by year (effectively continuously), reflecting changes in building costs and motor repair costs respectively. Other claim costs reflect the impact of court judgments in quantifying liability or assessing damages. Thus different types of business will experience different inflationary pressures. We use the term **claim inflation** to include the component due to general price inflation and an additional amount (which may in some circumstances be negative) in respect of factors specific to the particular line of insurance. In some classes of business, e.g. liability, this additional component is sometimes referred to as **social inflation**, since it reflects the changing values of society.

(c) Inflation and technical reserves. In some lines of business only a relatively small proportion of claims will be settled in the same accounting year as the event giving rise to the claim. At the end of the year provisions must be established for the cost of meeting all outstanding claim payments, including payments in respect of claims which have been incurred but not reported (section 9.5.1). It will usually not be known how long it will be before any particular claim is settled, although experience with a line of business may enable a reasonable estimate to be made of the pattern of claim settlement in respect of each year of underwriting.

Where claim payments are directly affected by inflation to the date of settlement, for example because of court awards of damages, it is prudent to allow for future claim inflation when assessing the appropriate outstanding claim reserve, and to provide for the ultimate cost of settlement. This approach is envisaged by the European Communities (EC) Insurance Accounts Directive, which requires provisions to be established in respect of the full expected cost of settling the claims. Implicit discounting, i.e. allowance for less than the full expected future cost of settling claims, in anticipation of the receipt of investment income from the assets backing the technical reserves, is not permitted under the EC rules.

Thus the inflation of claim costs during the period of delay until the claims are settled should be taken fully into account in assessing technical reserves. The EC Insurance Accounts Directive does not rule out the possibility, under certain conditions, of discounting to allow for future investment income in establishing technical reserves for long-tailed classes of business. This issue is discussed more fully in section 9.5.2.

(d) Inflation and premiums.   The inflation of claim costs should also be taken into account in setting premium rates. The claim experience which is available to an insurer will inevitably reflect a period prior to that for which premiums need to be set, often several years earlier, so that estimates will need to be made of the impact of subsequent inflation on claim costs in arriving at a suitable premium basis. In many classes of business the recent claim experience can only be analysed using estimates of the effect of future inflation on the ultimate settlement cost of claims. Such estimates may be implicit if the methodology effectively assumes that future inflationary trends will be similar to those in the past. Unless inflation is properly taken into account, premiums may be set at an inappropriate level. This could have serious consequences for the viability of the insurer, as inadequate premium rates can soon endanger solvency, particularly if their inadequacy is not recognized quickly.

Owing to the feature mentioned above whereby premium rates have to be estimated from information which is not fully up-to-date, sudden and unexpected changes in claim inflation are reflected in premium rates only after a certain lag. This is an important practical feature and one which contributes to cyclical fluctuations in the profitability of the insurance business (section 12.3).

(e) Inflation and investment return.   It has been observed in many countries that there is some correlation between investment return and inflation. It is not surprising that investors demand some compensation for the effect of inflation and that the expectation of future inflation has a strong influence on security prices and hence on the rate of return available to investors. If the rate of inflation is reasonably stable, bond prices tend to adjust so as to enable investors to obtain a reasonable positive real rate of return, i.e. a return over and above that necessary to compensate for the fall in the purchasing power of money due to inflation. When inflation is changing rapidly the full potential impact may not be reflected very quickly in prices and substantial fluctuations can occur in the real rate of return obtained by investors.

The income from real assets, such as equities (ordinary shares or common stock) and property (real estate), will itself reflect inflation, since both shareholder dividends and property rack rentals will tend to increase in an inflationary environment, not necessarily exactly in line with consumer price inflation but in a broadly similar way. (The rack rental is the appropriate underlying market rental. The actual rental may differ from this because of the incidence of rent reviews.) The market values of these types of investment will also tend to grow over time as a result (or fall if there is negative inflation), although the trend may be disguised by short-term fluctuations in the market, and there may be some lags as the effects work through.

Daykin (1976, 1987) has studied the real rates of return available in the UK since the end of the Second World War on investment in equities and irredeemable gilt-edged securities. These investigations confirmed the relative stability of real rates of return as compared to nominal rates of return (before adjustment for inflation).

(f) Inflation and claim frequency. Apart from the direct impact of inflation on claim size, as discussed in section 7.2(b), there may also be an indirect relationship with claim intensity. This could arise, for example, because of an association between economic cycles and changes in the inflation rate, giving rise to an increased level of danger for particular classes of insurance in certain phases of the economic cycle. The effects will differ for different classes of business, with some being most at risk during boom periods and others during periods of recession. In practical applications it is desirable to consider possible correlations of this nature, since they can greatly increase the potential dangers for an insurer.

The sharing of claim costs between the insurer and the insured will depend on the amount of any deductibles. These should in principle be increased in line with inflation. Otherwise there may be significant gearing of the insurer's claim costs. Claim frequency may also increase in an unpredictable way as claims of a size previously too small to involve a payment by the insurer come into the picture. Similar features can arise with excess of loss reinsurance, unless the excess limits are fully indexed (sections 3.4.2(d) and 3.4.4(c)).

7.3 Modelling inflation

(a) Characteristics of inflation. As was seen in section 7.1(b), the rate of inflation differs substantially from country to country and from time to time. Over the period illustrated in Figure 7.1.2, none of the countries shown experienced an inflation rate of more than 25%. However, much higher rates of inflation, sometimes many hundreds of per cent a year, and generally referred to as **hyper-inflation**, have been experienced in some countries in recent years (Israel, Argentina, etc.) and, for example, in Germany in the years following the First World War.

Even within the relatively modest range of values shown in Figure 7.1.2 it is easy to distinguish some periods of relatively slow change in inflation rates and other periods (for example 1971–74 and 1977–80) when inflation suddenly increased much more rapidly. These phenomena seem to have been common to all the countries illustrated (and many others), although the inflation rate rose much higher at its peak in some countries than in others. Such bursts of inflation can arise because of characteristics of the national economy or, as was the case in the 1970s, because of factors which affect a large number of economies around the world, such as a dramatic rise in the price of oil.

The relatively modest fluctuation in rates of inflation at other times may be explicable in terms of local economic conditions, and in the short term can sometimes be forecast, with greater or lesser success, using econometric models. Taking a somewhat longer view, however, there appears to be a substantial random element in the development of rates of inflation and they can be viewed as a stochastic process. The Finnish Working Parties (1982, 1989) proposed that a distinction should be drawn between **steady** inflation and **shock** inflation, where a relatively large sudden change in the rate of inflation takes place, and that this essential characteristic should be reflected in practical models. In view of some of the lags which are present in insurance systems, such as have already been noted in

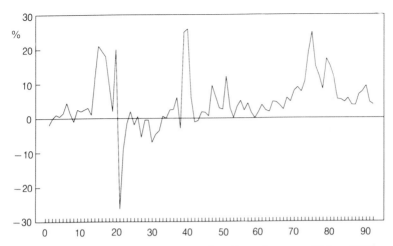

Figure 7.3.1   *The rate of inflation in the UK in the years 1900 to 1992.*

sections 7.2(d, e), the impact of a sudden change of inflation can be
very different from that of gradual but small changes. Clarkson, in an
appendix to Geoghegan *et al.* (1992), has proposed that a distinction
should also be drawn between high variability when inflation is high
and low variability when it is low.

It should be noted that negative inflation (a fall in consumer prices)
can occur, but such phenomena have not been much in evidence in
free-market industrialized economies since the Second World War.
Figure 7.3.1 illustrates an estimate of the rate of inflation in the UK
over a longer historical period, from which it will be seen that periods
of negative inflation have indeed occurred quite regularly. Many of
these periods reflected particular circumstances, such as the after-
math of a major war. Recent experience in industrialized economies
suggests that there is now more often upward pressure on inflation
than tendencies which would lead to negative inflation.

(b) Purpose of the inflation model.   As has been outlined in the
preceding sections, the effects of inflation can be seen in many areas
of the insurance business. For many practical applications of risk
theory, therefore, in modelling the behaviour of an insurer, or a part
of the business of an insurer, it is essential to have regard to the
potential impact of inflation. Allowing for a constant future rate of

inflation may be sufficient in some applications, but is essentially unrealistic and does nothing to expose the vulnerabilities of the business to fluctuations in the rate of inflation. For some purposes it may be appropriate to use a range of deterministic scenarios. Use of a stochastic model provides a generalization of this process. It will be seen in Chapter 8 that a stochastic model of inflation may be necessary in order to develop an adequate model of asset variability.

A stochastic model aims to simulate the behaviour of inflation so as to reproduce realistic future inflation scenarios across the whole range of what might be regarded as possible, but reflecting their respective probabilities of occurrence. The fluctuations should correspond in broad terms with those observed empirically. However, the inflation model does not forecast actual future inflation outcomes. Furthermore, the models discussed here do not attempt to relate inflation to exogenous predictor variables, such as might be the case with some types of econometric model.

(c) Autoregressive inflation process. Although different types of inflation index may be appropriate for different lines of business and for different purposes (for example in regard to claims and expenses), a useful starting point is a model of general price inflation, based on the historic development of the consumer price index. Wilkie (1984, 1986) made use of a first order autoregressive process to model the future changes in the consumer price index. In its simplest form the rate of inflation can be expressed as in (5.5.3), such that the divergence in one year of the ratio of inflation from a mean value is expressed as a constant times the same quantity in the previous year, together with a random white noise term.

With a non-zero autoregression coefficient a $(0 < a < 1)$ to control the tendency to regress to the mean, the rate of inflation $i(t)$, over the period $t - 1$ to t, is given by

$$i(t) - \bar{i} = a \cdot [i(t-1) - \bar{i}] + \sigma_i \cdot \varepsilon(t) \qquad (7.3.1)$$

where

$i(t) = I(t)/I(t-1) - 1$ is the incremental rate of inflation;

$I(t)$ is the value of the consumer price index at time t;

$\bar{i}$ is the average rate of inflation;

σ_i is the standard deviation of the noise term; and

$\varepsilon(t)$ is a random variable, independent but identically distributed for different values of t, standardized with zero mean, unit standard deviation.

Time series models of this type are discussed in more detail in Appendix G.

Equation (7.3.1) is intuitively appealing since it can be expected in general that the rate of inflation will not change very rapidly and so the level of inflation in the previous year will exert a strong influence. There may also be a general level of inflation which it seems reasonable to assume over the long term, rather than allowing the modelled future rate of inflation to fluctuate completely at random.

(d) The Wilkie model. As part of a set of autoregressive models designed to simulate the behaviour of UK $2\frac{1}{2}\%$ Consols (irredeemable government bonds) and quoted UK equities (common stock), Wilkie (1984, 1986, 1986a) proposed a modification of (7.3.1) as his model for the UK Retail Prices Index.

Wilkie argued that for all index variables it is proportional changes which are important, not absolute changes. A 10% increase in prices has the same relevance, whether the price level is 10, 100 or 1000. It is therefore appropriate first to take logarithms and to work on $\ln I(t)$. Furthermore, there is no 'natural' level of retail prices. If prices rise by 10%, there is no mechanism (at least nowadays, when gold is not the monetary base) to cause them to revert to their original level. It is therefore reasonable to assume that $\ln I(t)$ is not stationary, and to look instead at the annual differences, $\Delta(t) = \ln I(t) - \ln I(t-1)$.

The range of $\Delta(t)$ is unlimited in both directions. Empirical observations show that it is distributed roughly symmetrically and roughly normally (see, however, section 7.3(e, h)). For long periods in the past in the United Kingdom the mean was close to zero. During this century the mean has been generally positive. If successive values of $\Delta(t)$ were independent and identically distributed, then $\ln I(t)$ would perform a standard random walk. However, observation shows that successive values of $\Delta(t)$ are correlated, with a correlation coefficient of about 0.6, both in the UK and in other countries. Higher order autocorrelation coefficients are of little significance.

Why should this be? One reason is that nowadays the Retail Prices Index is calculated and published monthly, and the annual rate over the preceding year is emphasized. Firms tend to review their prices annually and also to review their wages annually. An element of argument in the wage negotiations is the increase in prices in the preceding year. But an increase in wages one year often requires a corresponding increase in prices. At the level of inflation experienced in Britain there is therefore a tendency for high inflation in one year to be followed by high inflation in the following year. This justifies a first-order autocorrelation. But there is no tendency for the effect to be carried forward for two years except through the medium of the one-year step.

There is some seasonable fluctuation in inflation in the UK. Certain foodstuffs, mainly fruit and vegetables, are cheaper in the summer months than in the winter months and increases in indirect taxation are often introduced by the Government around April; but these seasonal effects are not required in a model that considers only observations at annual intervals.

The Wilkie model assumes that the noise term is normally distributed with zero mean and expresses it in terms of the product of a constant, representing the standard deviation, and a unit normal random variable. Thus, using the same notation as in the previous section, the price index $I(t)$ is given by

$$\ln\left(\frac{I(t)}{I(t-1)}\right) = \bar{i} + a\cdot\left[\ln\left(\frac{I(t-1)}{I(t-2)}\right) - \bar{i}\right] + \sigma_i\cdot\varepsilon(t) \quad (7.3.2)$$

where σ_i is the standard deviation of the white noise term and $\varepsilon(t)$ are independent, identically distributed unit normal random variables. The Wilkie model is closely related to the model at (7.3.1).

Noting that

$$\ln\left[\frac{I(t)}{I(t-1)}\right] = \ln[1 + i(t)] \quad (7.3.3)$$

and, substituting (7.3.3) into (7.3.2), we obtain

$$\ln[1 + i(t)] - \bar{i} = a\cdot\{\ln[1 + i(t-1)] - \bar{i}\} + \sigma_i\cdot\varepsilon(t) \quad (7.3.4)$$

Since

$$\ln[1 + i(t)] = i(t) + \sigma(i(t)) \quad (7.3.5)$$

the models (7.3.1) and (7.3.2) can be seen to be equivalent, to the first order in $i(t)$. Nearly full numerical correspondence is achieved if $\bar{i}$ in (7.3.2) is taken as $E[\ln(1 + \varepsilon(t)]$. (See Appendix G, section G.4(c).)

Wilkie's (1986) model for price inflation was fitted from UK data from the period 1919 to 1982. This resulted in the following parameters:

$$\bar{i} = 0.05; \quad a = 0.6; \quad \sigma_i = 0.05. \quad (7.3.6)$$

More recently a similar model has been fitted by Wilkie to the data for the period 1945 to 1989 (Geoghegan *et al.*, 1992) and the conclusion was again drawn that a first order autoregressive model provided at least as good a fit to the data as any higher order linear model. The choice of parameters, in the light of past experience and expectations for the future, requires considerable judgement, having regard to the purpose for which the model is to be used.

(e) Comments on the applicability of the Wilkie model. The model described by Wilkie (1986) represented an important step forward in the stochastic modelling of asset values and investment returns. The investment aspects of the model will be considered in Chapter 8, but it is important to note the distinctive cascade structure of the model, whereby the sub-models for the yield on irredeemable government securities, the dividends and the dividend yield on equity shares, are all driven to a significant extent by the model for the retail price index.

In a review of the Wilkie model, Geoghegan *et al.* (1992) identified three areas of concern regarding the suitability of the model

(1) The existence of bursts of inflation, indicating that once an upward trend in inflation is established, there is a tendency for it to continue.
(2) The existence of large, irregular shocks, such as those in the mid-1970s.
(3) The possible skewness of residuals (white noise).

(f) Non-constant variance of residuals. The first area of concern referred to in section 7.3(e) may be addressed by allowing the variance of the white noise term to depend on the magnitude of previous residuals. This leads to what was described in Engle (1982) as an auto-regressive conditional heteroscedastic (ARCH) model. Engle showed, in an analysis of UK quarterly inflation data, that the estimated variance of the residuals increased significantly during the 1970s, as can be seen intuitively from Figure 7.1.3. Similar conclusions were reached by S. Taylor (1986) in a study of 40 daily financial time series. In Appendix B of Geoghegan *et al.* (1992), Wilkie demonstrates how his model could be adapted to incorporate ARCH effects.

(g) Random shock inflation. It has already been observed in section 7.3(a) that periods can be identified where the rate of inflation receives a sudden impetus, either from an event of international significance such as a major increase in the price of oil, or from local economic circumstances. It is possible to envisage a model for inflation along the lines of that described in section 7.3(c) but with an additional residual term, with relatively high standard deviation (and skewness), activated by a random number generator according to a specified probability distribution. Such a model may be difficult to fit on the basis of the past data since the different elements are hard to disentangle, in particular the probability of switching from one model to the other. An alternative procedure, which may be satisfactory for some applications, would be to apply specific deterministic shocks with a view to studying the system's response (for example, sections 8.5(i), 12.3(e) and 12.3(f), (3)).

(h) Skew distribution of residuals. Wilkie (1984) observed that the residuals were negatively skewed and fatter-tailed than under a normal distribution. However, his model proposed the use of a normal distribution for the residuals, adopting a somewhat higher standard deviation than might otherwise have been thought appropriate. Pentikäinen et al. (1989) suggested that it would be easy to specify the skewness as well as the mean and standard deviation of the residual term. The three-parameter gamma function (section 3.3.5 and the WH-generator of section 5.4(b)) may be a suitable candidate but a number of others would be possible. The exact shape is not so important but the skewness should be available as a parameter. Geoghegan et al. (1992) suggest the possibility of using a Pearson Type IV distribution, a t-distribution or a stable Paretian. We will use the generalized model of (7.3.1), allowing for a skew distribution of the white noise term. This is also an effective way of reducing the incidence of negative inflation in the projections.

Figure 7.3.2 demonstrates the effect of the choice of the noise term. Although the standard deviation of the noise term is smaller in the case with non-zero skewness, the upwards variability is similar. Negative outcomes are, however, largely avoided.

(i) A non-linear inflation model. Clarkson (1991) has suggested a more general model for inflation with several components. The principle on which it is based is that there are three main mechanisms which affect the year-on-year inflation rate:

(1) A tendency for the rate to return to some intrinsic value.
(2) A random error component which operates every year.
(3) A random shock component which operates infrequently.

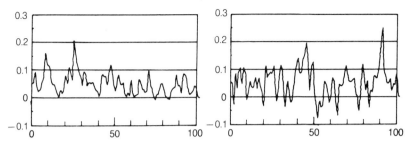

Figure 7.3.2 *The rate of inflation simulated using skew (left-hand side) and symmetric (right-hand side) noise.*

The model is postulated in the form (using the notation of Clarkson)

$$i(t) = i(t-1) - A \cdot (i(t-1) - B) + C \cdot \text{Trend}^{+}\{i(t-1)\}$$
$$+ D(t) \cdot \varepsilon(t) + \rho(t) \cdot E(t) \qquad (7.3.7)$$

where

A and C are positive constants,

B is the intrinsic rate of inflation,

$D(t)$ is a positive function, monotonically increasing with the recent average rate of inflation,

$\varepsilon(t)$ is a standardized normal random variable,

$\rho(t)$ takes the value 0 or 1 according to a defined rule,

$E(t)$ is the value of the shock impulse to inflation,

$\text{Trend}^{+}\{i(t)\}$ is the recent trend of inflation when this is positive and zero otherwise.

This formula contains one linear and three non-linear components. The non-linearity of the model presents difficulties in fitting the model to past data by standard statistical techniques and the choice of parameters relies heavily on actuarial judgement. Clarkson has suggested a set of parameters which provide a reasonable representation of UK inflation over the past 30 years.

This model is described briefly here as an example of a variety of more complex models which might be considered as suitable for modelling inflation.

CHAPTER 8

Investment

8.1 Investment as part of the insurance business

(a) The investment process. A key feature of insurance is that premiums are received in advance of the risk being borne and claim payments follow even later. In the meantime, provisions need to be established in respect of the expected future liabilities and additional solvency margins have to be maintained. These provisions and reserves should be backed by appropriate assets, having regard to the liabilities, and with a reasonable balance between security, liquidity and good investment return (including interest, dividend or rental income and also capital gains). Thus the investment process is an integral part of the financial management of an insurer.

(b) Assets as generators of cash flows. The balance sheet (see Figure 1.2.1) focuses on a capitalized value of the assets. From the point of view of an insurer, however, and of many other financial institutions, immediate realization of a significant quantity of assets is not a realistic prospect. The assets are held because of the stream of future income which they can be expected to generate. In looking at the financial strength of the insurer, therefore, it is necessary to look at the flow both of premiums and of income from the assets, some of which may arise from dividends or interest income and some from the maturity of fixed-term investments. Total income will exceed outgo on claims, expenses, etc., in certain periods, with the reverse applying in other periods. Account needs to be taken of the investment of surplus income and the realization of assets when there is a shortfall. However, bearing in mind the tendency of insurance liabilities to grow, and the long settlement periods which are associated with certain lines of business, the cash flow can often be expected to remain positive as long as the business is a going concern, with only a slight chance of any appreciable shortfall. If this is the case, liquidity may not be much of an issue, permitting greater freedom in investment strategy.

Assets may also be bought or sold as part of an **active tactical investment policy**, in order to maximize the return obtained on the assets. However, it is not usually necessary to take this into account specifically in any modelling of the assets, since one asset will simply be replaced by another. Such tactical changes are only of importance to the modeller if they result in a significant adjustment to the nature of the portfolio of assets, in terms of its risk characteristics or its suitability *vis-à-vis* the liabilities, or if the transaction costs (taxes, commissions, etc.) are significant.

(c) Market value.   In principle the market value can be seen as the market's assessment of the discounted value of future cash flows. Market value is the value at which buyers and sellers are in equilibrium. If buyers of a stock predominate at a particular market price, the price will rise until equilibrium is reached, and vice versa if sellers predominate. However, although the equilibrium of supply and demand is usually expressed in terms of price, the price that a buyer is willing to pay reflects his perception of a satisfactory yield on the investment from the future income flows and future sale value.

If the accounting practice is to adjust the book values of assets to reflect changes in market values, considerable volatility may ensue in the balance sheet, with direct consequences for the profit and loss account. The overall volatility of the results may be reduced if the basis for valuing the liabilities also reacts flexibly to market changes. Immunization techniques have been proposed to improve the linkage between assets and liabilities so that changes in value on the two sides of the balance sheet offset each other. These are discussed further in section 8.6.

Alternatively, changes in the value of the assets may be absorbed by a special fund, called, for example, an **investment reserve** (see Figure 15.2.2) or a fund for future appropriations, so that they do not flow directly into the profit and loss account unless a specific transfer is made. This practice is not so far removed from that in countries where book values, based on the purchase price of the assets, are permitted. In that case the book values are often not changed, at least not fully, when the market values increase (although they may be reduced when market values fall below book values). Unless the market values are disclosed, the difference between the market value and the book value may be held as a **hidden reserve**.

It serves as a buffer against potential future falls in asset values (section 1.2(e)).

(d) Selection of investment types. The problems of constructing an appropriate investment portfolio will be dealt with in section 8.6. Here we consider only some general ideas.

General insurance companies often have a predominance of relatively short-term liabilities and investment tends to be mainly in

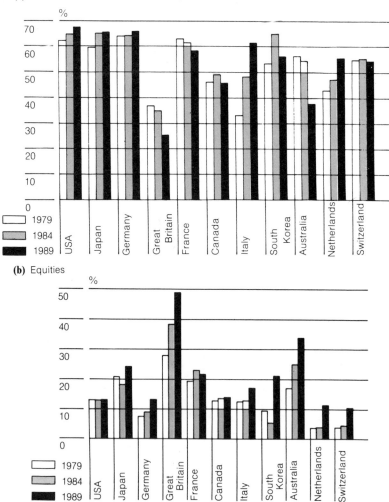

(a) Fixed interest securities and loans

(b) Equities

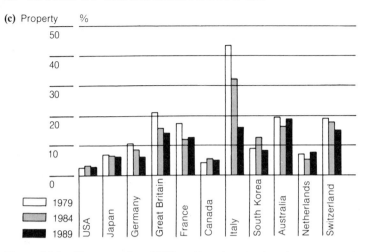

Figure 8.1.1   *The proportion of different investment categories as a percentage of insurance portfolios (life and non-life combined) in certain countries. Source: Sigma 5/91.*

cash, short-term deposits and government or corporate bonds of fairly short duration. However, if the cash flow is expected to be positive, a more diversified investment policy could be beneficial, including property and equity shares. These are known as real estate and common stock respectively in North America but we will use the British terminology of property and equities. Such investments may generate higher rates of return, and in particular better real rates of return relative to inflation, but they are usually associated with greater volatility and possibly also with a risk of significant loss of value.

REMARK   Following the British terminology **investment income** is used to refer to the cash income earned on the assets. **Return on investment** is the investment income together with changes ($\pm$) in the value of the assets.

   In some jurisdictions life insurance companies have a greater degree of investment freedom than non-life companies, in particular if they are writing substantial volumes of participating (with-profits) life business. In this case only a part of the sum assured may be guaranteed and the balance depends on the performance of the investments (and other aspects of the experience). Different bonus systems, discussed further in Chapter 15, help to explain some of the differences between investment policy in different countries, illustrated for life and non-life insurance companies together in Figure 8.1.1.

8.2 Investment returns

The investment manager needs, self-evidently, close familiarity with the investment markets and an idea of the expected returns and risks associated with alternative forms of investment. This is not, however, a topic which can be adequately discussed in this book. Considerations are, therefore, restricted to those features which are most relevant for risk analysis and which are useful for model-building.

(a) Testing, not forecasting. It is important to appreciate that the models which are considered in this chapter are not intended to be used for forecasting actual asset movements, or the times when the return on investments will be high or low. This is an important difference as compared to some models in the economics literature. Our models aim to explore how capital markets may influence the insurance process in quite general terms, for instance by testing insurers' capacity to overcome various adverse events whenever they might occur. For this purpose it is sufficient to find models which generate high and low periods of return on investments in similar patterns to those which can be expected to appear in the future, having regard to past experience and the expected values of key factors, and employing realistic basic characteristics, expressed in terms of the mean, standard deviation, skewness, autocorrelations and the linkage with inflation.

(b) Economic cycles. It is useful to have a view of the movements of and mutual interrelationships between the key indicators of the financial market. Figure 8.2.1 exhibits a simplified representation based on an article by Oppenheimer (1991).

The general economic or business cycle, for instance in terms of the growth rate of GNP, is the driving force in Figure 8.2.1. When the business cycle is falling, the resulting reduction in demand for commodities induces falls in both the rate of interest and the rate of inflation. Falling interest rates lead to an increase in bond prices. Furthermore, equity prices are typically already falling when the economic cycle turns towards a recession. When the cycle has reached its trough and begins its recovery, the reverse movements occur, as shown in the diagram. Inflation might also be deemed to show a direct correlation with interest rates, although not necessarily precisely in phase. Although not included in Oppenheimer's configuration, an inflation cycle has been incorporated in Figure 8.2.1.

Note the shifts in timing. These are caused in part by the inertia of the system and in part by the fact that it is expectations rather than actual data which drive the markets. Investors may begin

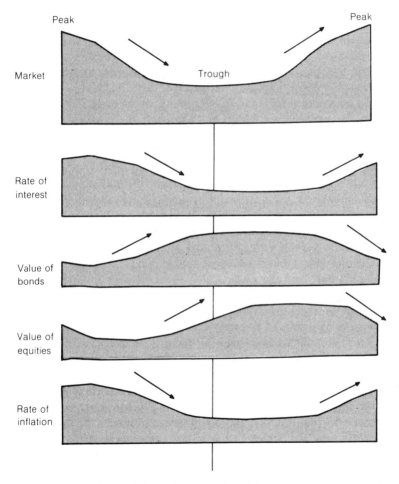

Figure 8.2.1 *Timing of the cycles of the financial market as compared to the rate of interest and inflation and the growth rates of bond and equity prices.*

purchasing equities before the economic cycle has reached its deepest trough and sell them before the peak.

(c) Aspects of the construction of an investment model. If the above reasoning is believed to be relevant, there would be no overwhelming problem in modelling a scheme to implement these properties, whilst at the same time incorporating stochastic noise terms. However, unfortunately, real life phenomena are much more complicated than this.

Notwithstanding the fact that the relationships shown in Figure 8.2.1 certainly exist, there are also other factors and events affecting the whole process, such as the oil crisis of the mid-1970s. The political and financial actions of governments and central banks, labour market negotiations, etc., can significantly influence the financial market. Moreover, international linkages disturb the cycles which would arise within each national economy. In the early 1990s many countries experienced a period of low growth and low inflation but high rates of interest, because of attempts to protect currencies against speculation by increasing prime rates.

Against this background, it is not unexpected to see that inflation has not behaved in the way that the theory would suppose. It is often determined to a large extent by political and random impulses, rather than by market forces which would be susceptible to modelling. If the model was intended to have a forecasting capacity, the driving force should probably be a set of two or more economic indicators. But having regard to our particular application of investment models (section 8.2(a)), and the uncertainties involved with comprehensive econometric models, it is sufficient for our purposes to use only one indicator as a general driving force, controlling the other relevant indicators by means of assumed interacting correlations. Following Wilkie (1984), inflation is taken as this driving factor, without any reference to general economic cycles.

One should appreciate that there need not be any very major difference in the eventual structure of the process according to which of the time series is chosen as the driving factor. The cycles and random fluctuations in inflation induce movements in the other variables in a way which can be much the same as if the general economic cycles had been generated first and the others as subsidiary variables. Assumptions can be made about the relationship between inflation and equity prices, as for example in the Wilkie model. The same is the case in regard to the influence of inflation on the general rate of investment income and its impact on bond prices, as will be discussed in sections 8.5(f, g). Another reason for using inflation as the driving force in the investment model is that it is a major factor in determining the liabilities. This approach ensures greater consistency in the approach to assets and liabilities.

It is for future research to ascertain whether the investment models can be made more suitable for the insurance environment by incorporating other exogenous variables to drive the process in parallel with inflation, for example, the growth of GNP or the prime rate of interest.

(d) Cash and short-term deposits. In an analysis of returns available on different types of investment in the US market over the period from 1926 to 1981, Ibbotson and Sinquefield (1977, 1982) showed that investment in Treasury bills (short-term fixed maturity value investments with no default risk) would have produced a return closely in line with the inflation rate, both for the period as a whole and for almost all sub-periods from 1947 onwards.

Figure 8.2.2 shows the real returns obtainable in the UK each

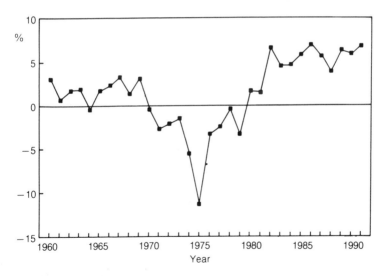

Figure 8.2.2 *Annual real returns on investment in UK Treasury bills, 1960 to 1991.*

year since 1960 on investment in Treasury bills. It can be seen that real returns were negative throughout the 1970s (averaging -3.3% a year) but that substantial positive real returns were available in the 1980s (averaging $+4.8\%$ a year). Over the whole period from the beginning of 1953 to the end of 1991 the average real return from investment in Treasury bills would have been $+1.3\%$ a year.

Some have proposed short-term deposits as a suitable hedge against inflation in the UK, but other investments, such as equities, tend to produce a higher real rate of return in the long-term and to match better the increases in the retail price index. In practice the short-term bill markets are dominated by a wide variety of economic factors and the UK experience has been that returns on such investments do not always match inflation, particularly in times of high inflation.

(e) Fixed interest investments. An ideal medium for matching liabilities for fixed payments, for example in respect of annuity payments, would be high grade bonds (fixed interest securities). In the UK this particular market is dominated by **gilt-edged** securities, fixed interest redeemable securities (and a small number of irredeemable securities) with regular (usually half-yearly) interest payments, issued by and guaranteed by the government. Such securities can be regarded as being without any measurable default risk. In North America there is a much more extensive corporate bond market, ranging from top-rated bonds (denoted by AAA or similar marks) with negligible risk of default to so-called **junk bonds** with a very significant risk of default and a correspondingly higher rate of return to compensate for the underlying risk. In some countries investment in personal and corporate mortgages is also common for insurance companies and pension funds.

In the study by Ibbotson and Sinquefield (1982) the return on long-term government bonds matched inflation over the period from 1926 to 1981. However, this was the result of returns in excess of inflation in the period up to 1950 and returns falling behind inflation over the next 30 years. They interpreted this as a failure by the market to foresee the higher inflation rates of the post-war period, resulting in underestimation of the maturity premium required on longer dated stocks as compared with short-term Treasury bills.

Daykin (1976, 1987) studied the rates of return in the UK from investment in Consols. This is an irredeemable government security with a $2\frac{1}{2}\%$ coupon (i.e. interest is paid at an annual rate of $2\frac{1}{2}\%$ of the nominal value of the stock, in fact in four quarterly instalments) for which a long time series of market prices and yields is available. Since there is no maturity value, there is a simple inverse relationship between yield and price, and this facilitates the calculation of the returns available over any required period.

The relationship between the yield on irredeemable securities such as Consols and the redemption yield on redeemable securities depends on the shape of the so-called **yield curve**, which relates redemption yields of gilt-edged securities to their term to maturity (in North America this is usually referred to as the **term structure of interest rates**). Although the shape of the yield curve has varied from time to time, the yield on irredeemable securities such as Consols has usually been close to yields on long-dated securities.

Daykin's study showed that in the UK, over the period since 1952, it has rarely been possible to achieve a positive real rate of return, relative to price inflation, by investing in Consols. To some extent this has been due to falling market values of irredeemable securities

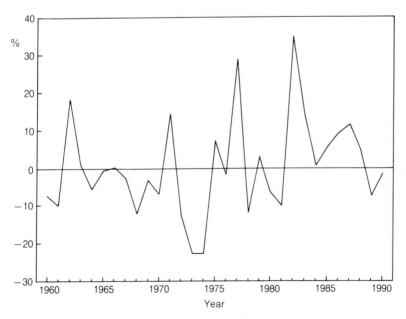

Figure 8.2.3   *Annual real returns from investment in $2\frac{1}{2}\%$ Consols, 1960 to 1990.*

over much of the period, due to the increase in market yields. However, even when this effect is controlled for, the real returns are low and often negative, so that it can generally be deduced that investment in government fixed interest securities, in the UK at least, would not have yielded positive real rates of return relative to price inflation over much of the post-Second World War period. Figure 8.2.3 shows the annual real rates of return from investing in Consols over the period 1960 to 1990.

The situation may be different in future if plans proceed for monetary union in the European Community. A stable Euromarket in an environment of generally fairly low inflation could favour bond investors and enable positive real rates of return to be obtained.

(f) Equities. The prices of equity shares fluctuate considerably, although in the long-term they can be expected to go up at least as fast as price inflation. Figure 8.2.4 shows the changes in the monthly averages of the Financial Times-Actuaries 500 share index in the

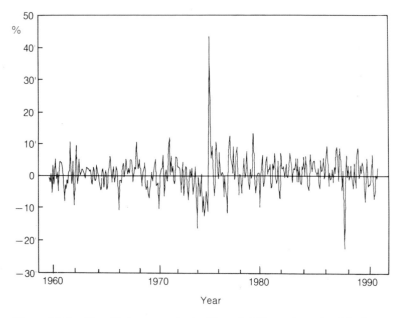

Figure 8.2.4   *Monthly increases in the Financial Times-Actuaries 500 share price index, 1960 to 1990.*

UK (which was published daily) from 1960 to 1990 until replaced by a new series in 1993. Although the market value of equities can go down instead of up, and frequently does, the total amount of dividends from a portfolio of equities tends to increase, although the rate of growth can vary considerably.

It is clear from the study of the US market by Ibbotson and Sinquefield (1982) that returns on investment in equities there have greatly exceeded price inflation over almost all periods of 10 years or more since 1926. There have been periods, such as from 1928 to 1932, and 1972 to 1974, when the total return on a portfolio was negative. Nevertheless, over the 55-year period the real return relative to price inflation on a portfolio of equity shares (with reinvestment of dividend income) was 5.9% a year.

Similar conclusions can be drawn from Daykin (1976, 1987) in relation to the UK equity market since 1952. With the exception of a period from 1972 to 1974 and other brief interludes, the return on

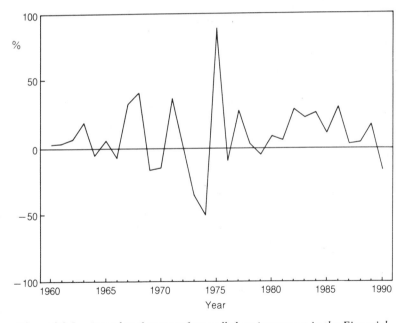

Figure 8.2.5 *Annual real returns from rolled-up investment in the Financial Times-Actuaries 500 share index 1960 to 1990 – net of increases in the consumer price index.*

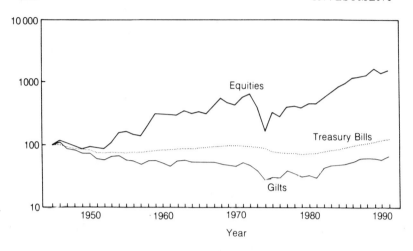

Figure 8.2.6   *Real values of UK equity, gilt and Treasury bill funds with gross income reinvested, 1945 to 1991. Source: Barclays de Zoete Wedd.*

equity investment has significantly exceeded the rate of inflation. On average, over the period from 1950 to 1990, the real rate of return on investment in a broad portfolio of UK equity shares (including both dividends and capital gains and assuming reinvestment of dividend income) was 7.7% a year, but for the periods before 1973 and after 1974 it was significantly higher. Figure 8.2.5 shows the real annual rates of return net of price inflation from investment in the Financial Times-Actuaries 500 share index from 1960 to 1990 (including both dividend income and capital gains).

Figure 8.2.6 shows the real growth (net of consumer prices) of a notional fund invested in equities, with gross income reinvested, from 1945 to 1991. Corresponding real growth patterns are shown for funds invested in UK gilts and in Treasury bills.

Equity investments may not only be in quoted shares. Many insurance companies own subsidiaries or have substantial interests in businesses which are not quoted on the stock market. Investments of this sort will need to be given special attention in any overall modelling.

(g) Mixed investment portfolios.   An insurer's investment portfolio is as a rule made up of a broad mix of the types of investments specified in the foregoing sections. The challenge for investment managers is to find an optimal mixture of the various sorts of assets so as to maximize the return, subject to the constraint that the potential capital losses and other adverse developments should be limited according to the insurer's resources and his willingness to accept risk.

In the applications to be presented subsequently, several categories of investments will be used, to demonstrate how varying mixtures affect the overall business outcomes and the financial strength of the insurer. Further consideration is given to the portfolio selection problem in section 8.6.

(h) Investment characteristics. From the point of view of modelling the behaviour of the assets it may not be necessary to treat each type of investment separately. Investments can be characterized according to

- whether the return they produce is primarily in the form of income or substantially in the form of capital gains;
- whether the income they produce is fixed or variable;
- whether their underlying value is stable or variable;
- whether they offer a higher than average or lower than average rate of return;
- whether they can be expected to provide a hedge against inflation; and
- whether there is any risk of default.

There may also be differences in tax treatment which need to be taken into account in a practical modelling situation, as well as the transaction costs of buying and selling if they are significant.

8.3 Modelling investment prices and returns

Models for projecting investment prices and returns have been developed in recent years, for example by Ibbotson and Sinquefield (1977, 1982) in the US and by Wilkie (1984, 1986) in the UK. The properties and practical value of the Wilkie model were investigated by a Financial Management Group Working Party of the Institute and the Faculty of Actuaries (Geoghegan *et al.*, 1992). The Working Party also reviewed econometric and time series modelling procedures other than the Wilkie model, which is based on Box–Jenkins methods, and made some proposals with regard to possibilities for developing and improving the Wilkie model. However, the Working Party concluded that it is not yet possible to give a definitive statement as to which model is most appropriate for general actuarial use.

It is clear that the development of this topic is still in its infancy. In this chapter, therefore, several alternative approaches are presented to readers for experimentation. Some earlier models are reviewed in this section, the Wilkie model is dealt with in section 8.4 and other approaches are considered in section 8.5.

It is worth noticing that the procedures, business strategies and regulatory conditions for investment practices and asset management differ more between countries than the claim process and the treatment of liabilities. This also suggests the need for alternative approaches.

(a) Historical deterministic approaches.   From the earliest days of actuarial science (De Wit, 1671; Halley, 1693; De Moivre, 1725) deterministic assumptions have been made regarding the future rate of interest in order to model the investment process. In the early days this related particularly to life insurance, but was subsequently applied for pension funding, discounting of general insurance outstanding claim liabilities and other applications. In some cases there may be a simple relationship between liabilities and corresponding assets, whereby a fixed rate of interest or redemption yield may be secured on appropriate assets, so that the asset returns correspond to the incidence of the liability payments.

In practice, however, there are many applications where assumptions have to be made regarding the rate of return on future investments. In such cases the traditional actuarial approach was to make a conservative assumption about the returns which would be available in the future on the relevant investments. For many life insurance products the exact nature of this assumption has not been too important, since long-term policies of this type frequently involve some degree of participation by the policyholder in the profits arising from the business. If a conservative assumption is made, and investment returns subsequently turn out to be much higher than assumed, the excess return can be passed back to the policyholder through bonuses, dividends or other profit-sharing arrangements.

In the 1960s unit-linked life insurance products began to emerge in the UK and in North America. Under such an arrangement the policyholder participates directly in the investment performance of the underlying assets and the insurer does not generate any investment profits as such, but is permitted to make certain defined deductions from the funds to cover expenses and to yield some profit.

Unit-linked products caused a major shift in actuarial emphasis, from the use of present value equations of equilibrium based on a fixed set of assumptions to a cash flow modelling approach first developed by Anderson (1959) and soon to become known as **profit-testing** (Smart, 1977; Lee, 1985). It was also the particular problems of unit-linked products with fixed maturity value guarantees which provided the impetus for the development of stochastic asset models in the UK (Maturity Guarantees Working Party, 1980), as described briefly in section 8.3(b).

Cash flow testing is being developed as a tool for solvency control in the USA and in Canada, as part of the professional actuarial requirements for **dynamic solvency testing**. Following the enactment of a new Canadian insurance law in 1992, dynamic solvency testing is now a professional requirement for the appointed actuaries of both life and non-life insurers (Brender, 1988). The assumptions that are made are, however, deterministic, rather than stochastic, and hence take into account in only a limited way the essential variability of assets and asset returns.

A similar deterministic approach to the problem of asset variability has been adopted in the UK through the **resilience test** proposed in November 1985 by the Government Actuary as a working rule for testing reserve adequacy (Fine *et al.*, 1988; Purchase *et al.*, 1989).

(b) Maturity Guarantees Working Party model. As indicated in section 8.3(e), early work on stochastic asset models in the UK was stimulated by the particular problems of reserving for maturity guarantees on unit-linked life insurance products. The Maturity Guarantees Working Party (1980) fitted autoregressive moving average models (ARIMA models) to an index of unit prices of an equity portfolio, based on the De Zoete and Bevan Equity Index with net reinvested dividend income. This led to a rather complex ARIMA $(7, 1, 0)$ model for yearly prices. Following Godolphin (1980) the Working Party analysed the following basic variables (following their notation)

- **the equity price index** $I_e(t)$, giving the price at time t for a unit of equities invested in the index (without any reinvested income) by reference to an initial time point $t = 0$ for which $I_e(0) = 100$;
- **the equity dividend index** $d_e(t)$, which is the gross dividend received during the period $t - 1$ to t on a unit of equities;

- the prospective **equity dividend yield** $y_e(t)$, being the gross dividend in the subsequent period on a unit of equities, expressed as a ratio of the equity price index at time t, i.e. $y_e(t) = d_e(t + 1)/I_e(t)$.

The Working Party decided to recommend an approach based on modelling first the gross dividends $d_e(t)$ and the dividend yield $y_e(t)$. The unit price $I_e(t)$ could then be obtained as the quotient of the gross dividend per unit and the dividend yield. The **roll-up index for a unitized fund with reinvested income** could easily be derived from this.

The model proposed for the equity dividend index $d_e(t)$ was

$$\ln d_e(t) = \ln d_e(t - 1) + \mu_d + \varepsilon_t \qquad (8.3.1)$$

where $\mu_d = 0.04$ and ε_t is a normally distributed random variable with zero mean and standard deviation $\sigma_\varepsilon = 0.13$.

The model proposed for the equity dividend yield $y_e(t)$ was

$$\ln y_e(t) = \mu_y + \alpha \cdot [\ln y_e(t - 1) - \mu_y] + \eta_t \qquad (8.3.2)$$

where $\mu_y = \ln(0.05) \approx -2.99573$, $\alpha = 0.60$ and η_t is a normally distributed random variable with zero mean and standard deviation $\sigma_\eta = 0.20$.

The unit price $R(t)$, with reinvested income, is given by

$$R(t) = R(t - 1) \frac{I_e(t) + k \cdot d_e(t)}{I_e(t - 1)} \qquad (8.3.3)$$

where k is the complement of the tax rate, taken as 0.625 (with the then current tax rate for life insurance funds of $37\frac{1}{2}\%$ of the excess of investment income over expenses).

This model assumed no correlation between the dividend model and the dividend yield model. No attempt was made to link either model to a separate driving force such as inflation. The considerations focused on UK equity funds. It was suggested that a similar model could be used for property funds and managed funds with a reduced standard deviation σ_η.

8.4   The Wilkie model

(a) The structure of the model.   Following on from the work of the Maturity Guarantees Working Party, Wilkie (1984, 1986, 1986a) developed a set of integrated autoregressive stochastic models to project retail price inflation, equity share yields, equity share dividends,

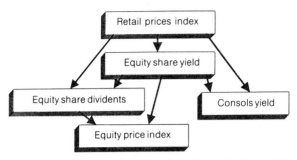

Figure 8.4.1 *Interrelationship between the submodels of the Wilkie model.*

and yields on $2\frac{1}{2}\%$ Consols (section 8.2(e)), as a measure of the general level of fixed interest yields in the UK market. The primary process is that relating to retail price inflation, described in section 7.3(d). This drives the other processes, as indicated by the cascade structure of Figure 8.4.1, where the relationships between the various models are illustrated diagrammatically, with arrows indicating how the different models feed into each other.

(b) Model for equity yields.   The primary variables are the same as were introduced in section 8.3(b), except that the equity price index is now defined as it is in practice, based on the latest dividend declarations. $d_e(t)$ is an index of equity share dividends at the end of year t, $I_e(t)$ is the corresponding equity share price index, and $y_e(t)$ is the dividend yield on the equity share index at the end of year t. These quantities are connected by

$$y_e(t) = \frac{d_e(t)}{I_e(t)}. \tag{8.4.1}$$

In other words, the dividend index is the product of the price index and the dividend yield. Wilkie linked together a number of historical series of UK equity prices and dividend yields from 1919 to 1982, culminating, for the most recent period, in information relating to the Financial Times-Actuaries All-Share Index. Indices are usually published in terms of a price index and a corresponding dividend yield (the weighted average of current dividend rates). The dividend rates can be constructed from the product of the price index and the yield.

Wilkie observed that the dividend yield over quite long periods appears to be stationary. This seemed reasonable, as investors expect a reasonably constant return in real terms on their investment. A very long history of interest rates on money shows that 4% to 5% a year in real terms is fairly normal. The dividend yield is necessarily positive, but it can be observed that its logarithm $\ln y_e(t)$ is approximately symmetrically distributed. This is a convenience in that the logarithm of the yield $\ln y_e(t) = \ln d_e(t) - \ln I_e(t)$ is the negative of the logarithm of the price/dividend ratio, which gives a certain symmetry to the structure.

Observation also shows that the logarithm of the yield is approximately normally distributed. However, it can be seen to change relatively slowly from day to day, and it can also be seen that values of $\ln y_e(t)$ at annual intervals are correlated with a correlation coefficient of about 0.6. Monthly values are correlated with a correlation coefficient of about 0.95 which is roughly $0.6^{1/12}$. There is no apparent significant seasonal fluctuation in dividend yields, although there are various legends in the Stock Exchange of such seasonal effects.

A stationary series with first order autocorrelation proportional to the power of the differencing interval is consistent with discrete observations of a particular continuous stochastic process, the Ornstein–Uhlenbeck process. However, observations in the UK show that there is an additional small effect of inflation on share prices. It seems that higher inflation disturbs the stock market and causes prices to fall, at least temporarily, even though they rise later as higher dividends come through. Possible reasons for this are as follows: high inflation has often been followed by government controls on prices, which hold back company profits for a period; a period of rising prices has often been associated with rising wages, so that employees are taking a higher proportion of the available funds, leaving rather less for shareholders; high inflation is often associated with high short-term interest rates, which means that shareholders have to pay more for bank loans and other sources of working capital. The overall effect is that, when inflation is high, share prices fall, or share yields rise. This is expressed in the first term on the right-hand side of (8.4.2) which has a relatively small effect, putting the average yield level up by about one quarter per cent.

The model proposed by Wilkie for the dividend yield $y_e(t)$ was

$$\ln [y_e(t)] = \omega_{ye} \cdot \ln [I(t)/I(t-1)] + n_{ye}(t) \qquad (8.4.2)$$

where the first term depends on the current rate of inflation and the second introduces a first order autoregressive process

$$n_{ye}(t) = \ln(\mu_{ye}) + \alpha_{ye} \cdot [n_{ye}(t-1) - \ln(\mu_{ye})] + \sigma_{ye} \cdot \varepsilon_{ye}(t), \qquad (8.4.3)$$

$\varepsilon_{ye}(t)$ is a sequence of independent, identically distributed, unit normal random variables. The values proposed by Wilkie for the model were

$$\mu_{ye} = 0.04, \ \alpha_{ye} = 0.6, \ \omega_{ye} = 1.35, \ \sigma_{ye} = 0.175.$$

(c) Model for share dividends. The next variable is the index of dividends on ordinary shares. Dividends depend on the earnings of companies, which in turn depend on their trading profits, which in turn depend on their turnover, which is measured in money terms. An increase in the general level of retail prices, therefore, flows through into dividends in due course, possibly with a time lag, and generally with 'unit gain', i.e. a 1% increase in prices in due course results in a 1% increase in dividends.

The variable used is an index of dividends $d_e(t)$. This is an index for which the average level is of no significance, but proportional changes are. It is therefore appropriate to study $\ln d_e(t)$ and, in particular, changes in $\ln d_e(t)$. Annual changes in dividends depend on annual changes in the Retail Price Index, with a distributed lag whose coefficients sum to unity. This is expressed in the model in two parts: an exponentially weighted moving average of inflation, denoted $m_{de}(t)$, and the immediate effect of current inflation, $\Delta(t)$. The total effect of inflation is taken as the weighted sum of these two effects.

As a first approximation the remainder of the change in the dividend can be modelled as an independent series, approximately normally distributed and in fact with roughly zero mean, which implies no upwards or downwards drift in the real value of dividends. However, two extra terms are introduced into the model. One is a 'moving average' term, denoted $\gamma_{de} \cdot \sigma_{de} \cdot \varepsilon_{de}$ $(t-1)$, i.e. last year's residual multiplied by a coefficient. A possible explanation for this term is as follows: company earnings are variable but directors like to smooth their dividend payouts to some extent; if there is a rise in earnings, not all of this is paid out in dividends in one year, but part is carried forward and paid out the next year.

A further feature is that share prices anticipate changes in share dividends. Dividends are paid in one year based on the profits of the preceding year. As that year passes, investors are aware of the developing economic situation and also of the special factors affecting particular companies. They are therefore able to make reasonably good forecasts of the dividends that might be paid in the next few months. If dividends are likely to rise, share prices also rise and dividend yields (being based on the previous dividend) fall.

However, although this is the true sequence, it is easier to express this in the model by reversing the connection, and making dividends appear to depend on changes in the dividend yield in the preceding year. This is done through the term $\beta_{de} \cdot \sigma_{de} \cdot \varepsilon_{ye}(t-1)$. But it is not changes in share yields that cause changes in dividends; instead it is expected changes of future dividends that cause changes in share yields.

The index of share dividends is made to depend in part on the rate of inflation and in part on a lagged autoregressive term, with a mixture of residuals, including a share of the residuals of the yield model:

$$\ln[d_e(t)/d_e(t-1)] = \omega_{de} \cdot m_{de}(t) + \alpha_{de} \ln[I(t)/I(t-1)] + \mu_{de}$$
$$+ \beta_{de} \cdot \sigma_{ye} \cdot \varepsilon_{ye}(t-1) + \gamma_{de} \cdot \sigma_{de} \cdot \varepsilon_{de}(t-1)$$
$$+ \sigma_{de} \cdot \varepsilon_{de}(t) \tag{8.4.4}$$

where

$$m_{de}(t) = \delta_{de} \cdot \ln[I(t)/I(t-1)] + (1 - \delta_{de}) \cdot m_{de}(t-1)$$

and $\varepsilon_{de}(t)$ are independent, identically distributed standardized normal random variables. From this we see that the amount which enters the dividend model each year is δ_{de} times the logarithm of the increase in inflation (i.e. approximately δ_{de} times the current rate of inflation), plus $(1 - \delta_{de})$ times the amount brought forward from the previous year – an exponential smoothing formula. The values proposed by Wilkie for the parameters were

$$\omega_{de} = 0.8,\ \delta_{de} = 0.2,\ \alpha_{de} = 0.2,\ \beta_{de} = -0.2,\ \mu_{de} = 0,$$
$$\gamma_{de} = 0.375,\ \sigma_{de} = 0.075.$$

(d) Model for the yield on Consols.   The final series modelled is that for long-term interest rates, represented by the yield on Consols, almost irredeemable stock of great antiquity.

During the nineteenth century the average inflation rate was approximately zero, and it was reasonable to assume that investors ignored the possibility of steady changes in the value of money. Investment in government stocks was therefore equivalent to investment in a guaranteed real asset. The running yield on Consols was typically between 3% and 4%, dipping briefly below 3% for a short time, and more frequently rising above 4%, but not higher than 6%. It is therefore reasonable to assume that investors are happy with a guaranteed real yield of approximately $3\frac{1}{2}\%$. Index-linked stocks in the UK have generally yielded between 3% and 4%, though when first issued their yield was below 3%, and more recently has been above 4%.

The American economist Irving Fisher suggested early this century that money interest rates could be seen as the sum of a real interest rate and a compensation for inflation, the decline in the purchasing power of the investment. It is therefore reasonable to model long-term money interest rates as a real rate of roughly $3\frac{1}{2}$% plus an allowance for expected inflation in the future. There is no way of measuring what the market's estimate of future inflation has been (though it is now possible to make comparisons between conventional fixed interest stock and index-linked stock), but it is plausible that investors look back over the past 20 years or so when estimating inflation over the future of a long-term loan. It is convenient to model this by means of an exponentially weighted moving average, and the parameter chosen for the UK of 0.045 corresponds roughly to an average lag of 20 to 25 years.

The money yield on Consols, $c(t)$, is therefore modelled as two parts: an allowance for inflation, modelled simply as the exponentially weighted moving average of past inflation, and a real yield, which has a mean of about $3\frac{1}{2}$%, and changes rather slowly.

Investigations of the yield on Consols during the nineteenth century show that successive values of the yield have a correlation coefficient of about 0.9, and can therefore be modelled by a first order auto-regressive model (analogous to an Ornstein–Uhlenbeck model) with an autoregressive parameter of about 0.9. It is reasonable to assume that the real part of the Consols yield can be modelled similarly nowadays. Investigations show that additional second and third order terms could be justified in the UK, but these are relatively small in their effect, and do not apply in other countries. These terms could be omitted, using only a first order parameter of about 0.91.

It can also be observed that share yields and fixed interest yields move to some extent in sympathy, rising or falling together. There are good reasons for this: a rise in long-term interest rates is often associated with a rise in short-term interest rates, which has a bad effect on the profits of companies, reducing the prospects for future dividend increases, causing share prices to drop and share dividend yields to rise. More directly, shares and fixed interest stocks are alternative investments and a rise in the expected yield of one makes the other relatively less attractive unless its yield also rises.

This effect is most easily represented in the model by including the term $\phi_c \cdot \sigma_{ye} \cdot \varepsilon_{ye}(t)$... into the Consols model. It could equally have been represented by a comparable term in the share yield model.

The projected yield on Consols is made up of an estimate of the rate of inflation, derived from an exponentially weighted moving average of past rates of inflation, and a third order autoregressive term, which introduces the residuals from the equity yield model.

The model proposed by Wilkie for $c(t)$, the yield on $2\frac{1}{2}\%$ Consols, is

$$c(t) = \omega_c m_c(t) + n_c(t) \qquad (8.4.5)$$

where

$$\ln[n_c(t)] = \ln(\mu_c) + \alpha_c \cdot \ln[n_c(t-1)/\mu_c] + \beta_c \cdot \ln[n_c(t-2)/\mu_c]$$
$$+ \gamma_c \cdot \ln[n_c(t-3)/\mu_c] + \varphi_c \cdot \sigma_{ye} \cdot \varepsilon_{ye}(t) + \sigma_c \cdot \varepsilon_c(t)$$

and

$$m_c(t) = \delta_c \cdot \ln[I(t)/I(t-1)] + (1 - \delta_c) \cdot m_c(t-1)$$

where $\varepsilon_e(t)$ is a sequence of independent, identically distributed unit normal variables. The values proposed for the parameters were

$\omega_c = 1.0$, $\delta_c = 0.045$, $\mu_c = 0.035$, $\alpha_c = 1.20$, $\beta_c = -0.48$, $\gamma_c = 0.20$, $\phi_c = 0.06$, $\sigma_c = 0.14$.

When the model is fitted to data from the United Kingdom, or any other economy, it is appropriate to check that the four series of residuals do not have significant autocorrelations, nor significant cross-correlations. If they had, this would justify additional terms in the model; but not only are additional terms not statistically justifiable, those terms that have been included, and have been described above, all have some sort of economic or market rationale.

Further connections might be discovered if an index of wages or salaries were included in the model, or a measure of short-term interest rates.

(e) Model for property rental yields. Wilkie's original paper did not include a model for property rentals and yields, but a supplement to the model was proposed by Wilkie for use by Daykin and Hey (1990). By analogy with section 8.4(b) we define $d_p(t)$ as an index of property rents at the end of year t and $I_p(t)$ as the corresponding index of property prices. The rental yield on the property price index is then given by

$$y_p(t) = \frac{d_p(t)}{I_p(t)}. \tag{8.4.6}$$

In other words, the rental index is the product of the property price index and the rental yield. The complications of different rental review periods are ignored.

The model proposed by Wilkie for rental yields is

$$\ln\left[\frac{y_p(t)}{\mu_{yp}}\right] = \alpha_{yp} \cdot \ln\left[\frac{y_p(t-1)}{\mu_{yp}}\right] + \sigma_{yp}\varepsilon_{yp}(t) \tag{8.4.7}$$

where $\varepsilon_{yp}(t)$ is a sequence of independent, identically distributed, unit normal random variables, assumed to be correlated with $\varepsilon_{ye}(t)$ because of many common influences on rental yields and dividend yields.

Historic series of property prices and rental yields are unfortunately not as readily available as corresponding share indices, but in Daykin and Hey (1990) the values proposed, somewhat arbitrarily, for the parameters were

$$\mu_{yp} = 0.05, \ \alpha_{yp} = 0.6, \ \sigma_{yp} = 0.075 \text{ and}$$
$$\text{correlation coefficient } \varrho(\varepsilon_{yp}(t), \ \varepsilon_{ye}(t)) = 0.4.$$

(f) Model for property rents. As with the dividend series, the model considers the annual change in the index of property rents (absolute amounts, as contrasted with the rental yields of section 8.4(e)) and relates this to the rate of inflation and also to an autoregressive term depending on previous values and the rate of inflation in previous years.

The model proposed is

$$\ln[d_p(t)/d_p(t-1)] = \omega_{dp}m_{dp}(t) + \alpha_{dp}\ln[I(t)/I(t-1)] + \mu_{dp} + \sigma_{dp} \cdot \varepsilon_{dp}(t) \tag{8.4.8}$$

where

$$m_{dp}(t) = \delta_{dp} \cdot \ln[I(t)/I(t-1)] + (1 - \delta_{dp}) \cdot m_{dp}(t-1)$$

and $\varepsilon_{dp}(t)$ is a sequence of independent, identically distributed, unit normal random variables.

The values proposed for the parameters were

$$\omega_{dp} = 1.0, \ \delta_{dp} = 0.1, \ \alpha_{dp} = 0, \ \mu_{dp} = -0.01, \ \sigma_{dp} = 0.05 \text{ and}$$
$$\varrho(\varepsilon_{dp}(t), \varepsilon_{de}(t)) = 0.4.$$

With these particular values (in particular with $\alpha_{dp} = 0$) it is only the lagged effect of inflation which affects the index, and not the current rate of inflation.

(g) Cash and redeemable bonds. The model proposed by Wilkie generates projected values for the yield on $2\frac{1}{2}\%$ Consols. This was taken as a proxy for yields on fixed interest securities more generally. As mentioned in section 8.2(e), the yield on Consols has in recent years not been far different from the yields on longer dated securities. In the UK yields on corporate bonds, local authority bonds, etc. would generally be a little higher than yields on comparable redeemable government securities.

For some analyses it may be necessary to superimpose on the Wilkie model a term structure of interest rates. Models for this have been much discussed in the US (for example, Richard, 1976, Brennan and Schwartz, 1977, Cox *et al.*, 1985). An alternative approach is outlined in section 8.5(e), with a link to the general yield formula presented in section 8.5(f).

8.5 Other model structures

(a) A simpler model structure. While following the general spirit of the Wilkie model, and in particular using inflation as the primary driving factor for the investment models, Pentikäinen *et al.* (1989) have suggested a rather simpler and more homogeneous model structure to apply to a variety of types of investment. It was modified by a Finnish Life Insurance Solvency Working Party (1992). The model has been supplemented by proposals for further development, along the lines discussed, among others, by Geoghegan *et al.* (1992). Several approaches are described, providing users with the opportunity to experiment with alternatives, as suggested in the introduction to section 8.3.

(b) The general structure of the modified model. We propose here an alternative set of investment models to those proposed by Wilkie. The idea is to create a pair of investment models for each investment category, one concerned with the value of the assets, and the other with the cash income which they generate. The entry $J(t)$ in the basic equations (1.1.1) and (1.2.5), representing the return on investments, is decomposed into a cash income element $J'(t)$ and a change in value element ΔA. Each of these, in addition to the total asset value $A(t)$, is further subdivided, as in the Wilkie model, into investment categories specified by the index k:

$$J(t) = \sum_k J'_k(t) + \sum_k \Delta A_k(t). \qquad (8.5.1)$$

It is not necessary to use a different model for each investment category. Assets of similar character can be combined into reasonably homogeneous groups. The differing behaviour of different investment types within the group can be controlled by adjusting the model parameters (section 8.2(h)).

The returns on the different investment categories are assumed to be correlated to the extent that they depend on the rate of inflation, but not otherwise. This is not likely to cause any major shortcomings in the model, having regard to its specified use and the other features discussed in section 8.2(c) above. Note that the model generates a correlation between the $J'(t)$ terms and the $\Delta A(t)$ terms in (8.5.1) because the income is calculated from the amount of the assets, which is built up from increments ΔA from current and earlier years.

(c) Equities. We consider first how to model changes of value and the income from equities as an example of how the relevant terms in (8.5.1) can be generated. A look at the general index of equity prices for several countries shows that in the long term this index follows fairly closely the index which is arrived at by multiplying the consumer price index $I(t)$ by a multiplicative factor $(1 + j_e)$ representing the real growth of the equity values, as illustrated in Figure 8.5.1. In other words, equity prices on average not only retain their real value but show positive real growth, although the extent of this may differ significantly for different time periods. Care should be taken to distinguish between this graph of the share price index, with no reinvested income, and the growth of an equity fund, with reinvestment of gross dividends, discussed in section 8.2(f) and illustrated in Figure 8.2.6.

This might suggest that the equity price index $I_e(t)$ should be modelled first from the inflation index, and the price increments and dividend yields derived from it. (This is the reverse order to that in the Wilkie model, where the cash dividends and dividend yields are modelled first and the price index is derived from them using (8.4.1).) Making the simplified assumption that the real growth rate of shares is constant, the price index can be presented as a product of three

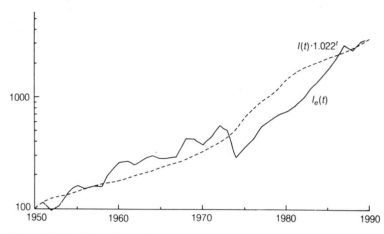

Figure 8.5.1   *The enhanced price index* $I(t)\cdot(1+j_e)^t$, *where* $j = 0.022$, *and the index of equity prices* $I_e(t)$ *in the UK for the years 1950 to 1990. Logarithmic vertical scale.*

factors as follows.

$$I_e(t) = I(t)\cdot(1 + j_e)^t\cdot(1 + \delta(t)) \qquad (8.5.2)$$

where j_e is the constant real growth rate, $I(t)$ the inflation index and the factor $1 + \delta(t)$ represents the inflation-independent variability of $I_e(t)$.

The factor $1 + \delta(t)$ is modelled so that an attraction towards $I(t)$ will arise, which implies that the real value of shares fluctuates around $(1 + j_e)^t$. This can be done by defining the logarithm of $1 + \delta(t)$.

$$d(t) = \ln[1 + \delta(t)], \qquad (8.5.3)$$

by a second order autoregressive process ((G.3.1) of Appendix G)

$$d(t) = \alpha_1\cdot d(t-1) + \alpha_2\cdot d(t-2) + \sigma_\varepsilon\cdot\varepsilon(t,\gamma_\varepsilon) \qquad (8.5.4)$$

where $\varepsilon(t,\gamma_\varepsilon)$ is the standardized noise variable having mean 0, standard deviation 1 and skewness γ_ε (section 5.4(c) and Appendix G, (G.2.2), (G.2.3) and (G.3.1)).

The coefficients should, of course, be estimated to fit the application, ensuring, among other things, the stationarity of the time series (8.5.4) (Appendix G, section G.3). A set of parameter values which correspond approximately to the UK curves is as follows

$$j_e = 0.022, \alpha_1 = 0.333, \alpha_2 = -0.111, \sigma_\varepsilon = 0.237 \text{ and } \gamma_\varepsilon = 0. \quad (8.5.5)$$

Using these parameters a fluctuation of $d(t)$ is generated which has cycles of average length 6 years and standard deviation 0.25, as can be obtained from (G.3.5) and (G.3.6) of Appendix G.

The equity element of the second term of (8.5.1) ($k = e$) is

$$\Delta A_e(t) = i_e(t) \cdot A_e(t-1) \quad (8.5.6)$$

where

$$i_e(t) = \frac{I_e(t)}{I_e(t-1)} - 1.$$

Figure 8.5.2 exhibits a simulated realization of $I_e(t)$. When compared with the empirical diagram Figure 8.5.1, one can see that the modelled outcome does not contain any abrupt fall, followed by a lengthy recovery, such as that which occurred in 1974 in the UK and in many other countries. In order to generate fluctuations which would reproduce the 1974 effect, the parameter σ_ε would have to be rather large. This creates a degree of volatility which is excessive for normal periods and suggests a further amendment to the model, as will be discussed in section 8.5(i) below.

REMARK The idea of a close relationship between the price index and the equity index is the same as in the Wilkie model. However, as a result of

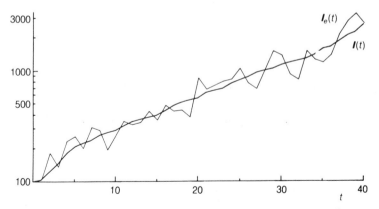

Figure 8.5.2 *A realization of the price index $I(t)$ and the associated equity price index $I_e(t)$ (8.5.2) applying the parameters (8.5.5). Logarithmic vertical scale.*

simulating the growth rates, (8.4.2) results in a slight but growing random divergence between the simulated price and equity indices over long time periods, whereas by choosing $\delta(t)$ stationary in formula (8.5.2) this divergence can be avoided. In the short term the two approaches are not significantly different.

For the income term $J'_e(t)$ in (8.5.1) a simple formula will be presented following a proposal by Bonsdorff (1991). It is derived from the observation that dividends constitute a fairly stable process, increasing broadly in line with movements in equity prices. The formula is analogous to (8.5.2) as follows

$$J'_e(t) = J'_e(0) \cdot I(t) \cdot (1 + \bar{j}_e)^t \cdot (1 + \zeta(t)), \qquad (8.5.7)$$

where $J'_e(0)$ is an initial value and $I(t)$ and $\bar{j}_e$ are as in (8.5.2). The last factor introduces a noise effect, generated from

$$1 + \zeta(t) = \beta \cdot (1 + \delta(t)) + (1 - \beta) \cdot (1 + \zeta(t - 1)). \qquad (8.5.8)$$

Here β is a smoothing factor subject to the constraint $0 < \beta < 1$ and $1 + \delta(t)$ is as in (8.5.2).

The exponential smoothing (section 6.5.3) in (8.5.8) forces the income to follow the development of the corresponding equity prices, with some delay and smoothing out the fluctuations. Optionally, a noise term $\sigma_z \cdot \varepsilon(t, \gamma_z)$ can be added to the right-hand side of (8.5.8). An alternative formula, based on a general income formula, will be given in section 8.5(g) below.

(d) Property can be dealt with by using the same formulae as for equities. However, the volatility might be smaller and the long-term real gain j_p lower than for equities. Instead of (8.5.5), the following set of illustrative values are suggested, though in practice the data should be derived from the relevant environment for each planned application:

$$j_p = 0, \ \alpha_1 = 0.3, \ \alpha_2 = -0.1, \ \sigma_\varepsilon = 0.1 \text{ and } \gamma_\varepsilon = 0. \qquad (8.5.9)$$

REMARK Experience from some countries has shown that the types of property which are the principal objects of investments for the insurance industry, for instance office or factory buildings, have their own booms and downturns depending on movements in general economic conditions and also on the demand for and supply of these types of buildings. Hence, the dependence on inflation may be different from that provided for by the proposed model. For example, during a period of low growth of GNP and low inflation property prices can have a tendency to decrease. This could be adjusted for by allowing a larger range of variation for the control variable

$\delta(t)$ or allowing it to be manipulated along the lines discussed later in section 8.5(i). Another approach would be to use an index of construction work (if one is available) as the reference basis. Then the real growth j_p by reference to this index might be slightly negative, having regard to the depreciation of the value of old buildings as compared to new ones.

(e) Fixed interest assets such as cash, redeemable bonds and direct loans can be modelled by basing both price movements (value in the case of non-marketable assets) and future income flows on the market rate of interest $j_k(t)$. The latter will be considered in section 8.5(f). In principle, market rates of interest (or, more precisely, redemption yields) may be expected to vary by term to redemption of the bonds. There will also be somewhat higher yields for less marketable bonds and for bonds where there is a material risk of default. Here we assume a single series of interest rates to be available and outline briefly how the series of prices of fixed interest assets could be deduced from it.

The market value of a redeemable bond (and of similar assets) is the discounted present value, at the appropriate market rate of interest, of the promised cash flows, over the remaining life of the bond. Hence, when the current market interest rate rises (based on future expectations), bond prices fall, and vice versa when rates decline. To model these movements the market rate of interest should be generated first, whereupon the discounted present values for different types of fixed interest assets can be calculated in a straightforward way from information about the redemption value and the promised interest payments. The general income model proposed in section 8.5(f) below might be a candidate for modelling the interest rate.

The term $\Delta A_k(t)$ of (8.5.1) in respect of fixed-interest assets is obtained directly from the generated time series for market values. The income term $J'_k(t)$ for (8.5.1) should not present any particular problems, as the income expected from each asset in each time period is well-defined and can simply be aggregated for each time period. A more detailed consideration can be found in Elton and Gruber (1987), Chapter 18 (see also Exercise 8.5.1).

(f) A general formula for investment cash income will now be set out as an alternative to the approaches of the preceding sections and also to serve in cases where no special formula is available. The terms of the first summation in (8.5.1) can be written

$$J'_k(t) = j_k(t) \cdot \frac{A_k(t-1) + A_k(t)}{2}, \tag{8.5.10}$$

where the market values A_k of the assets are generated, for example, in the way outlined in the previous sections, taking into account new investments and any disinvestment (sections 8.6(b, c)). The coefficients j_k indicate average running yields over year t, for example, dividends, rents or interest, calculated by reference to these market values. In the case of equities and property, both the total income J_k and the market values A_k go up broadly in line with inflation. The running yields j_k have proved to be relatively stable, and might be represented as varying around a constant average level $\bar{j}_k$. A simple first order autoregressive process, similar to one proposed by Ibbotson and Sinquefield (1977, 1982), could be used as a first approximation, formulated (Appendix G, section G.2) as follows

$$j_k(t) - \bar{j}_k = \alpha_{jk} \cdot (j_k(t-1) - \bar{j}_k) + \sigma_j \cdot \varepsilon(t; \gamma_{jk}). \tag{8.5.11}$$

A logarithmic version of this could be used ((G.4.2) of Appendix G). If there were no autocorrelations, i.e. the successive values for the running yield could be expected to be mutually independent, the coefficient α_j would be put equal to zero. It would then be an independent, identically distributed series, whereas (8.5.11) with $0 < \alpha_{jk} < 1$ defines a stationary time series.

It might be useful to be able to take into account a tendency for the running yield $j_k(t)$ to react to changes in the rate of inflation $i(t)$ with a time lag, owing to inertia in the system. For example, if the rate of inflation increases suddenly, the market value A_e of equities may follow suit, in anticipation of future inflationary increases in dividends, but the actual growth in the dividend amount J'_e may lag behind, because dividend declarations are made only once a year, and reflect business activity over the previous year. The result of this lag in dividend growth is a sudden drop in the running yield. If account is to be taken of this phenomenon, the total dividends J'_e should be modelled directly as a lagged function of inflation, as in the Wilkie model, or the rate $j_k(t)$ will have to be made to reduce for a period in these circumstances, gradually increasing again thereafter. Then the cash income rate earned on investment category k can be described by a formula as follows

$$1 + j_k(t) = (1 + \bar{j}_k) \cdot \left[1 + \sum_{s \geqslant 0} a_{ks} \cdot (i(t-s) - \bar{i}) \right] \cdot [1 + v_k(t)] \tag{8.5.12}$$

where the term $v_k(t)$ introduces an autoregressively modified noise, being generated, for example, by a first order stationary time series (see Appendix G, section G.2)

$$v_k(t) = b_k\, v_k(t-1) + \sigma_{vk} \cdot \varepsilon_{vk}(t; \gamma_{vk}). \tag{8.5.13}$$

By taking several terms into the summation in (8.5.12), a suitably long time delay can be achieved for the inflation impact, providing the possibility of flexibly adjusting the formula to fit different circumstances. The first non-zero coefficient a_{ks} in (8.5.12) should be negative in order to create a fall in the running yield when the rate of inflation increases. This same feature can also be observed in the Wilkie model.

The above general formulae can be adapted for different asset categories, as will be proposed for equities and property in the sections which follow. Formulae (8.5.11) and (8.5.12) can also serve to generate the total rate of return from investment in asset category k if required, for example to calculate the market values of bonds, etc., as discussed in section 8.5(e). Then put

$$\bar{j}_k = (1 + \bar{i}) \cdot (1 + \bar{j}_{rk}) - 1 \approx \bar{i} + \bar{j}_{rk}, \tag{8.5.14}$$

where i is the average rate of inflation and $\bar{j}_{rk}$ indicates the average overall real rate of return on the asset category in the long term.

The fact that financial markets are periodically subject to substantial changes can be expected to lead to changes in the values of the background parameters from time to time, including the mean running yields $\bar{j}_k$. Therefore, the idea of splitting the period of application into sub-periods, as outlined in section 8.5(i), might also be appropriate in this context.

(g) Cash income from equities. The general formula (8.5.11), or (8.5.12), could be used as an alternative to (8.5.7). The inflation increment $i(t)$ could be replaced by the increment of the equity price index (8.5.6), replacing $i(t)$ in (8.5.12) by $i_e(t)$.

The formulae in section 8.5(c) could be further enhanced by an amendment to take into account the observation that the cash income from investments generally fluctuates less than the market value of the investments. In fact, this is the same feature which led to (8.5.12) and these can be regarded as alternative approaches. The stream of income (dividends) can be related to a specially constructed notional asset value, which, regardless of projected market values, grows steadily at the average growth rate. For this purpose, we define the stabilized current value at time t, either by putting the attraction factor $\delta(t)$ in (8.5.2) equal to zero, or alternatively by

$$\bar{A}_e(t) = (1 + g_e)^{t - t_0} \cdot A_e(t_0) \tag{8.5.15}$$

where g_e is an anticipated growth factor and t_0 is the selected base year. The base year should be chosen so that the stabilized current value represents a reasonable central trend value for the category, rather than taking a base year which is either a peak or a trough of market value. The cash income (8.5.10) is then replaced by

$$J'_e(t) = j_e(t) \cdot \frac{\bar{A}_e(t-1) + \bar{A}_e(t)}{2}. \tag{8.5.16}$$

An example of a set of relevant parameters for (8.5.12), (8.5.13) and (8.5.15) proposed by the Finnish Life Insurance Solvency Working Party is

$$j_e = 0.03, \; a_{e0} = -0.04, \; a_{es} = 0 \text{ for } s > 0, \; b_e = 0.7, \; \sigma_{ve} = 0.005,$$
$$\gamma_{ve} = 0, \; g_e = 0.08. \tag{8.5.17}$$

This set generates a relatively smooth flow of income corresponding to the experience that the level of dividends is fairly stable in practice. The rate g_e is the sum of the mean rate of inflation and the real growth rate: $0.05 + 0.03$. The influence of inflation is mainly through the changing value $A_e(t)$ in (8.5.10) or (8.5.16). In fact, this approach is close to that defined in a more straightforward way by (8.5.7).

(h) The income (rent) earned from properties can be simulated by using the some formulae as for equities (changing the subscript k into p) but the parameters should be different, for instance:

$$\bar{j}_p = 0, \; a_{ps} = 0 \text{ for all } s, \; b_p = 0.7, \; \sigma_{vp} = 0.005, \; \gamma_{vp} = 0, \; g_p = 0.05. \tag{8.5.18}$$

(i) Piecewise alternating variants.   Figure 8.5.3 shows the UK equity price index from 1950 to 1990. A striking feature is the abrupt crash in 1974. The data from other countries reveal similar events from time to time, for example the great depression of the 1930s. They are caused by strong exogenous impulses such as an international oil crisis. Another source, or at least a contributory element, may be a boom in equity markets driven by speculation, which ends in the 'bursting of the bubble'. This can lead to a rapid and severe drop in equity prices. The crash of October 1987 may have been of this type, possibly exacerbated by investment managers operating computer-driven portfolio insurance schemes. On the other hand the recovery of prices is typically slower and the return to the average level in terms of the equity index may take years.

Because sudden falls in equity prices may be quite significant, possibly up to 50% or more as in Figure 8.5.3, it is an important risk factor to incorporate in cases where the investment portfolio has a substantial proportion of equities. Neither the co-integrated formula (8.5.2) nor the Wilkie model deal satisfactorily with this crash risk, still less the conventional models of economists, which usually measure risk and volatility in terms of standard deviations. If such a model is fitted to data which includes occasional crashes, it will result in rather large values for the standard deviation and skewness and produce a general variability which may be too great to represent the normal course of events satisfactorily. One way to amend the model is to incorporate a sub-module which can generate crashes of (randomly) varying magnitude. This permits the general variability to be specified at a lower level. This same idea was considered in sections 7.3(g, i) in the context of inflation.

Figure 8.5.3 indicates how the model of section 8.5(c) might be adapted. If the periods before and after 1974 are considered separately, the real growth factors for these periods turn out to be significantly larger than for the years 1950–1990 together, which was evaluated as 2.2% in Figure 8.5.2 (the parameter values (8.5.5) and the Figure 8.5.3). Moreover, the deviations of the $I_e(t)$ curve from the separate period curves are significantly smaller than those from the all-period curve.

If the equity index significantly exceeds the presumed long-term reference index for a lengthy period, the probability of the bubble bursting is intensified and, when the burst has occurred, a recovery period follows with an increased growth rate. Figure 8.5.4 demonstrates an example obtained by amending the formula (8.5.2) with an auxiliary factor, which could be of the type

$$[1 - \varphi(t) \cdot d_c \cdot b_c^{(t - t_c)}] \tag{8.5.19}$$

where t_c, d_c and b_c (< 1) are the time point of a crash, its depth and recovery control coefficient, respectively, and $\varphi(t)$ is a trigger function defined to be $= 0$ for $t < t_c$ and $= 1$ for $t \geq t_c$ (Exercise 8.5.2). The parameters were given the values $25, 0.6$ and 0.95. The recovery period can be terminated after a suitable duration Δ, i.e. at $t = t_c + \Delta$, after which $\varphi(t)$ is again zero and the system is ready for a new crash to be triggered.

There are, of course, many ways of constructing a model which mimics these features. It may be appropriate to insert a crash into the

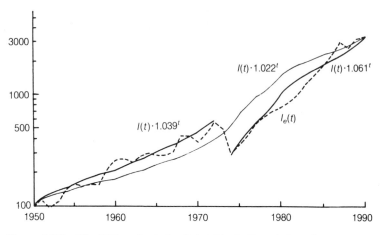

Figure 8.5.3   *The UK equity index $I_e(t)$ (dashed) and the reference curves $I(t)\cdot(1+j)^t$ for the period 1950 to 1990 and separately for the sub-periods before and after the crash of 1974 (continuous lines). $I(t)$ is the retail price index. Note the differing real growth rates $j = 2.2\%$, 3.9% and 6.1%, respectively. Time t is counted from the beginning of each period. Logarithmic vertical scale.*

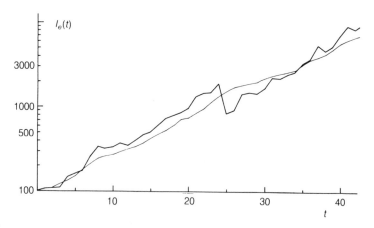

Figure 8.5.4   *A simulation where the time range is split into three parts. For $t = 0$ to 24 the real growth j in (8.5.2) is 0.039 and for $t = 26$ to 40 j is 0.061 (as in Figure 8.5.3). At $t = 25$ a crash is activated, as described in the main text. The reference curve based on the retail price index (thin) is also plotted and calibrated for the whole period with $j = 0.022$. The standard deviation σ_ε is about 40% less than in Figure 8.5.2. Logarithmic vertical scale.*

process deterministically as in Figure 8.5.4. In this way the capacity of the insurer to overcome various crashes can be tested.

A stochastically triggered model can be devised by incorporating the trigger time and step into the $d(t)$ in (8.5.4) providing the expression with a factor of type (8.5.19). $\phi(t)$ can be binomially distributed, generating crash time t_c, and d_c is a random variable defined by a d.f. supplied by the user of model. The recovery can be achieved by the attraction effect. (Exercise 8.5.2 gives more details.)

Another approach would be to replace the constant coefficients in the relevant time series, for instance in (8.5.4), by coefficients which depend on the randomly varying state of the process. Priestley (1980, 1981) has considered a large class of state dependent time series and how they can be estimated from observed data. Feigin and Tweedie (1985) have investigated the asymptotic behaviour of autoregressive processes with randomly changing coefficients.

(j) A formula for the overall return on investments. In some applications the main focus of analysis is on some aspect other than the investments, for example, on claim expenditure. The investments might then be assumed to follow the past experience, as characterized in a relatively simple way. As a first approximation, it may be convenient to simulate the total return on the investments, the term $J(t)$ in (1.1.1) and (1.2.5), as a single item, including both the cash income and changes in value. For this purpose, first simulate the total amount of assets, for example, from (1.1.1) or from one of its variants. The rate of return $j(t)$ can then be simulated from (8.5.11) or (8.5.12), fitting the coefficients so that a formula of type (8.5.10) produces values of $J(t)$ that fluctuate in a similar way to that which might be expected on the basis of experience, having regard to any special relevant features in the particular application.

Having regard to the fact that in this context a fairly rough formula may be sufficient, the multiplicative shape of (8.5.12) can be replaced by an additive form as follows

$$j(t) - \bar{j} = a \cdot [\, j(t-1) - \bar{j}\,] + b_1 \cdot [\, i(t - t_{\text{lag}}) - \bar{i}\,]$$
$$+ b_2 \cdot [\, i(t - t_{\text{lag}} - 1) - \bar{i}\,] + \sigma_j \cdot \varepsilon(t) \qquad (8.5.20)$$

where t_{lag} is the time lag for the impact of inflation. Furthermore, it may be useful to provide for the rate j to be kept within limits

$$j_{\min} \leqslant j(t) \leqslant j_{\max}. \qquad (8.5.21)$$

Finally

$$J(t) = j(t) \cdot A(t-1) \qquad (8.5.22)$$

in (1.1.1) and (1.2.5).

(k) Validity and parametrization. As stated above, markets and practices differ to such an extent that it is not feasible to propose any single asset model of general applicability. Research into actuarial applications of these ideas is still developing rapidly and is far from having reached maturity. In practical applications several alternative models should be investigated in order to find the one which seems to fit best in the circumstances.

When a suitable model has been found, perhaps after trial and error, the parameters will need to be assigned numerical values. The **estimation of parameters** can be treated as a standard exercise in mathematical statistics and is dealt with in text-books on time series, for example, Chatfield (1989). However, judgement is required to arrive at a suitable overall set of parameters. Further consideration is beyond the scope of this book, other than a few general comments and the short Exercise G.2 of Appendix G.

When parameter values are to be derived from observed data, it is very useful to plot the relevant time series first as diagrams (Appendix G). Figures 8.5.1 and 8.5.2 are examples (note the appropriateness of the logarithmic vertical scale for index type variables). A visual investigation usually gives a good idea of a phenomena to be analysed, such as whether the relevant time series are homogeneous over the whole observation period or whether significant changes are evident. Unfortunately standard methods of parameter estimation usually require rather long homogeneous series of observed data and may give erroneous results if heterogeneity is ignored.

It is important to find values which can be expected to be relevant in the future period which is to be investigated (rather than the past). There may be reasons why the future can be expected to differ from the past. For example, the movement towards a single European market and currency union can be expected to affect the national markets significantly. Estimates obtained by coarse, less sophisticated methods and by visual evaluation, supported by commonsense amendments and judgement, are likely to give better results than blind calculation using text-book formulae. The simulation outcomes must, of course, be compared with observed outcomes. The comparison may lead to adjustments to the parameter values, or may suggest rejection of the model in favour of another. The procedure was exemplified by comparing the simulation of Figure 8.5.2 against the observed data of Figure 8.5.3. The obvious discrepancy gave rise to amendments to the model, as seen in Figure 8.5.4. This shows index patterns which conform very satisfactorily with the observed data.

Methods for testing the validity of models are also standard exercises in the literature, for example, residual analysis (Chatfield, 1989). The same problems of the applicability of the strict conditions which were referred to above in connection with parameter estimation apply also to these methods. A visual investigation is, therefore, also necessary in this context. In fact, it may often be sufficient for practical purposes. Furthermore, it must be appreciated that, even if a particular model passes a residual analysis test, it does not prove that there is not some other model which would give an even better fit if tested similarly.

Exercise 8.5.1 Write down the flow chart for the changes in the market values of redeemable bonds can be obtained following the idea in the third paragraph of section 8.5(f). The amortization amount to be paid in year τ is $a(\tau)$ for years $\tau = t + 1, t + 2, \ldots, T$.

Exercise 8.5.2 Give an outline of a flow chart for a model which simulates the changes in equity values described in section 8.5(i) and specified in the caption of Figure 8.5.4. The 'crash' should be assessed (*i*) deterministically, and (*ii*) stochastically. In the latter case the probability that a crash is triggered should be proportional to the elapsed time from the previous crash and also proportional to the rate by which the simulated index exceeds the reference index. Furthermore, it should be assumed that there is a minimum time distance T_{min} during which a further crash cannot occur.

8.6 Asset/liability considerations

An important step forward in risk theoretical considerations has been to introduce stochastic models for inflation, interest rates, share prices, etc. However, we must now look more closely at what is involved in modelling a portfolio of assets. Recall that the analyst requires projections both of profits and losses (accounting model, section 1.2) and of cash flow (section 1.1). The latter is the main focus of this section, since the investment process is determined by actual cash flows, rather than by accounting quantities.

(a) Modelling a portfolio.   In principle an existing portfolio can be modelled by subdividing it into the different classes of investment and applying appropriate stochastic models to each part. In the case of cash the capital value will be regarded as fixed and the models will be used to generate estimated interest flows. For redeemable and irredeemable bonds the yields generated by the models can be translated directly into prices by means of normal bond-pricing formulae (section 8.5(f)). In the Wilkie model for equities and property the ratio of the dividend (rental) index to the dividend (rental) yield provides a price index which can be used to assess changes in market value and project actual dividend flows. In the variant models of section 8.5 the price index is generated first and the yields and changes are deduced therefrom.

(b) New investments.   In modelling an insurance process some decision rule is necessary for handling excess income (or shortfalls). The amount $A^+(t)$ available for new investment in year t is obtained from the basic cash flow equation (1.1.1) modified as follows

$$A^+(t) = B'(t) + J'(t) + A^-(t) - X'(t) - E(t) - D(t) + U_{new}(t) + W_{new}(t)$$
$$(8.6.1)$$

where

$A^-(t)$ represents the income from sold or matured assets, for instance from redeemable bonds.

$J'(t)$ is the net cash flow received from investments.

The definitions of other variables of (8.6.1) were given in section 1.1(b). However, for simplicity here we have taken premiums and claims to be net of reinsurance.

In practice care needs to be taken with debtor items (in particular amounts due from agents, and possibly from reinsurers) since they represent a further class of assets, remaining invariable in amount and providing (normally) no income. It may be easier in the cash flow analysis to bring in such items only when cash in actually received by the insurer.

The amount available for investment could be held as cash, or used to purchase further investments at their current market price. For most purposes it may be sufficient to assume that net additions are held in cash until the end of the financial year before being invested. At the end of the financial year it may be appropriate to operate some rebalancing of the investment portfolio, in order, for example, to maintain some criterion of asset allocation.

Asset allocation rules can be set for the portfolio as a whole (for example, rebalancing to maintain 25% in equities, 55% in bonds and 20% in cash) or separately for assets backing the technical reserves – or technical reserves for particular classes of business. Rules of this kind play an important part in simulations which extend for a number of years. However, care may be needed to ensure that assumptions are made which are not obviously unrealistic and which do not take advantage of changes in asset value to an extent which would clearly not be possible in practice.

(c) Disinvestment. Similar problems may arise when outgo exceeds income, with a possible requirement to realize investments. Provided that a reasonable amount is maintained in cash, this can be drawn on to meet net outgo, with any rebalancing of the portfolio left to the end of the year. If the cash becomes exhausted there needs to be a mechanism for selecting which investments to realize, or alternatively, an arrangement for temporary borrowing to generate the cash required, with an appropriate interest charge.

A particular problem is to evaluate the requirement for liquid cash, so that it is sufficient but not unnecessarily large. This is akin to the storage problem of Operational Research, and risk theory methods, such as those used in the evaluation of capital needs, may find useful application.

(d) The portfolio selection problem. A central task in investment management is to find a proper – preferably optimal – composition for the asset portfolio. The problem can be described as follows.

Different sorts of investments are available in the capital market: bonds, various kinds of equities, property, etc. It is assumed that the investment manager has a view of the expected average rate of return $r_k = E(r_k)$ for each of them $(k = 1, 2, ...)$, as well as the variance σ_k^2 and (possibly) covariances $Cov(r_i, r_k)$. (We are using the notation which is conventional in capital pricing theory: in fact r_k is a total rate of return from interest and capital growth, equal to $j_k + \Delta A_k / A_k$ in terms of section 8.5 (8.5.1).) The variance terms serve as a measure of the risk involved with the category (section 6.6.1 discusses the idea of using variance as the measure of risk). The manager wishes to maximize the return, but subject to an acceptable degree of risk.

One way of specifying the problem for a single period is to fix a certain upper limit σ^2 for the total variance of the asset portfolio, reflecting the resources available and the willingness to accept risk. If the proportional share to be invested in assets of type k is denoted by w_k, the problem is to find a set (w_k) so that the expected total return for the whole portfolio

$$r_p = \sum_k w_k \cdot r_k \qquad (8.6.2)$$

will be maximized subject to the constraints

$$\sum_k w_k^2 \cdot \sigma_k^2 + \sum_{i \neq k} w_i w_k \operatorname{Cov}(r_i, r_k) = \sigma^2; \quad \sum_k w_k = 1; \quad 0 \leqslant w_k \leqslant 1, \quad (8.6.3)$$

where the quantities r_k, σ_k and $\operatorname{Cov}(r_i, r_k)$ are assumed to be known, and σ^2 is given. The equations have to be solved for the distribution (w_k).

This is the maximization problem which is well known in mathe-

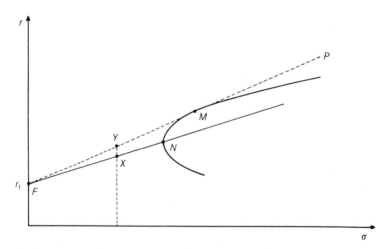

Figure 8.6.1 *The efficient frontier showing the risk and return trade-off for optimal combinations of a given set of risky assets. F represents the risk-free portfolio, N an efficient investment mix totally in risky assets and X and Y combinations of these. M is related to the so-called market portfolio.*

matical statistics and in operational research in the context of quadratic programming. Standard text-books can be consulted for methods of finding the solution (using a Lagrangian multiplier or by other methods; Exercise 8.6.1). Assuming that a unique solution (w_k) can be found, the expected return r can be calculated from (8.6.2).

It is usual in financial theory to plot the pair (r, σ), for a given combination of risky assets, on a plane diagram, as illustrated in Figure 8.6.1. Allowing the standard deviation σ to vary, a locus is obtained to describe the optimal or **efficient** solutions. It is called the **efficient frontier**. An efficient portfolio is a feasible portfolio which is not **dominated** by any other. A portfolio is dominated by another if for the same total expected return r the other has a lower variance σ^2, or if for the same variance σ^2 the other has a higher expected return r. A portfolio which is not efficient will be represented by a point (r, σ) to the right of the efficient frontier. It is always possible, by adjusting the composition (w_k) of the portfolio, to increase the rate of return without increasing the variance, or to reduce the variance without reducing the return, until a portfolio is obtained which lies on the efficient frontier.

For the mathematics of the construction of the efficient frontier, reference may be made to the monograph by Huang and Litzenberger (1988).

(e) Discussion on assumptions. A suggested investment mix for a single period can be obtained straightforwardly by the method which was outlined in the previous section, provided that reliable estimates of the quantities r_k, σ_k and the covariances in (8.6.3) can be made. However, it is not clear that the rather simplified model defined by equations (8.6.2) and (8.6.3) provides sufficient insight into the real-world behaviour of investment markets for practical applications, especially if a longer period is in view. Even if it is appropriate for describing ordinary movements of the yield and changes of value of individual investment objects by reference to average market levels, it is almost certainly inadequate to describe what happens when the whole market, or a major class of investment, for example, equities or real estate (not to speak of junk bonds), is hit by a depression, causing a simultaneous drop in values.

There is an extensive literature on financial risks and capital markets. Only a few key concepts will be noted in the following sections without any derivations or in-depth explanations. The capital market models have been subjected to critical discussion at the colloquia of the AFIR-section of the International Actuarial Association (Wilkie, 1990, 1991).

(f) The capital asset pricing model introduces the concept of a **risk-free rate of interest** available, for example, on government fixed interest securities, where there is no measurable risk of default (in other words with zero variance for the single period in question). The investor is assumed to invest a fraction w of his portfolio in an efficient portfolio of risky assets corresponding to point N on the efficient frontier (Figure 8.6.1), and the remaining portion $(1 - w)$ in the risk-free asset, with risk-free rate of return r_f. The expected return r_p is then given by

$$r_p = (1 - w) \cdot r_f + w \cdot r_N \qquad (8.6.4)$$

and the variance of the return σ_p^2 on the portfolio

$$\sigma_p^2 = w^2 \sigma_N^2 \qquad (8.6.5)$$

since the return r_f on the risk-free asset is certain.

From (8.6.4) and (8.6.5) we can deduce that

$$r_p = r_f + \frac{r_N - r_f}{\sigma_N} \cdot \sigma_p. \qquad (8.6.6)$$

This set of possible portfolios of risky assets, together with some investment in the risk-free asset, can be represented by the line FN in Figure 8.6.1. Point F corresponds to investing wholly in the risk-free asset and point N to investing wholly in the portfolio N of risky securities. The points in between represent mixed portfolios determined by the choice of w.

Given a risk-free asset and a portfolio of risky investments N, it is possible to increase the return on a mixed portfolio, at the expense of an increase in riskiness, by taking combinations nearer to N than to F. However, it is possible to further increase the expected return, without any increase in riskiness, by moving from a point on the line FN (X say) to the point Y with corresponding value of σ on the line FMP which is tangential to the efficient frontier for the portfolio of risky assets. This line gives the set of the most efficient combinations of risk-free assets and a portfolio of risky assets, corresponding to weighted mixes of the risk-free asset F with the risky portfolio M. It is known as the **capital market line** and is the new efficient frontier, now that a risk-free asset has been introduced. Points on the line in the section MP reflect values of the weighting factor w greater than 1, i.e. assumed borrowing at the risk-free rate to finance additional investment in the risky portfolio M. By analogy with equation (8.6.6) the equation for the capital market line is

$$r = r_f + \frac{r_M - r_f}{\sigma_M} \sigma. \tag{8.6.7}$$

On the basis of this derivation, one would expect investors always to select the portfolio M of risky assets, and to adjust their risk/return trade-off by investing in the risk-free asset alongside the risky assets M, or by borrowing at the risk-free rate in order to gear up their holding of the risky portfolio M (a **leveraged** or borrowing strategy). From this it is deduced that the best investment strategy (i.e. the choice of the portfolio of risky assets) is independent of the choice of the degree of gearing (i.e. whether and to what extent to invest in or borrow at the risk-free rate of interest). This is known in financial theory as the **mutual fund separation theorem**.

A further property of the capital market line is that it is tangential to the efficient frontier at the point M. Since any rational investor should in principle invest in this portfolio, together with a suitable portion of risk-free assets, it should reflect the composition of the market in terms of types of security, weighting etc. This explains the term **market portfolio**, which is, according to this reasoning, common to all investors, i.e. characteristic of the market.

REMARK   It should be appreciated that the practice of gearing up by borrowing risk-free money and investing it in potentially lucrative but risky assets (segment MP in Figure 8.6.1) is associated with the major risk that the capital market will be hit by falling market values and defaults on mortgages. Then the speculating investor might run into difficulties, because he is liable under the original terms of the loan. The dangers of such gearing have been highlighted by specific cases in the insurance industry (and even more so in the banking sector). It is arguable whether speculative practices of this nature can be justified at all for insurers, given their commitments to claimants and policyholders. As was discussed in section 8.6(e), it is not clear that the simple capital asset pricing model exposes this kind of risk.

(g) Beta coefficients.   It is useful to employ the market portfolio M, characterized by σ_M, r_M, to calibrate the component asset items, with characteristics σ_k, r_k (and $w_k > 0$ in the composition of the market portfolio) in relation to the market characteristics. This is achieved as follows.

Consider a mixed portfolio composed of a fraction w invested in asset k and $(1 - w)$ in the market portfolio (or in any other portfolio on the efficient frontier). The expected return and variance of return on this portfolio are given by:

$$r(w) = w \cdot r_k + (1 - w) \cdot r_M \tag{8.6.8}$$

$$\sigma^2(w) = w^2 \cdot \sigma_k^2 + (1 - w)^2 \cdot \sigma_M^2 + 2w(1 - w) \cdot \text{Cov}(r_k, r_M). \tag{8.6.9}$$

As w varies, the point $\sigma(w), r(w)$ describes a locus inside the area to the right of the efficient frontier and approaches M as $w \to 0$. The slope of the tangent of the locus curve is

$$\frac{dr}{d\sigma} = \frac{r'(w)}{\sigma'(w)} = \frac{2(r_k - r_M)}{\sigma^{-1} \cdot [2w\sigma_k^2 - 2(1 - w)\sigma_M^2 + 2(1 - 2w)\text{Cov}(r_k, r_M)]}.$$

At the end point $w = 0$, where $\sigma = \sigma_M$, this expression becomes

$$\frac{\sigma_M \cdot (r_k - r_M)}{\text{Cov}(r_k, r_M) - \sigma_M^2}. \tag{8.6.10}$$

It can be deduced that this tangent should at the limit coincide with the tangent of the efficient frontier curve at point M. If this were not the case, an infinitesimal step with $dw < 0$ would take the locus over the efficient frontier. This would contradict the property that every non-efficient portfolio is characterized by a point σ, r to the right of the efficient frontier. If we suppose that the tangent at M has slope K/σ_M, we obtain

$$r_k = r_M + K(\beta_k - 1) \tag{8.6.11}$$

where

$$\beta_k = \frac{\text{Cov}(r_k, r_M)}{\sigma_M^2}.$$

This is the security market line corresponding to point M on the efficient frontier. The derivation is quite general and applies to any point on the efficient frontier and securities included in that particular portfolio. In the case of the market portfolio M, however, if there is a risk-free security, we can write $K = r_M - r_f$, so that,

$$r_k = r_f + \beta_k \cdot (r_M - r_f) \tag{8.6.12}$$

The excess return on asset k over the risk-free return is expressed in terms of the product of the **beta** for asset k and the **market risk premium**, i.e. the excess return on the market portfolio as compared to the return on the risk-free asset.

The actual return on asset k can be redefined as:

$$r_k' = r_f + \beta_k \cdot (r_M - r_f) + \text{Cov}(r_k, r_M) \cdot \varepsilon_k \tag{8.6.13}$$

where the last term introduces uncorrelated random variation, representing diversifiable or non-systematic risk, and the second term represents systematic or non-diversifiable risk arising from correlation between the return on asset k and the return on the market portfolio.

The β_k and $\mathrm{Cov}(r_k, r_M)$ are often referred to in the financial economic literature, and numerical values have been estimated and published for them. Unfortunately, there are many problems with such estimates because of instability in the observed behaviour of particular securities. In order to be able to apply the formulae (8.6.2) and (8.6.3), the relationships between the relevant coefficients are required. For the calculation of r_k from a given β_k in (8.6.13), values are needed for r_f and r_M, and in the case of simulation also $\mathrm{Cov}(r_k, r_M)$. For long-term simulation, separate sub-models are necessary for the generation of r_f and r_M.

The connecting equation between the sigmas is

$$\sigma_k^2 = \beta_k^2 \cdot \sigma_M^2 + \mathrm{Cov}(r_k, r_M) \cdot \sigma_{\varepsilon k}^2. \tag{8.6.14}$$

Another possibility would be to make use of the distribution patterns of the asset risks, rather than to rely simply on the variance as a risk indicator. This would complicate the considerations and reduce the possibility of getting results in a concise, often analytic, form.

(h) Matching and immunization. In its purest form matching of assets and liabilities involves structuring the flow of income and maturity proceeds from the assets to coincide precisely with the outgo in respect of the liabilities. This requires both the asset proceeds and the liability outgo to be known deterministically as regards both timing and amount. It is rarely possible to achieve pure matching, although a close approximation may be possible with certain life insurance products, such as guaranteed income bonds, if matching investments are made in government fixed interest securities.

Redington (1952) proposed a generalization of the matching concept, based on the idea of immunizing the liabilities by investing in such a way that a change in the value of the assets can be reflected by an exactly similar change in the value of the liabilities, in other words that the balance of the value of the assets and the liabilities is immune to changes in interest rates.

Redington used Taylor's expansion to show that the financial balance of assets and liabilities would not be disturbed by a small change in the force of interest if

$$\frac{dV_J}{d\delta} = \frac{dV_X}{d\delta}; \quad \frac{d^2V_J}{d\delta^2} > \frac{d^2V_X}{d\delta^2} \tag{8.6.15}$$

where δ is the force of interest, V_J is the present value of the asset proceeds and V_X is the present value of the liability outgo, both at the ruling force of interest δ.

Now

$$V_J = \sum v^t J_t; \quad V_X = \sum v^t X_t \tag{8.6.16}$$

where J_t is the expected proceeds from the assets in year t, X_t is the expected net outgo in respect of liabilities in year t, and $v_t = e^{-\delta t}$.

It follows that the conditions for immunization can be restated as

$$\sum t \cdot v^t \cdot J_t = \sum t \cdot v^t \cdot X_t; \quad \sum t^2 \cdot v^t \cdot J_t > \sum t^2 \cdot v^t \cdot X_t. \tag{8.6.17}$$

These can be expressed in words as follows

(1) The mean term of the value of the asset-proceeds must equal the mean term of the value of the liability-outgo.
(2) The spread of the value of the asset proceeds about the mean term should be greater than the spread of the value of the liability-outgo.

(i) A generalization of immunization. As presented by Redington, the formulation of immunization involved a number of simplifying assumptions. In particular it assumed (1) a flat yield curve, (2) that asset proceeds are received and liability payments made at a single fixed time each year, and (3) only infinitesimal changes occur in the force of interest.

However, an important step forward had been taken in ensuring that the proceeds from the assets, including interest income, were considered as a single set of cash flows, to be balanced by another set of cash flows incorporating all the elements of the liability outgo.

Boyle (1978) established conditions for immunization with a stochastic model of the term structure of interest rates, generalizing Redington's work in the light of models of interest rates published in the USA by financial economists such as Cox *et al.* (1977). The subject has been extensively developed by financial economists and actuaries in the US, with a particular focus on stochastic models of the term structure of interest rates.

A further step was taken by Wise (1984), who generalized the matching problem to pension fund portfolios with inflation-linked liabilities and considered stochastic variation of inflation and investment returns. He defined matching in terms of a closed fund or run-off principle, where the ultimate surplus can be measured in terms of the realizable market value of the assets remaining when all the liabilities have been extinguished.

Wise defined a **positive unbiased match** as a portfolio of assets which gives rise to a mean ultimate surplus of zero and which minimizes the mean square ultimate surplus. *Positive* refers to the exclusion of negative asset holdings, i.e. borrowing, from the admissible portfolios. *Unbiased* refers to the condition of mean ultimate surplus. A **positive match** is one which minimizes the mean square ultimate surplus but without necessarily having mean ultimate surplus of zero. Wise went on to show that, given

(1) Any suitable actuarial model of future conditions in which the demographic factors are fixed and the factors of interest and inflation follow a specified statistical behaviour.
(2) A suitable model of the investments available in the market and of the future cash flows which each is expected to generate.
(3) Any pattern of liability cash flows,

then there exists a unique positive match to the given liabilities. Further details of this analysis are given in Appendix H.

(j) Stochastic modelling of assets and liabilities. In the previous section we introduced the principle of relating riskiness to the liability outgo. An insurance company or pension fund is concerned with the risk that the proceeds from the assets will be insufficient to meet the required liability outflows. There may also be a need to demonstrate a sound financial position in terms of a balance sheet position at a particular point in time, possibly with an additional solvency margin requirement, but the ultimate test is whether the actual cash liability outflows can be met. Variability of the investment return as such is not the issue.

In practice, therefore, portfolio selection in the presence of defined liabilities, which may themselves be uncertain, involves an attempt to maximize some criterion which measures success in meeting the liabilities over a given number of years, or perhaps over the period until all the liabilities have arisen. One approach is to assume notional discontinuance and allow the business to run off as a closed fund (sections 9.5(c), 14.6(d, e) and 16.3(c) for further discussion of closed fund approaches).

Applying the modified basic cash flow equation (8.6.1) successively for $t = 1, 2, ..., T$, we obtain an estimate of the remaining assets at the end of year T. Simulating the process a large number of times will generate a distribution A_T of assets at time T. For each simulation the run-off of the liabilities can also be monitored and a procedure defined for setting a value on the outstanding liabilities L_t at the end of each year t. This will enable a distribution to be obtained of the surplus R_T $(= A_T - L_T)$ at the end of year T.

A variety of optimization criteria can be considered, such as

- maximize the expected surplus at time T
- maximize the expected surplus at time T subject to a specified maximum variance of the surplus at time T
- minimize the variance of surplus at time T
- maximize the expected surplus at time T subject to Prob$\{R_T < 0\} \leqslant \varepsilon$

If T' is such that $L_t = 0$ for all $t > T'$, but $L_t > 0$ for all $t < T'$, the above criteria relate to the distribution of ultimate surplus at the point at which all the liabilities have been paid off. The probability that $A_{T'} < 0$ is defined as the probability of ruin.

Maximization would generally be over the available portfolios of assets which are of market value equal to that of the portfolio of assets currently held (at $t = 0$). In practice there may be practical difficulties in testing all possible portfolios, or indeed of defining suitable stochastic models for each asset which might be part of an optimal portfolio. A realistic approach is to define a limited number of asset categories for which a suitable investment model is available and optimize the strategic allocation between these investment categories. The choice of individual assets within each category would then be made on traditional investment criteria, or having regard to the portfolio selection techniques referred to in section 8.6(d).

(k) Example. Daykin *et al.* (1993) give a case study of the use of stochastic asset-liability modelling techniques with a funded occupational pension scheme which is closed to new members.

Figure 8.6.2 shows the results for four alternative asset distributions, expressed in terms of the projected funding level 15 years from the date to which the assessment relates. The funding level is defined as the ratio of the asset values to the residual liabilities (A_{15}/L_{15}). The four alternative portfolios are

A. A mixed portfolio corresponding to the existing holdings.
B. 100% in UK equities.

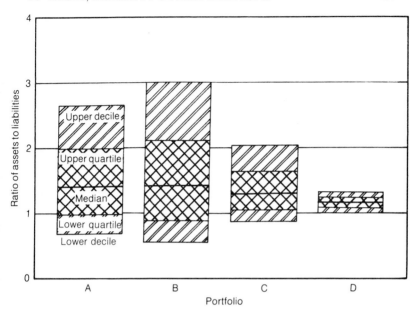

Figure 8.6.2 *The distribution of funding levels after 15 years for four alternative portfolios of assets.*

C. 50% in UK equities and 50% in index-linked government securities.
D. 100% in index-linked government securities.

In Figure 8.6.2 the bold line in each column represents the median 15-year funding level for that portfolio. The cross-hatched portion above and below the median represents 50% of the probability

Table 8.6.1 *Results at 15-year point from asset/liability study*

Portfolio	Median funding level	Mean funding level	Standard deviation of funding level	Proportion of simulations with 15-year funding level below 1 (%)
A	1.42	1.56	0.80	25
B	1.41	1.62	1.00	29
C	1.30	1.39	0.48	20
D	1.17	1.17	0.12	8

distribution, lying between the lower and upper quartiles. The diagonally hatched band takes the distribution out to the lower and upper deciles.

Table 8.6.1 shows some key results from the analysis for the four portfolios. Since the distributions of ultimate surplus are skew, the mean funding level is significantly higher than the median shown in Figure 8.6.2. The simulations which produced a 15-year funding level of less than one do not necessarily represent ruins in the sense defined earlier. At this stage the business has still some years to run and the valuation basis used for the test is a conservative one.

The results shown in Figure 8.6.2 are consistent with intuition. The median funding level of portfolio B (100% equities) is high, but so is the spread around the median. Portfolio D has a lower median funding level, but the variation is considerably reduced. The median 15-year funding level is in all cases well above the initial funding level. This arises because of the conservative assumptions underlying the funding level assessment, contrasted with the more realistic expected actual assumptions for the progress of the fund over the 15-year period.

CHAPTER 9

Claims with an extended time horizon

9.1 Description of the problem

In Part One of this book the time-span was limited to one year. This restriction will now be relaxed and the considerations extended to cover an arbitrary finite time T.

The financial status of insurers is usually reported at the end of each calendar (accounting) year, and possibly also at other time points, for instance at the end of each quarter. So a discrete treatment of claim and other processes is assumed. According to the convention established in section 1.3(d), the time variable t will in most contexts indicate the count of successive periods. The state variables, such as the assets $A(t)$, solvency margin $U(t)$, etc. refer to the status of the business (process) at the end of year (period) t. We make use of the accounting version of the fundamental transition equation (1.2.5), so that the flow variables, such as the amount of aggregate claims $X(t)$, premium income $B(t)$, etc., indicate the amount accrued or incurred during the period t, rather than cash amounts paid or received.

The aggregate incurred claims for year t will be denoted by $X(t)$ and the accumulated amount in years $t_1,...,t_2$ by

$$X(t_1, t_2) = \sum_{t=t_1}^{t_2} X(t). \tag{9.1.1}$$

As specified in sections 1.2(c) and 3.1(c), $X(t)$ includes both paid and outstanding claims. We assume at this stage that the outstanding claim amounts can be accurately estimated at the end of the year in which the claims have been incurred. In practice this is not usually the case, and the problem of estimating outstanding claims, and dealing with uncertainties in the pattern of settlement of claims, will be considered in section 9.5.

It was stated in section 2.4(a), and illustrated in Figures 2.4.1 and 2.4.2, that the risk propensity and other determinants of the claim process may be subject to continual change. This must be taken into account in analysing periods of more than a single year. Appropriate changes can be made to the assumptions of the mixed Poisson process; in section 9.2, we consider the claim number process, going on to claim size in section 9.3. The properties of these processes and further applications are looked at in section 9.4. The claim settlement process is considered in more detail in section 9.5 and catastrophe claims in section 9.6.

9.2   Claim number process

(a) The structure of the claim number process.   As stated above, the risk process is often subject to continual changes in risk propensity. As was demonstrated in section 2.4, these affect the number of claims, partly as (1) long-term systematic slow-changing trends, partly as (2) more or less irregular cyclical up- and downturns and partly as (3) short-term random variation. Each of these three categories will be discussed in the sections which follow. The model to be employed will be constructed by superimposing the different effects, making the relevant amendments to the mixed Poisson formulae. For this purpose the Poisson parameter n will be designated as a suitably defined time-dependent function $n(t)$ or a stochastic process $n(t)$.

(b) Trends in claim numbers are caused by a number of background factors, such as when new policies are added to the portfolio, policy conditions are extended to cover new risks, or the risk propensity changes, for example as a result of new construction methods and materials (fire insurance) or changes in traffic conditions (motor-car insurance).

Because the trends are usually fairly slow and stable, a simple deterministic formula will often be adequate for introducing the relevant effect into the model. This will be assumed to be the case. The formula can be either multiplicative

$$n_g(t) = n \cdot r_g^t \tag{9.2.1}$$

or linear

$$n_g(t) = n + d_g \cdot t, \tag{9.2.2}$$

where $n_g(t)$ is the trend adjusted expected number of claims in year t and $n = n_g(0)$.

The model parameters n, r_g or d_g should be derived from relevant experience. They are normally different for different lines of business.

As background it might be helpful to observe that the total volume of demand for insurance in each country grows to a large extent in line with the Gross National Product (GNP). De Wit (1978, amendments 1992 in a private letter) has investigated how GNP growth is reflected in the demand for insurance. He estimates that a 10% growth in GNP induces an increase in the total demand for general insurance of about 15% in the Netherlands and a somewhat higher increase for life insurance. However, these numbers vary over time and with the type of insurance, and are, so far as scattered information is available, very different in different countries.

Figure 9.2.1 exhibits the general development in the OECD area, confirming the feature that the growth rate of premium income varies over time. The penetration depicted in the figure is premium income as a ratio of Gross Domestic Product.

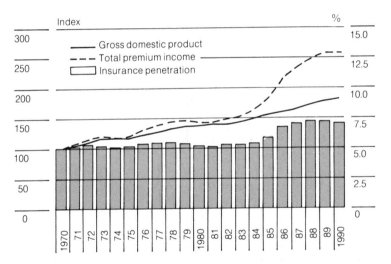

Figure 9.2.1 *Development of the Gross Domestic Product and total premium income in the OECD in the life and non-life sectors. Source: Sigma 4/92.*

(c) **Deterministic cycles.** Claim curves typically exhibit deviations from the trend line. There may often be periods of several years with more or less systematic upwards or downwards movements relative to the trend (Figures 2.4.1 and 2.4.2 and the remark about the term 'cycle' in section 12.2(b)). These longer term movements are generally

associated, moreover, with a considerable degree of short-term variation of the risk propensity, again varying to a large extent by line.
A formula of the type

$$n(t) = n_g(t) \cdot (1 + d(t)) \cdot q(t) \qquad (9.2.3)$$

might be appropriate to describe the effects on the Poisson parameter
for year t. In this simple model the deterministic function $d(t)$ introduces deviations from the normal trend which can extend over
several years. The factor $q(t)$ introduces short-term variation of the
risk propensity. It is conceptually the same as the mixing variable
in section 2.4. The values for consecutive years are assumed to be
(stochastically) independent.

The function $d(t)$ should be specified to correspond to the experience
in the relevant environment. A convenient formula may be obtained
by using a sine-curve

$$d(t) = \alpha_d \cdot \sin(\omega t + v) \qquad (9.2.4)$$

where α_d is the amplitude and ω the (angular) frequency

$$\omega = 2\pi / T, \qquad (9.2.5)$$

T being the length of the postulated cycle. The parameter v determines the phase of the cycle.

If the formula is to be used for projecting the future course of the
process, the phase v should be set to reflect the anticipated phase
of the current cycle. Since the future is generally very uncertain,
several alternatives can be examined in order to provide scenarios
for business planning. In practice it is not possible to make reliable
forecasts for more than a couple of years ahead, if that, and a more
useful application of formula (9.2.4) is in testing how well the insurer
could cope with the cycles which might occur. Realistic but hypothetical cycles can be postulated purely for the purpose of the test. For
example, the adverse phase of the cycle can be programmed to begin
immediately, or at any given point of time, regardless of what is
seen as the best estimate of the phase of the current cycle. The
purpose of the model is to explore the potential consequences of a
given set of assumptions and, in particular, whether the insurer has
the resources to meet them.

(d) Cycles, time series approach. Another way of handling cycles
is to make use of time series. They generate upward and downward
movements which are not so regular as a sine curve, so being in better

conformity with experience. A summary of some of the most relevant features of time series is given in Appendix G. As an example of this approach (9.2.4) will be replaced by

$$n(t) = n_g(t) \cdot (1 + d(t)) \tag{9.2.6}$$

where the variable $d(t)$ is generated by a second order autoregressive time series (Appendix G, section G.3)

$$d(t) = \alpha_1 \cdot d(t-1) + \alpha_2 \cdot d(t-2) + \varepsilon(t) \tag{9.2.7}$$

with mean equal to zero.

The short-term random variable in (9.2.3) is incorporated into the factor $1 + d(t)$. Note that it will now affect the subsequent flow of the simulated process, whilst in (9.2.3) this was not the case.

With the coefficients α_1 and α_2 suitably defined (Appendix G), this algorithm generates a sequence of values for $d(t)$ which may satisfactorily simulate the observed cyclical course of the claim number process.

The time series approach is very handy for practical simulation applications. It gives effect to the idea that cycles can occur irregularly

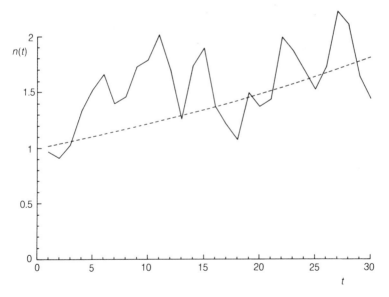

Figure 9.2.2 *A realization of the time series (9.2.6) with coefficients* $r_g = 1.02$, $\alpha_1 = 0.438$, $\alpha_2 = -0.096$, $\sigma_\varepsilon = 0.182$ *and* $\gamma_\varepsilon = 0.2$, *involving a superimposition of a trend, cycling (T = 8 years) and short-term fluctuation.*

and that they can be adverse or profitable, large or small (Figure 9.2.2). The phase of a present cycle can be taken into account by specifying the initial values of $d(t)$ (Appendix G, Exercise G.4).

(e) Combined notation. It is convenient to introduce a notation which covers both the deterministic case of section 9.2(c) and the time series approach of section 9.2(d). For this purpose let

$$q_c(t) = (1 + d(t)) \cdot q(t) \quad \text{(deterministic case)}$$
$$q_c(t) = (1 + d(t)) \qquad \text{(time series case)}. \tag{9.2.8}$$

The variable $q_c(t)$ represents the combined effect of cycles and short-term variation, and

$$n(t) = n_g(t) \cdot q_c(t) \tag{9.2.9}$$

is valid for both cases, giving the expected number of claims in year t, including a correction factor which incorporates the effects of trends, cycles and short-term mixing variation.

(f) Exogenous impacts. The validity of both the deterministic and the time series approaches can sometimes be improved if exogenous impulses are incorporated into the system. For example, the frequency of employment accidents is clearly correlated with the level of activity of the national economy (Pentikäinen et al., 1989, p. 33; see also Figure 2.4.2) and third-party motor claims with the total volume of traffic and with sales of new cars (Becker, 1981, p. 190).

9.3 Claim amounts

(a) Claim size. Apart from the risk propensity, the claim size distribution also changes so much over time that the variation cannot be ignored when looking at longer time periods. The main cause is inflation, which was considered in general terms in Chapter 7. The claim inflation discussed in section 7.2(b) is relevant in this context. We have

$$I(t) = I(0) \cdot \prod_{\tau=1}^{t} (1 + i(\tau)) \tag{9.3.1}$$

where I is the index of claim inflation and i the incremental rate.
It is convenient to scale the index so that $I(0) = 1$.

Our assumption now is that **claim size changes is proportional to the index of inflation.** Hence the moments about zero of $Z(\tau)$ ((3.2.9) and (1.4.10)) are changed as follows

$$a_j(\tau) = I(\tau)^j \cdot a_j(0) \qquad (9.3.2)$$

and consequently

$$\mathrm{E}[Z(\tau)] = \mu_Z(\tau) = m \cdot I(\tau)$$
$$\sigma_Z(\tau) = \sigma_Z \cdot I(\tau) \qquad (9.3.3)$$
$$\gamma_Z(\tau) = \gamma_Z$$

where the symbols without the argument (τ) indicate the initial values associated with the period $\tau = 0$ and $m = \mathrm{E}[Z(0)] = \mu_Z$.

The reader is referred to paragraph 2 of section 3.1(c) for discussion about the assumption that large and small claims change in the same proportion.

(b) Aggregate claims, cross-section. We are now in a position to consider the aggregate claim amount emerging in a single year τ and the amount accumulated in the period $1,\dots,t$. A major complication is that, according to the assumptions made in this and previous sections, the distribution of the claim amount variable X may be conditional on both the combined cycle-mixing variable q_c and the inflation index I. This makes rigorous analytical calculation of the d.f. so difficult that we will mainly resort to simulation. It is necessary to derive some basic characteristics as primary input quantities for the methods which were discussed in Chapter 5.

REMARK Rigorous analytical handling of the long-term development of the claims distribution is usually based on the **convolution** of the one-year distributions. Numerous suggestions for solutions can be found in the literature (for instance Beard *et al.*, 1984, section 6.7). Regrettably, these methods are mostly based on restrictive premises and are highly inconvenient for practical application.

The main characteristics of $X(\tau)$ are first presented subject to the conditions that the values q_c and I of the combined cycle-mixing variables q_c and the inflation index I are given. The values to be assigned to these primary variables at future time points should be first evaluated or simulated, as outlined above and considered further in section 9.4. Then the process reduces to the Poisson case and formulae (3.2.13) can be applied.

$$\mu(X(\tau)|q_c, I) = n(\tau) \cdot m(\tau)$$
$$\sigma^2(X(\tau)|q_c, I) = n(\tau) \cdot a_2(\tau) \qquad (9.3.4)$$
$$\gamma(X(\tau)|q_c, I) = \frac{a_3}{a_2^{3/2} \cdot n(\tau)^{1/2}}$$

where

$$m(\tau) = m \cdot I(\tau) \qquad \text{(see (9.3.3))}$$
$$a_2(\tau) = a_2 \cdot I(\tau)^2 \qquad \text{(see (9.3.2))}$$
$$n(\tau) = n_g(\tau) \cdot q_c(\tau) \qquad \text{(see (9.2.9))}.$$

For brevity, we drop the argument 0; hence $m = m(0)$ and $a_j = a_j(0)$, using the notation introduced in preceding chapters (3.2.9).

The above formulae are immediately available for the calculation of the d.f. of the cross-sectional variable $X(\tau)$ or approximations to it, using the techniques which were introduced in the previous chapters. The background variables I and q_c are first separately simulated or calculated and then substituted into these equations for further processing.

(c) Cumulative aggregate claims, mean value. As was stated in section 9.3(b), we do not aim to derive any rigorous formulae for the d.f. of the cumulative claim variable $X(1, t_2)$ defined by (9.1.1), owing to the complications which arise from the specification of the model. However, some characteristics of this variable can be obtained easily and these are presented here.

The mean value can be derived immediately from (9.1.1) and (9.3.4)

$$\mu(X(1, t)) = m \cdot \sum_{\tau=1}^{t} n(\tau) \cdot I(\tau). \qquad (9.3.5)$$

In the particular case where the growth rate $r_g(t)$ (9.2.1) for the portfolio and $i(t)$ for the rate of inflation are both assumed to be constant and no cycles are present, formula (9.3.5) reduces to a geometric series, the sum of which is

$$\mu(X(1, t)) = nm \frac{r_{gi}^t - 1}{r_{gi} - 1} r_{gi} \qquad (9.3.6)$$

where, briefly

$$r_{gi} = r_g \cdot (1 + \bar{i}). \qquad (9.3.7)$$

This factor represents the average growth of nominal volume of the claim expenditure.

(d) The variance of the accumulated total amount of claims is obtained in a similar way

$$\text{Var } X(1, t) = \sum_{\tau=1}^{t} \text{Var}(X(\tau)) + 2 \sum_{\tau_1 < \tau_2} \text{Cov}(X(\tau_1), X(\tau_2)). \qquad (9.3.8)$$

The calculation of the relevant covariances depends on the conditions assumed.

In the particular case where the inflation rate is constant (or otherwise deterministic) the rule

$$\text{Cov}(X(\tau_1), X(\tau_2)) = \text{Cov}(q_c(\tau_1), q_c(\tau_2)) \cdot E(X(\tau_1)) \cdot E(X(\tau_2)) \qquad (9.3.9)$$

can be derived ((G.1.1) of Appendix G).

In addition, if the cycles are deterministic or non-existent (section 9.3(c)) and the mixing variables $q(t)$ are independent and identically distributed for all t, then the covariance terms drop away and (9.3.8) reduces (see also (3.2.14)) to

$$\text{Var } X(1, t) = \sum_{\tau=1}^{t} [n(\tau)a_2(\tau) + \mu_X(\tau)^2 \sigma_q^2]. \qquad (9.3.10)$$

(e) The skewness can be obtained under the same special conditions as (9.3.10) by first calculating the third cumulant (3.2.14)

$$\kappa_3(X(1, t)) = \sum_{\tau=1}^{t} [r_3/n(\tau)^2 + 3r_2\sigma_q^2/n(\tau) + \sigma_q^3\gamma_q] \cdot \mu_X(\tau)^3 \qquad (9.3.11)$$

and then deriving the skewness from (1.4.25)

$$\gamma(X(1, t)) = \frac{\kappa_3(X(1, t))}{\sigma^3(X(1, t))}. \qquad (9.3.12)$$

9.4 Simulation of the claim process

There are several methods available for the calculation of the d.f. of the claim variable $X(1, t)$, most of them analytic (Beard *et al.*, 1984, Chapter 6). However, when the structure of the process is extended as far as was indicated in section 9.2, particularly introducing autoregressive background variables, the analytic approach becomes cumbersome, indeed virtually intractable. Therefore, in this and in subsequent chapters most calculations are based on simulation.

(a) The basic algorithm. The formulae for the characteristics of the claim process which were derived above, and the technique which

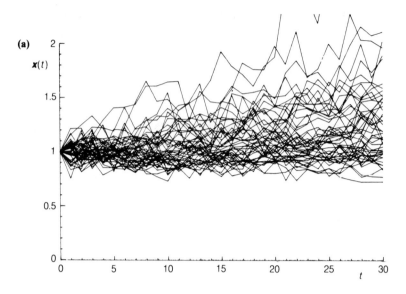

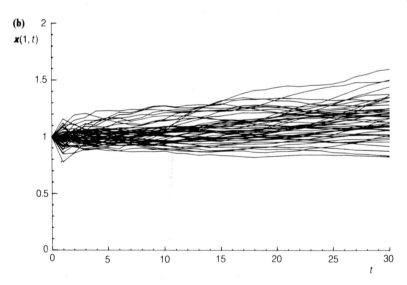

Figure 9.4.1   *Realizations of the relative cross-sectional variable x(t) ((a) and (c)) and the relative cumulative variable x(1,t) ((b) and (d)). In (c) and (d) the standard deviations of the noise terms for the mixing variable and for inflation were taken as zero.*

(c)

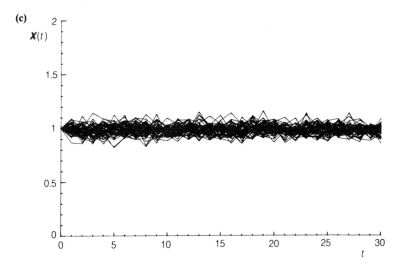

(d)

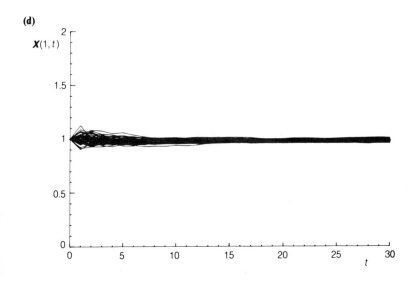

was introduced in section 5.4, now enable a straightforward simulation to be carried out, both of the cross-sectional variables $X(t)$ and of the cumulative claim variable $X(1, t)$.

As was outlined in section 5.5(a), the simulation progresses as an algorithm from year $t - 1$ to t, resulting in a sample of the cross-sectional variables $X(t)$ and the accumulated claim amount $X(1, t)$ therefrom.

(1) Generate the primary variables by means of an appropriate stochastic process, or calculate them from deterministic formulae:

- The trend adjusted $n_g(t)$ ((9.2.1) and (9.2.2)).
- The mixing state $q_c(t)$ ((9.2.3), (9.2.6) and (9.2.7)) and by means of its $n(t)$ ((9.2.8) and (9.2.9)).
- The index of claim inflation $I(t)$ ((7.3.1) and (7.3.2), also (9.3.1)).

(2) Calculate the characteristics of the cross-sectional aggregate claim $X(t)$, using (9.3.4).
(3) Generate $X(t)$; note that when the values of the mixing variable and claim inflation are given in the previous steps, the conditional X variable at this step is of the simple compound Poisson type, so that the procedure of section 5.4(b) applies.
(4) Accumulate the claims for year t with the claims up to year $t - 1$ using

$$X(1, t) = X(1, t - 1) + X(t).$$

The above algorithm is given in more detail as a part of the flow chart in section F.5 of Appendix F.

Note that the above approach might require modification and additional sub-modules in particular situations. For example, claims of catastrophic size may require separate treatment. Amendments may, therefore, be necessary, as will be proposed in section 9.6.

(b) Examples. Figure 9.4.1 exhibits realizations of the relative one-year claim amount $x(t) = X(t)/E(X(t))$ and the cumulative claim amount variable $x(1, t) = X(1, t)/E(X(1, t))$.

The outcomes of the simulation represent a superimposition of (1) a trend, (2) variation of the risk propensity over a period of several years (cycles), (3) short-term mixing variation and pure Poisson fluctuation. The data used in these diagrams are as follows

Claim variable: $n = 10\,000$, $m = £6160$, $r_2 = 37.3$,
$r_3 = 3832$ (these are from the line $M = £1$ million in Table 3.4.1),

$r_g = 1.02$; Mixing variable: second order time series (G.3.1) of Appendix G with coefficients corresponding to 8-year average cycle length (see (G.3.5)) and with standard deviation 0.05 and skewness 0.2. Inflation: formula (7.3.1) with $\bar{\imath} = 0.04$, $a = 0.6$ and $\sigma_i = 0.03$.

$$(9.4.1)$$

As might be expected the amplitude of the fluctuating waves of $X(t)/E(X(t))$ are larger than that of the cumulative variable $X(1,t)/E(X(1,t))$, where the fluctuation is smoothed over a number of years. The range of fluctuation in diagrams (a) and (b) is mainly due to the variation inherent from both the mixing and inflation. To illustrate this, the standard deviations of the noise terms of these variables were taken as zero in diagrams (c) and (d). As can be seen, the amplitude of fluctuation is considerably reduced.

A visual scan of a diagram such as Figure 9.4.1 can provide a very good idea of the properties of the process. Methods of adding outlined confidence boundaries to the diagram are dealt with in section F.4 of Appendix F.

Note the divergence of the bundle of realizations in (a) and (b). In terms of time series theory (Appendix G, section G.1(c)) the process is not stationary. In some applications special control arrangements may be needed in order to make the outcome more realistic.

The claim process will be a component of the more general processes which will be introduced in subsequent chapters. It is advisable to defer further consideration until then.

9.5 The settlement of claims

We will now relax the assumption that claims are paid as soon as they occur. In fact there is always a time delay, whilst claims are notified to the insurer, handled by the claim manager and eventually settled. The indemnity may be paid in one or more stages or, if the compensation is given in the form of an annuity, by establishing an appropriate provision for future payments. In what follows we will use paying and settling as synonyms, notwithstanding that the latter is, strictly speaking, a more general concept.

The time lags from the occurrence of the event causing the claim to the final payment can be quite long, often years and in extreme cases decades, and may commonly be still longer in reinsurance than in direct business. This can give rise to considerable uncertainty which warrants careful attention.

Reserving for outstanding claims is one of the central topics of actuarial theory and practice. However, stochastic evaluation of the uncertainties involved in the settlement process has gained little attention until recent years. The importance of the outstanding claim reserve in solvency analyses was dealt with by the British Solvency Working Party in a series of reports (Daykin and Hey, 1990, Daykin *et al.*, 1984, 1987). Particular articles on this topic, among others, have been presented by Stanard (1986) and by Pentikäinen and Rantala (1986, 1992).

9.5.1   Outstanding and paid claims

(a) The run-off pattern.   For the analysis of the claim settlement process it is necessary to distinguish the year u when the event causing the claim occurred and the year t when the claim payment is made. This approach is known as **year of accident** analysis and will be used consistently in what follows. Some actuarial methodology for estimating outstanding claims is based on the **year of underwriting**, which focuses on when the relevant premium was paid. Let $X_p(u, t)$ be the amount of claims which originate from year u and are paid in year t. The term **cohort** is used to describe all the claims which originate from the same year.

The claim settlement procedure and the development of cohorts is illustrated in Figure 9.5.1, using the current time t and the development time $d = t - u$ as coordinates, i.e. placing the payment $X_p(u, t) = X_p(t - d, t)$ in the cell (t, d). The cohort starts in the cell $(u, 0)$ in the year u of its origin. The next year, when the current time and the cohort's development time have both increased by one, the cohort is moved diagonally up a step and to the right, to cell $(u + 1, 1)$. After d years it appears in the cell $(u + d, d) = (t, d)$. Hence, the payments relating to the cohort of accident year $u = t - d$ are located in the ascending diagonal which is indicated by heavier shading in Figure 9.5.1.

The settlement process is also referred to as the **run-off** of the cohort. This term emphasizes the feature that the amount of claim payments outstanding is decreasing, and falls ultimately to zero, as the development time d runs its course. For applications it is convenient to find an upper limit d_{max} for the development time d, within which all the claims of the cohort can be regarded as paid. Then, in Figure 9.5.1, the claim history of all the cohorts which contribute

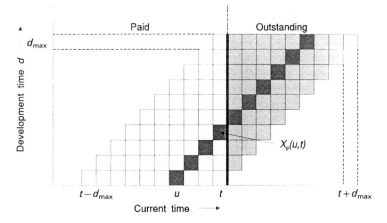

Figure 9.5.1 *Claims which occurred before the end of year t arranged according to time of settlement and development time.*

to claim payments and reserving in year t is constrained to the pattern of cells plotted in the diagram. Year t is usually taken to indicate the current accounting year.

(b) The claims paid in year t are

$$X_p(t) = \sum_{u \leqslant t} X_p(u, t). \tag{9.5.1}$$

This is the sum of the $X_p(u, t)$ located in cells on the vertical pillar at current time t.

(c) The reserve for outstanding claims at the end of year t is

$$C(t) = \text{estimate for } \sum_{u,d} X_p(u, t + d) \cdot v(t, t + d) \tag{9.5.2}$$

where $v(t, t+d)$ is the discounting factor to be specified below and the sum covers all the $X_p(u, t)$ located in Figure 9.5.1 in the cells to the right of the end of year t, in the so-called **run-off triangle** (shaded in the diagram).

Care has to be taken to ensure that the estimates of $X_p(u, t)$ allow properly for claim inflation up to year t. The rules introduced in section 9.4(a) provide the possibility of simulating the effect of future inflation in generating the payment amounts positioned in the run-off triangle of Figure 9.5.1.

There are diverging practices as to whether the expected future payments should or should not be discounted (i.e. to keep or drop the factor v in (9.5.2)). It is common practice in general insurance not to discount, resulting in a certain (usually unquantified) margin in the reserves, providing full allowance has been made for future inflation. However, accounting practice, actuarial standards or regulatory requirements may limit the discount factors which can be used, for example to take into account only the income return on investments or to require a cautious, low assumption or not to permit discounting at all for public statements.

In cases where discounting is performed the rate $j(t)$ should be estimated or simulated for the average yield earned for the relevant years $t + 1, t + 2, \ldots$ for assets which are held to back the claim reserve (section 8.5(f)). The discounting factor for the accrued interest from the midpoint of year $t + d$ to the end of year t is obtained from

$$v(t + 1) \cdot v(t + 2) \cdots v(t + d - 1) \cdot v(t + d)^{1/2} \qquad (9.5.3)$$

where

$$v(t + \tau) = (1 + j(t + \tau))^{-1}. \qquad (9.5.4)$$

More conventionally, a constant rate of interest j may be assumed, or a series of such constants for different time periods. In the case of constant interest

$$v(t, t + d) = v^{d - 1/2} \qquad (9.5.5)$$

where $v = (1 + j)^{-1}$ is the conventional discounting factor.

(d) The terminology concerning the claim settlement process follows the recommendations laid down by the Institute of Actuaries in their Claim Reserving Manual (1991), Vol. I, section A5.

The term **outstanding claims** is used as an umbrella concept to include **open claims** and so-called **IBNR**. A claim is open if it has been notified to the insurer but has not been settled.

The IBNR provision is an estimate of the reserve required for all outstanding claims less the reserve held at the accounting date in respect of open claims. It includes an estimate of claims which have been incurred but not reported to the insurer by the accounting date. In addition it may include a reserve for adverse deviations, because of the uncertainty relating to the reserve for open claims, as well as a provision for claims which have already been settled but may be reopened.

Open claims may be assessed case by case or collectively using a statistical or actuarial approach. The IBNR provision is estimated by statistical or actuarial methods, as will be discussed in section 9.5.2.

(e) **Incurred claims** in year u, according to a definition based on the cohort structure, indicate the estimated total outgo in all development years in respect of the claim cohort which originated in year u, i.e. the sum of the $X_p(u, t)$ located in the shaded diagonal in Figure 9.5.1. This is one of the key entries in the accounting equation (1.2.5), being the counterbalance to the earned premiums on the income side. It consists of the amount paid in year u, together with the reserve at the end of year u for outstanding claims relating to this cohort. However, this definition is usually replaced by a more practical accounting rule, which is based on the quantities available in the conventional income statement (1.2.2)

$$X(t) = X_p(t) + C(t) - C(t-1). \qquad (9.5.6)$$

The payments relating to earlier cohorts are incorporated in this formulation. When a claim that was reserved in the claim reserve $C(t-1)$ at the beginning of the year is paid out in year t, the relevant reserve can be released. If the reserve was for exactly the correct amount, then the released reserve and the payment fully offset each other and the payment of the claim does not affect the financial result for year t. However, since the claim reserve is just an estimate, the two items do not in general cancel out. So formula (9.5.6) brings into the income statement any corrections to earlier reserves which emerge in year t, either when the claims are settled, or because claim reserves are revised.

9.5.2 Claim reserving rules

There is an extensive literature on claim reserving methods, including, for example, the *Claim Reserving Manual* issued by the Institute of Actuaries, London (1989), *Loss Reserving Methods* by van Eeghen *et al.* (1981) and *Claims Reserving in Non-life Insurance* by Taylor (1986). Claim reserving methodology constitutes a branch of actuarial mathematics which has traditionally been developed in isolation. However, in fact, it has links with risk theory in so far as some of the basic concepts have found application in both areas. In particular, analysis of the uncertainties involved with claim reserving can be carried out using the methods of risk theory. As examples, three different reserving methods will be briefly considered, and their effects investigated, in this section and those immediately following.

(a) Chain-ladder methods include a wide range of reserving rules. A common feature is that growth rates of the payments attributable to a cohort are assumed to be sufficiently stable that observed growth rates can be used to forecast corresponding growth rates for the future. A typical example is presented below.

The reserve is first calculated for each cohort and then the results are summed. The method makes use of the following growth rates as auxiliary coefficients

$$a(d) = A_1(d)/A_0(d). \qquad (9.5.7)$$

These are derived from the development triangle consisting of the sums of $X_p(u, t)$ in the cells relating to the accounting year t and earlier years, as shown in Figure 9.5.2 by the shaded parallelograms A_1 and A_0.

It is now assumed that the accumulated amount of claims of cohort $t - d$ paid in years $t - d, ..., t + 1$ can be expressed as the multiple of the growth rate $a(d)$ and the accumulated amount from the cohort up to year t

$$\sum_{B_1} X_p / \sum_{B_0} X_p = a(d)$$

where the sums are taken over the cells located on the heavier shaded

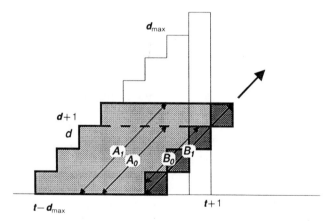

Figure 9.5.2 *Derivation of the chain-ladder rule. The parallelogram $A_0(d)$ is obtained from $A_1(d)$ by removing the top row. B_0 and B_1 are the chain of the developmemt cells of cohort $t - d$ up to t or $t + 1$ respectively.*

diagonals B_1 and B_0 respectively. The sum of the diagonal B_1 then grows in the ratio $a(d + 1)$ for the next year and so on. By applying this rule repeatedly for the rest of the run-off $t + 1, ..., t + d_{max}$ the following estimate is obtained for the cohort reserve at the end of year t.

$$C_c(t - d, t) = c_c(d) \cdot \sum_{\tau = 0}^{d} X_p(t - d, t - d + \tau) \qquad (9.5.8)$$

where

$$c_c(d) = \prod_{\tau = 0}^{d_{max} - d - 1} a(d + \tau) - 1. \qquad (9.5.9)$$

The sum in (9.5.8) stands for the claims paid from the cohort $t - d$ up to the end of accounting year t. Then the total claim reserve is the sum of the cohort reserves

$$C_c(t) = \sum_{d = 0}^{d_{max} - 1} C_c(t - d, t). \qquad (9.5.10)$$

If the reserve is to be discounted to allow for investment income, the formula (9.5.8) should be restructured so that the increments for years $t + 1, t + 2, ..., t + d_{max} - d$ are separately identified, with corresponding discounting factors $v^{1/2}, v^{3/2} ...$. Details of the procedure are given in Exercise 9.5.1.

(b) The premium-based method. The cohort reserves at the end of the accounting year t may be derived by assuming them to be proportional, with ratio $c_p(d)$, to the premium income $P(t - d)$ attributable to the cohort

$$C_p(t - d, t) = c_p(d) \cdot P(t - d). \qquad (9.5.11)$$

The coefficients $c_p(d)$ are a descending sequence as the development time d increases. A first approximation is to assume that the coefficients do not depend on the reserving year t. There are various ways to define the P and c_p in detail (section 4.3 in Pentikäinen and Rantala (1992) and Exercise 9.5.2).

Formula (9.5.11) leads to a fairly stable development of the reserve but it has a tendency to hide actual fluctuations, as will be illustrated in section 9.5.4.

(c) The mixed method is constructed as a combination of the chain-ladder and premium-based formulae

$$C(t) = \sum_{d = 0}^{d_{max} - 1} [z(t, d) \cdot C_p(t - d, t) + (1 - z(t, d)) \cdot C_c(t - d, d)]. \qquad (9.5.12)$$

The idea is to assign values to the coefficients $z(t, d)$ so that the premium-based C_p predominates at the beginning of the run-off, reflecting the prior experience in a stable way, whilst the chain-ladder component C_c is still volatile. As the cohort matures the weight should move to the latter term, because it takes into account the most recent experience available from the claim process. This formula is in principle similar to experience-rating; it 'learns' all the time from the actual data. The credibility coefficient $z(t, d)$ may be independent of t.

Formula (9.5.12) was proposed by Benktander (1976). The logic is analogous to that of Bornhuetter and Ferguson (1972), but it focuses on a different variable.

The objective can be realized, for instance, by taking z to be the same as the premium-based coefficient $c_p(d)$ in (9.5.11). An alternative derived from credibility considerations was presented by Pentikäinen and Rantala (1992) (Exercise 9.5.2).

9.5.3   Modelling the run-off process

(a) Evaluation of reserve uncertainty.   Conventional reserving rules provide only deterministic evaluations of the claim reserve. For risk theory considerations this needs to be supplemented by analysis of the uncertainty. This can be done, for instance, by evaluating the standard deviations of the relevant quantities or using other tools of mathematical statistics (references are given in section 9.5.4(f)). Unfortunately, if analytic techniques are to be used, some rather restrictive assumptions are necessary. In order to study the problem under more general conditions, as indicated in the first sections of this chapter, including allowing for stochastic mixing with cycles, stochastic inflation, and stochastic investment yield in cases where discounting (9.5.3) is involved, then it is necessary again to resort to simulation. The approach described in section 9.4 may be used.

First it is necessary to find methods for simulating the variables $X_p(u, t)$. If the reserves are to be discounted, then the interest variables can be generated in the way described in Chapter 8.

(b) Cell parameters.   In order to apply simulation methods of the type developed for the cross-sectional claim variable in section 9.4(a), the relevant input parameters, the Poisson n and the basic characteristics (9.3.4) of the claim size distribution must be specified for each cell in the configuration of Figure 9.5.1. When this has been done, more complex problems can be dealt with using the simulated claim amounts for individual cells as building blocks, for example testing the adequacy of the claim reserve to cover the range of possible outcomes for outstanding claims (section 9.5.4(c)).

The total estimated claims in respect of a cohort can be obtained as the sum of the simulated amounts for individual cells.

In order to obtain the Poisson parameter $n(u, t)$ for cell (u, t) it is necessary to postulate the distribution of the claim numbers within each cohort. Let $\rho(d)$ be the probability for the event that, if a claim occurs in year u, its settlement can be attributed to the run-off cell $(u, u + d)$. If the total number of claims in the cohort u is $n(u)$ then

$$n(u, u + d) = \rho(d) \cdot n(u). \qquad (9.5.13)$$

The probabilities $\rho(d)$ can be derived from the claim data using conventional statistical techniques. They are assumed to be independent of the cohort year u. They can also be derived from the c_c coefficients of the chain-ladder reserve (Exercise 9.5.3).

(c) The simulation proceeds according to the following steps:

(1) The expected number of claims $n(u)$ for the whole cohort is determined, as was described in section 9.4(a).
(2) The Poisson parameters for each of the cell claim amount variables $X(u, u + d)$ are obtained from (9.5.13).
(3) The distribution of claim size in each of the cohort cells is assumed to be the same as was postulated for the claim process as a whole. The claim sizes are assumed to increase with inflation in each successive settlement year as the cohort matures, in the way described in section 9.3(a). Hence, the index of inflation $I(t)$ has to be generated (Chapter 7) for each of the relevant years t and the claim size characteristics (9.3.4) transformed accordingly.
(4) The algorithm of section 9.4(a) for a cross-sectional claim variable can then be used (Exercise 9.5.4), resulting in $X_p(u, u + d)$.

REMARK Some authors have observed that the size of claims with delayed settlement may change for reasons other than inflation. For example, claims settled later may be significantly larger than those settled earlier. There is no particular difficulty in providing an auxiliary correction factor in the simulation if this phenomenon is considered relevant and sufficient data are available to evaluate the required parameters, as Pentikäinen and Rantala

(1986) have demonstrated. To avoid further complication of the formulae this aspect is not considered further here.

9.5.4   Applications

(a) Run-off error is a concept which will now be introduced to quantify the claim reserving uncertainty. It is defined as the difference between the claim reserve and the present value of the eventual amounts required to settle the claims in full, which the reserve is intended to estimate

$$R(t) = C(t) - \sum_{u,d} X_p(u, t + d) \cdot v(t, t + d). \tag{9.5.14}$$

The sum is extended over the entire run-off triangle in Figure 9.5.1 (9.5.2).

In practice the actual value of $R(t)$ cannot be obtained until all the cohorts with outstanding claims at time t have been completely settled. However, $R(t)$ can be used as an *a posteriori* check on the adequacy of the reserving method.

One of the advantages of stochastic modelling is that it is possible with simulation to evaluate in advance the variation range and other characteristics of the run-off error $R(t)$ for various reserving rules and also to evaluate the effect of the various types of uncertainties referred to in section 1.3(e).

(b) Closed fund evaluation.   The run-off error $R(t)$ can be simulated by assuming a closed fund situation where no further business is written and the outstanding claims are settled as they fall due, out of the initially assessed reserve.

The tools provided in section 9.5.3 make it possible to generate values for all the variables in (9.5.14). Repeating the computation with fresh simulations of the random variable (Chapter 5), the distribution of $R(t)$ can be constructed and the standard deviation and other characteristics obtained. Figure 9.5.3 illustrates the d.f.s of the ratio $r = R/P$ obtained from a simulated run-off with three different reserving rules.

Note from (9.5.14) that the fundamental unknown is the claim amounts $X_p(u, t + d)$ but the recognition of the run-off errors depends also on the calculation of the outstanding claim reserve $C(t)$, which is based on the development triangle.

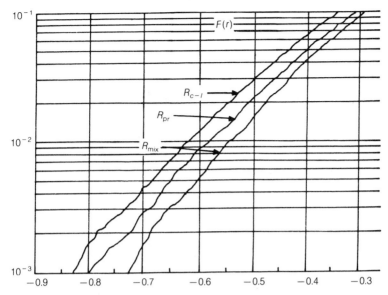

Figure 9.5.3 *The left-hand part of the cumulative distribution function F of r = R/P, computed in parallel for the three reserving rules of section 9.5.2 (Pentikäinen and Rantala, 1988), in the case of relatively long-tailed business with $d_{max} = 12$ years).*

A British Working Party (Daykin *et al.*, 1984, 1987) used the run-off simulation in a form which also utilized information about the distribution of the assets. The idea is to mimic the settlement process where claims are paid out step by step from the fund, which is initially made up from the reserve $C(t)$, but supplemented by the return on the residual fund as it runs down, including both interest and changes in asset values. Ideally the pattern of maturity of the asset items should be in conformity with the need for liquid money but the simulation assumes that assets are sold if necessary to meet the excess of outgo over income, according to a specified disinvestment policy. Reinvestment is required when assets mature before they are needed for payments or generally when there is an excess of income over outgo. At the end of the first development year $t + 1$ the residual fund $C_r(t + 1)$ is

$$C_r(t+1) = C(t) \cdot (1 + j(t+1)) - (1 + j(t+1))^{1/2} \cdot \sum_{d=0}^{d_{max}-1} X(t-d, t+1).$$
$$(9.5.15)$$

The residual funds $C_r(t + 2), C_r(t + 3), \ldots$ are obtained in a similar way until finally $C_r(t + d_{\max})$ shows whether the initial reserve was sufficient or not in the particular realization. Repeating the simulation a diagram such as Figure 9.5.4 is obtained. The terminal values represent the distribution of accumulated run-off error. The run-off error $R(t)$ defined in (9.5.14) is the equivalent of this accumulated run-off error discounted back to the starting date.

The closed fund (run-off) test offers a way of testing the solvency of the insurer or, taking a more limited target, of evaluating the adequacy of the claim reserves. If the reserve were calculated deterministically as a best actuarial estimate, it would frequently (in fact in about 50% of the cases if the distribution is not very skew) turn out to be inadequate as compared to the real (or simulated) outcome. If it is required that the run-off residual should be non-negative with a given confidence probability, it is necessary to incorporate a safety margin in the reserve, or hold an adequate solvency margin in terms of free assets to cover the risk of adverse deviation from the best estimate. In practice there will also be model error and parameter error in the calculation of the outstanding claim reserve, necessitating further margins (section 1.3(e)).

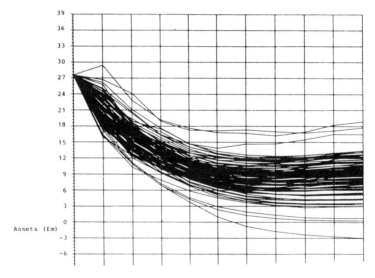

Figure 9.5.4   *Run-off of assets (100 simulations), Daykin et al. (1987).*

(c) The effect of run-off error in a going concern. In the case of a going concern the accounting formula for incurred claims (9.5.6) is what determines the emergence of profit. Each of the terms is subject to run-off inaccuracy and the technique of section 9.5.3 enables the errors to be simulated. However, some care is required in accounting consistently for the yield of interest in any case where the reserve is discounted. Formula (9.5.6) might need a corrective term which depends on how investment income flows are handled within the model. In the scheme outlined above the amended approximate formula is

$$X(t) = X_p(t) + C(t) - C(t-1) - j(t) \cdot \frac{C(t-1) + C'(t)}{2} \qquad (9.5.16)$$

where $C'(t)$ is that part of the claim reserve $C(t)$ which was included already in $C(t-1)$ (Exercise 9.5.6).

Figure 9.5.5 illustrates a simulated going-concern realization of a claim process where the run-off error is first omitted and the outstanding claims are assumed to be known accurately in advance (X_0/P) and then taken into account using the chain-ladder formula for establishing reserves (X/P). The proportional error arising from the run-off of a closed fund (R/P) is simulated in the same figure.

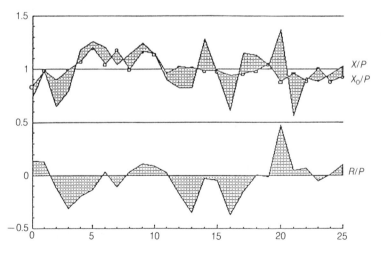

Figure 9.5.5   *Simulated ratios X_0/P, X/P and R/P where $P = n \cdot m$ is the net risk premium; X_0 is simulated assuming perfect knowledge of the outstanding claims and X assuming run-off errors from the claim reserving process, chain-ladder rule, long-tailed business (12 years). Source: Pentikäinen and Rantala (1992).*

The volatility of the outcome increases when the claim reserve is calculated by the chain-ladder formula.

(d) Adequacy measure for reserving rules. Figure 9.5.5 demonstrates how the features of different reserving rules can be revealed by simulating them in parallel. Standard deviations (and other characteristics) can be derived from the simulated outcomes to show the properties of different reserving rules. This permits comparison of the suitability of the different reserving methods. Pentikäinen and Rantala (1992) have explored this idea, which is illustrated in Table 9.5.1.

The test portfolio, which was the same in all cases, assumed a fairly long settlement period ($d_{max} = 12$ years). The mean values are set out in the table to verify the range of the simulation inaccuracy. Theoretically they should be equal to unity for the various ratios X/P and zero for R/P.

The standard deviation σ_0 relates to the case where there is no run-off uncertainty X_0/P. Hence comparison of the numbers in the fourth column shows the variability introduced by run-off uncertainty under the three reserving methods. In the going-concern case, which focuses on the ratio X/P, the chain-ladder method produces the highest level of volatility of the three. In the break-up test where the key measure is R/P, it does not perform as well as the mixed method. This suggests that the latter may be superior to the former. Tests on a variety of portfolios and under different conditions should, of course, be performed before any definitive conclusions can be reached. More rigorous testing is beyond the scope of this book, which is limited to a presentation of the testing methodology.

Table 9.5.1   *Standard deviations of the ratios X/P and R/P simulated in parallel using the chain-ladder method (c), the premium-based method (p) and the mixed method (m). To calibrate the results the same processes were also simulated without any run-off uncertainty (σ_0)*

Variable	Mean	Std. dev. σ	Rel. std. dev. σ/σ_0
X_0/P	1.003	0.087	1.000
X_c/P	1.001	0.240	2.759
X_p/P	0.980	0.065	0.745
X_m/P	0.993	0.125	1.431
R_c/P	−0.002	0.259	2.979
R_p/P	0.039	0.267	3.066
R_m/P	0.004	0.221	2.534

The relative standard deviations for the premium-based method show that it produces the lowest volatility of the three reserving methods on the going-concern basis relative to the baseline where the ultimate cost of claims is assumed to be known. On the other hand, on a closed fund run-off basis it gives the largest errors. In fact, this method is inherently inaccurate for estimating outstanding claims, as it smooths out the peaks and troughs of claim expenditure over several years.

A well-known experience is that the premium-based method is more advantageous when pure (Poisson) short-term fluctuation is predominant, whereas the chain-ladder approach can be expected to be better in cases where the long-term mixing is strong.

(e) Stability profiles offer an alternative way of displaying the outcomes of testing reserve methodology. Figure 9.5.6 shows the same test indicators ($\sigma_{X/P}$ and $\sigma_{R/P}$) as were introduced in section 9.5.4(d). The tests have been applied to four different portfolios. The run-off tail was assumed to be either short (3 years) or long (12 years) and the premium income P was assumed either to be constant in real terms or to be experience-rated according to the past claim experience (for more details see Pentikäinen and Rantala (1992)). In both cases the premiums are assumed to be adjusted in line with a simulated index of inflation.

The assumed portfolios are:

(1) Short-tail, varying rate of premiums;
(2) Short-tail, constant rate of premiums;
(3) Long-tail, varying rate of premiums;
(4) Long-tail, constant rate of premiums.

As would be expected, the volatility is in most cases less in the case of the short-tailed business than in the case of the long-tailed business. However, the premium-based system again shows higher level of stability for the reason given in section 9.5.4(d). Experience-rating of premiums reduces the volatility of the outcomes considerably. This suggests the possibility of improving the optimality properties of reserving methods by adopting better premium-rating rules.

(f) References. Besides the works mentioned in the third paragraph of the introduction to section 9.5 and in the introductory paragraph of 9.5.2, the papers of Ashe, 1986; Hewitt, 1986; Norberg, 1986; Renshaw, 1989; Sundt, 1990; and Verrall, 1989, 1990 are relevant to the analysis of claims reserving.

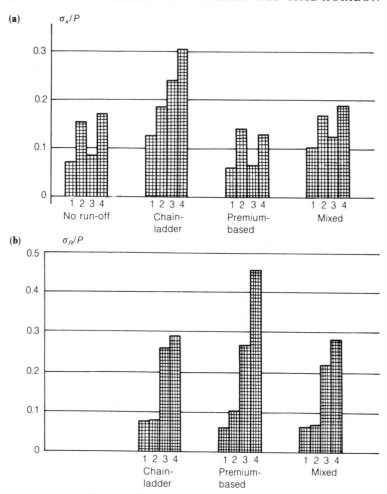

Figure 9.5.6 *Stability profiles obtained by displaying in parallel the test indicators for four different portfolios.*

Exercise 9.5.1 Write down the formula for the calculation of the chain-ladder reserve $C_p(t)$ where the reserve is discounted to allow for future investment income.

Exercise 9.5.2 Propose a rule to indicate how the coefficient $c_p(d)$ in (9.5.11) could be derived from the configuration of Figure 9.5.1. Note that both the premiums and the claims are continually affected by inflation and by the real growth of the portfolio.

Exercise 9.5.3 Find a connection between the cohort settlement distribution $\rho(d)$ and the development coefficient $c(d)$ of the chain-ladder method.

Exercise 9.5.4 Write down a flow chart for the simulation of the 'cell' claim amount $X(u, u + d)$.

Exercise 9.5.5 Do the same as in Exercise 9.5.4 for the run-off error $R(t)$.

Exercise 9.5.6 Derive formula (9.5.16).

9.6 Catastrophes

(a) Two sorts of catastrophes. The term catastrophe is generally used to refer to any very large claim caused by major damage to a single large object, such as a nuclear plant, an oil-drilling rig, a jumbo jet or a large building. These can be referred to as **single object risks**.

Another type of catastrophe arises when one primary event causes damage to a large number of objects. Natural catastrophes such as earthquakes and windstorms are risks of this sort, as are toxic tort (e.g. asbestosis) and other such calamities affecting liability business or workers' compensation insurance. These **multi-object risks** can simultaneously affect numerous insured objects that were treated as independent risk units when the insurance cover and reinsurance were arranged. This is not only a concern for primary insurers operating in sensitive areas, but, through the international reinsurance network, it can lead to major unexpected losses for reinsurers anywhere in the world who have accepted reinsurance (or retrocession) from them. A catastrophe can hit a significant number of insurers operating in the area, each of whom would then be claiming recoveries from their reinsurers. Simultaneous recovery on a number of treaties may trigger a variety of higher level protection covers and may flow round the international reinsurance network to give an ultimate liability to a reinsurer that is many times greater than the exposure that was taken into account when the reinsurer originally evaluated his risk-bearing capacity. This may be described as the **multi-channel effect**.

Note that even when the primary event is related to a single very large insured unit, for example an oil-drilling rig, the claim can spiral in the international reinsurance network in a complicated way and reach the reinsurer by several different channels.

Normal reinsurance arrangements can be expected to give protection against single-object risks along the lines indicated in previous chapters. The particular problems which arise with large claims were dealt with in section 3.3.8. This section can, therefore, be limited to considering the special features of multi-object and multi-channel risks.

It is clear that, if the risk portfolio of an insurance company is concentrated in an area where the catastrophe risk is exceptionally high, the financial position of such a company is particularly vulnerable. To survive natural catastrophes the insurer must have an adequate geographical spread of insured risks. The same can be said about its investment portfolio. No company should invest its assets in such a way that a natural catastrophe could affect the value of a major part of them.

(b) Some estimates of dimensions. It is useful to have an idea of the order of magnitude of potential catastrophes. For this purpose some examples will be given.

According to experts in continental plate theory, it is practically certain that earthquakes such as that in San Francisco in 1906 and that in Tokyo in 1923 will reoccur periodically. The total amount of losses may be enormous, at least theoretically, in areas like these where insured property and people are densely concentrated on the edge of the continental plates. However, a large part of the total potential loss is not insured (or is excluded from insurance coverage). Furthermore, improved modern building construction may significantly limit the damage.

The California Insurance Department publishes reports annually about the situation in California. A recent report (Roth, 1992) estimated that the total aggregate PML on earthquake insurance in Los Angeles/Orange County was about $13 000 million in case of a potential severe earthquake. In addition to that the insurer would also be liable to pay claims for fire, life, health, workers' compensation and other coverages.

The largest insured loss in the USA prior to 1992 was Hurricane Hugo (1989) with losses of about $7000 million, of which about $4000 million was covered by insurance. This was exceeded by Hurricane Andrew (1992), a primary evaluation of the total losses for which being over $30 000 million, of which about half were insured. The extent of the damage caused by hurricanes depends heavily on the wind velocity and on the locations that are affected.

The explosion on the oil-drilling rig Piper Alpha in 1988 is estimated to have cost about $2000 million and the Exxon Valdez (1989) oil spill in Alaska also some $2000 million.

The cost of **man-made catastrophes** such as asbestosis, hazardous waste sites and nuclear fall-out can reach similar dimensions to those of earthquakes and hurricanes.

(c) Modelling the multi-unit risk. The straightforward simulation approach outlined in section 5.4(a) can be used to evaluate the total expenditure in respect of a given period (year), because the number k of events is small. First k is generated by assuming, for example, either the Poisson or the Pólya distribution. Then the total amount of claims Z_i attributable to each event can be simulated to produce the total amount of all the catastrophic claims

$$X_c = \sum_{i=1}^{k} Z_i. \qquad (9.6.1)$$

(d) Distribution of the size Z_i of catastrophic claims can, according to the available experience, often be approximated by the Pareto law

$$\text{Prob}\{Z_i \leqslant Z\} = 1 - \left(\frac{Z_0}{Z}\right)^{\alpha}, \quad Z \geqslant Z_0, \alpha > 0, \qquad (9.6.2)$$

where Z_0 is the smallest claim that is still considered as catastrophic (section 3.3.7).

(e) Multi-channel effect. A particular feature of catastrophes is that settlement of claims, arising from business written by a large number of direct writers, spirals in a complicated way through the international reinsurance network, sometimes reaching the same reinsurer by several different channels, giving rise to an unpredictable share of the total claim cost. Landin (1980) has found empirically that the number of channels through which claims impinge on a reinsurer is strongly correlated with the total size of the catastrophe. In statistics from the years 1974–1978 he found a correlation coefficient as high as 0.88. The data, however, contained many claims that we would not consider to be catastrophes.

Using the Landin data, the Finnish Working Party (Pentikäinen and Rantala, 1982) approximated the relationship of dependency between the number of channels K and the size of the catastrophe Z by a linear equation:

$$K = a \cdot Z + \varepsilon \qquad (9.6.3)$$

where ε is a noise term to describe stochastic deviations from a linear course. The estimated value of coefficient a was about 0.04 [$ million^{-1}] (transformed into current money values).

If the average claim per channel is equal to M, then the reinsurers' share of the total loss Z_{tot} is on average

$$Z_i = K \cdot M = a \cdot Z_{tot} \cdot M. \qquad (9.6.4)$$

For instance, a catastrophe of $5000 million would result in 200 channels for the reinsurer. If the average loss per channel M is equal to $0.25 million, then the share of the particular reinsurer is $50 million. For a conservative evaluation it may be appropriate to assume that M is equal to the (average) maximum commitment that can fall due in respect of each contract.

It is clear that the appropriate formula for participation in a potential catastrophe will depend critically on the number, contents and coverage of the reinsurance contracts of each particular insurer and should be dealt with in accordance with local conditions. The formula (9.6.4) is only a suggestion for the model structure.

(f) A straightforward evaluation. The above problem can also be approached in the following way (Pentikäinen et al., 1989). Assume that the (re)insurer has a market share δ of the total premium income relating to the sensitive area of exposure. If a catastrophe of size Z_{tot} occurs, its effects will penetrate through the whole international reinsurance network, probably resulting eventually in a share

$$Z_i = \delta \cdot Z_{tot} \qquad (9.6.5)$$

for this particular insurer. For example, if the market share δ were 1% of the California business, then the PML $5000 million would again result in a $50 million share for this particular insurer.

When such a catastrophe occurs, it will also reach the upper layers of reinsurance cover. However, because the premiums for high layers are small, owing to very low claim frequencies, premiums are not an adequate basis for evaluating the share δ. Instead, quantities describing the size of the exposure should be used, or the premiums, if applied, should be modified to correspond to the risk.

(g) To complete the model, the Z_i simulated either from (9.6.2) or (9.6.5) are incorporated in (9.6.1). The amounts should be added to the ordinary claim expenditure evaluated by the technique discussed in the previous sections. Examples can be found in Pentikäinen and Rantala (1982), I, pp. 4.3–4 and II, p. 5.2, and Beard *et al.* (1984), p. 248. The models outlined in the previous sections for a normal insurance business are not sufficient to describe this phenomenon. A supplementary module is necessary, but care is required to avoid counting the same risk twice.

(h) Risk of failure of reinsurance recoveries. If a catastrophe achieved the dimensions of a supercatastrophe, as anticipated by some experts (section 9.6(b)), then numerous (re)insurers would become insolvent, causing problems for many insurers throughout the world, who may be unable to realize expected recoveries in respect of their ceded business. The risk can be reduced if both inward and outward reinsurance are carefully sub-divided and diversified adequately and if the contracts are properly worded. It is up to the model-builder whether or not to attempt to have regard to this kind of extreme risk.

It is clear that, if there is a risk of a significant loss of reinsurance cover, insurers should keep their accounts on a gross basis in order to be aware of the size of these risks. If the balance sheet and other accounts are written on a net basis, the company management and, *a fortiori,* outsiders may obtain a false idea of the real exposure.

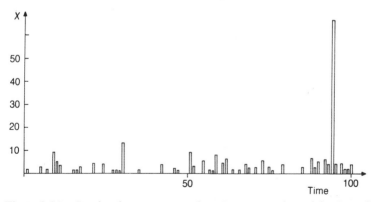

Figure 9.6.1 *Simulated supercatastrophes. Parameters derived by Rantala (1982) from data given by Landin (1980), total losses exceeding £15 million counted as catastrophes (wind storms), average number of events n = 0.5, claim sizes according to Pareto distribution with α = 1.76, multi-channel hypothesis.*

CHAPTER 10

Premiums

10.1 General framework

The next stage in the introduction of the basic features of practical risk theory is to find ways of modelling the premium income $B(t)$, which will be one of the main entries in the comprehensive model to be based on the fundamental transition equations (1.1.1) and (1.2.5).

The theory and practice of insurance premium-rating constitutes a wide branch of actuarial science and can only be considered briefly here, concentrating on the aspects which are most relevant for stochastic model-building.

In practice the rating of premiums depends not only on the past claim experience (X), but also on prospects concerning the return on investments (J), the financial position of the insurer, measured by the solvency margin U, and the current market situation. It is one of the key instruments in the realization and control of the insurer's business aims and strategies, as will be discussed in more detail in Chapter 14. The logic of premium-rating can be represented by the symbolic equation

$$\{X, J, U, \text{Market}, \text{Strategy}\} \Rightarrow \text{Premium } B\{P, E, \Lambda\}. \qquad (10.1.1)$$

Recapitulating section 6.1(a), the premium B_i is broken down into three components: the pure risk premium P_i, the safety loading Λ_i and the loading for expenses E_i:

$$B_i(t) = P_i(t) + \Lambda_i(t) + E_i(t). \qquad (10.1.2)$$

The subscript i refers to the risk unit or tariff class which is to be rated. The time variable t has been included in the notation as a reminder that all of the variables may be time-dependent.

Before turning to some practical problems in section 10.3, some theoretical principles are reviewed briefly in section 10.2, with comments on the components of (10.1.2).

10.2 Theoretical background

(a) **The pure risk premium** should in theory match the expected claim expenditure

$$P_i(t) = E[X_i(t)]$$ (10.2.1)

where $X_i(t)$ is the total expenditure in respect of claims incurred in year t in the risk unit or class i.

An estimate of the distribution of the variable $X_i(t)$ can be made on the basis of previous claim experience, allocated to underwriting years, and knowledge of the different features of the business. Clearly one should take into account inflation, by transforming the monetary amounts from different years to the same real level. Attention also needs to be paid to the run-off of claims and possibly to the discounting of delayed payments. Furthermore, the relevant distributions may have been changing and the distribution for business yet to be written cannot be known for certain in advance.

Tariff theory analyses how the premium should depend on different tariff factors, i.e. what kind of classification of policies would lead to a fair and practicable premium for each policyholder (e.g. Pitkänen, 1975; Lemaire, 1977; van Eeghen *et al.*, 1983; Goovaerts *et al.*, 1984, and Casualty Actuarial Society, 1990).

(b) **The effect of the return on investments** can be, and nowadays commonly is, taken into account by lowering the premiums so that the premiums and the return on investments together are still sufficient to cover claims and other expenditure. This can be seen in Figure 10.2.1 where the combined ratio, i.e. the sum of claims and expenses expressed as a ratio to premium income, is shown, together with the interest which could be earned on investments. The combined ratio tended to exceed 100%, hence the premiums have not fully covered the expenditures, but the interest income has substantially compensated for this.

There are several techniques for incorporating the return on investments into the model. It can be specified as a provision or a margin in the premium calculation formulae or it can simply be taken into account by allowing for a low, possibly negative, safety loading, without direct reference to the yield of interest.

REMARK   The common practice in the risk theory literature was to omit both the return on investment and the run-off tail, with the uncertainties involved,

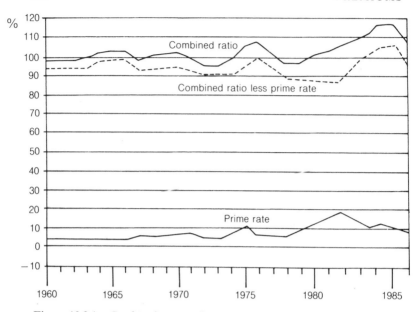

Figure 10.2.1   *Combined ratios of US property/casualty insurers and prime rates of interest. Source: BEST's Review, January 1989.*

as well as the question of discounting. The theory focused on those elements of the insurance process which were assumed to be most sensitive for stochastic behaviour and which proved to be most amenable for treatment by the techniques that were available before the era of computers. A pioneering step in incorporating these vital elements into practically orientated risk theory was taken by the British Solvency Working Party (Daykin *et al.*, 1987; Daykin and Hey, 1990).

(c) The expense loading *E* may conveniently be expressed in a ratio form

$$e(t) = \frac{E(t)}{B(t)}. \qquad (10.2.2)$$

This is one of the strategic decision variables in the comprehensive models and will be considered in more detail in Chapter 11.

As a first approximation $e(t)$ is often assumed to be constant.

(d) The safety loading, concepts and notations.   There are various principles and ideas concerning the determination of the safety loading for individual risks and tariff classes, as will be considered briefly in subsequent sections. Regardless of which method may be used, it is useful to employ the average level of the safety loading as a ratio to the pure risk premium (see also (6.1.3) and (6.1.4)), defined as follows:

$$\lambda(t) = \frac{\sum_i \Lambda_i(t)}{\sum_i P_i(t)}, \tag{10.2.3}$$

where the sums are taken over the whole portfolio or over that part of it which is under consideration. Then the total premium income can be written in the form

$$B(t) = E[X(t)] \cdot \frac{1 + \lambda(t)}{1 - e(t)}. \tag{10.2.4}$$

It is also useful to introduce the concept of **risk premium net of expenses** (see (6.1.3))

$$P_\lambda(t) = B(t) \cdot (1 - e(t)) = E[X(t)] \cdot (1 + \lambda(t)). \tag{10.2.5}$$

Notice that the risk structure, and even the size of the portfolio, affects the coefficients λ and e, and premiums may be significantly adjusted in practice to take into account the competitive market situation. The values of the coefficients may, as a result, be changing all the time, even though in many applications they can be kept constant as a first approximation.

Depending on the perspective, safety loadings are also known as **profit margins**.

(e) Safety loading, classical approaches. The examination of the proper size of the safety loading is one of the principal topics of traditional risk theory. A dozen or more approaches have been proposed by various authors, for instance Bühlmann (1970) and Goovaerts *et al.* (1984). The approach most often adopted is to suggest that the safety loading should be proportional either to the pure risk premium or to the standard deviation or variance of claims, or to some combination of them

$$\Lambda_i(t) = \lambda_1 P_i(t) + \lambda_2 \sigma_i(t) + \lambda_3 \sigma_i^2(t) \tag{10.2.6}$$

where one or two of the coefficients λ may be zero and they may be defined differently for different risk classes i. The justification for these formulae is that the loading should be larger the greater the uncertainty involved with the risk unit in question. The standard deviation or the variance can be taken as a measure for this purpose. On the other hand, a straightforward application of this formula may lead to bias, which can give rise to unsound antiselection. This might be modified by letting the λ depend on the volume of the risk class and its properties. Then the problem arises as to what is an equitable and fair allocation of the required safety loadings between the various groups of policyholders. This particular problem was discussed in section 6.6.6, applying the ideas of multiplayer games. It can be shown that there is no unique mathematically defined solution for the allocation problem. In fact, this leaves discretion to management so that aspects such as market pressures, bargaining between various interests, etc. can be taken into account within certain limits.

A fundamental question is what the level of the λ should be. Should each of the tariff classes be self-supporting or should it be permissible to equalize (random) profits and losses between the classes? In what follows we will assume the latter alternative. A sound theoretical principle, having regard, among other things, to equity between policyholders, is that the *a priori* **expectation** of the profit rendered by the safety loading in each class separately should be positive, notwithstanding that the actual outcome fluctuates randomly, sometimes giving profits and sometimes losses, and not-withstanding that it may be difficult, or even impossible, to follow such a theoretical principle strictly in a competitive market, as will be discussed in section 10.3.

(f) Safety loading, practical approaches. The safety loading Λ has several important functions: it provides protection against any uncertainty, it generates profit and it is a vital element (together with various risk reserves, reinsurance, etc.) in ensuring continuing financial strength. In fact, from the point of view of financial strength, the total amount of the safety loadings

$$\Lambda(t) = \sum_i \Lambda_i(t) = \lambda(t) \cdot E[X(t)] \tag{10.2.7}$$

is relevant, rather than how it is divided among the individual policies, notwithstanding that the latter aspect may have indirect effects which are discussed in section 10.3(c). This observation simplifies the analysis, because often only the overall coefficient λ needs to be considered. It is one of the key decision variables, as will be seen in the treatment of the comprehensive all-company models in Chapter 14.

The required safety loading depends, self-evidently, on the current position of the insurer (10.1.1). If the insurer is financially strong, then a lower λ may be sufficient than would be the case if it were weak.

(g) Dynamic control of premiums. The rating procedure is unavoidably subject to uncertainty. This is characteristic for all statistical inference, but there is also uncertainty due to the fact that the environment and relevant economic determinants such as inflation, the activity of national and international economies, etc. are changing all the time. A particular feature, called the **time lag effect**, is of importance for the insurance business, both at the level of individual insurers and for the insurance market as a whole. It arises from the fact that any bias in rates cannot be observed until some time afterwards and further time is needed for a potential correction to be made effective. During these delays the biased rates give rise to profits or losses which can be seen in the form of periodical (or cyclical) variation of the underwriting result. This crucially important phenomenon will be studied in Chapter 12 in the context of the underwriting process.

In practice, amendments to premium rates are under management control. Among other things regard is taken to the current market situation and to the implementation of possible changes to business strategy (more in section 14.5(b)). For long-term models rules are required to mimic this management function and to analyse the consequences of alternative strategies, in particular, how the premiums might behave in different situations. A simple rule was illustrated in Figure 5.5.3, assuming that, if the solvency ratio grows high, premiums will be lowered (or dividends and bonuses increased) and in the case of an adverse development the rates will be increased.

It is worth mentioning that it is the dynamics of premium control which keep the simulated outcomes stable. Without such control they tend to diverge, as was demonstrated in Figure 5.5.2. In fact, the interaction between the ever-developing business situation and premium-rating is one of the crucial aspects of making longer-term insurance models realistic.

Experience-rating and, in particular, exponential smoothing, reviewed in section 6.5.3, is much used as a control procedure. It is based on the idea of allowing the premium rate to be continually updated according to the accruing claim experience. The standard formula (6.5.7) might be modified by replacing the term X relating to the one-year claim amount with, for example, a weighted average derived from several previous years:

$$P(t) = Z \cdot P(t-1) + (1-Z) \cdot \sum_{\tau=1}^{T} \alpha_\tau \cdot X(t-\tau) \qquad (10.2.8)$$

where $\sum \alpha_\tau = 1$. This makes the premium development more stable.

The interaction between the financial position and premiums is treated in the context of the insurance process in Chapter 12.

The *bonus-malus* rules, according to which individual premiums are adjusted, particularly in motor-car insurance, are also worth mentioning in this context.

10.3   Premiums in practice

(a) Theories versus real world.   Even though the principles referred to above may be behind actual rating practices, they are unlikely to be followed strictly in practice. In fact, rating is only one aspect, though an important one, of the total conduct of business management. There are numerous reasons why the solutions suggested by theories may be ignored, in particular under competitive and bargaining pressures. With this in mind we have already commented, in section 10.2(f), that what is necessary is to maintain the total premium income at an adequate level in the long run. However, various uncertainties and market factors may have consequences which it is important to know about and to take into account. This section is devoted to a general discussion about them.

It is also worth noticing that, although the theoretical rates may not be followed strictly in practice, thorough analysis is still worthwhile. It is prudent to know the actuarially correct rates and to monitor the deviations of the actual rates from them, since biases may have harmful, and possibly fatal, long-term consequences. If management is not aware of such deviations, it cannot adequately carry out its task.

(b) Market.   When the considerations are extended to cope with the market environment, as indicated in (10.1.1), problems arise in accounting for the ensuing effects. In particular, a key feature is whether the market prices $B_M(t)$ deviate from the insurer's own premiums $B(t)$. If this is the case, particular techniques are required to evaluate the consequences.

The behaviour of customers and markets as a function of prices is examined extensively in the literature of general business economics, for example, in Kotler (1975). Unfortunately very few studies of insurance markets are available, nor are there many references to experience of how well the formulae used in general supply and demand analyses apply in this environment (Pentikäinen, 1979). We will limit our remarks, therefore, to a couple of examples. It will be necessary for the user of the model to find empirical bases for the behaviour of the market and clients in his own environment.

A suitable candidate may be the so-called **exponential price elasticity** function. It gives the increment ($\pm$) of the portfolio, measured here as an increment ΔB of the premium income B, if the level of company prices deviates from the level of market prices for a corresponding risk by $-100\pi\%$ (if π is positive, then the company's price is lower, and vice versa if it is negative)

$$\Delta B/B = (1 - \pi)^{-\alpha} - 1 \approx \alpha\pi, \qquad (10.3.1)$$

where α is an empirical **elasticity** parameter, also called a **gearing factor**.

A simpler approach, bearing in mind the paucity of the data, is to postulate directly the latter form of (10.3.1), i.e. assuming the change ($\pm$) in volume to be proportional to the difference in premiums (Daykin and Hey, 1990).

Another formula is based on input–output reasoning. It is assumed that the insurer invests an extra amount E_m for a sales campaign. The resulting return, in the form of an increment ΔB in premium income, is assumed to be proportional to the input:

$$\Delta B = \alpha_m \cdot E_m. \qquad (10.3.2)$$

The return coefficient α_m must, of course, be determined empirically. It can perhaps be derived by collecting statistics on the average sales cost required for one contact to be made with a potential client, and on the proportion of such contacts which are successful (Exercise 10.3.1).

As a particular example of an application of (10.3.1) let us assume that the company follows the market price level in year t and then lowers premiums relative to the market in year $t + 1$ (π positive). New business can be expected to increase but, on the other hand, all the new and old (if the reduction applies to them as well) policyholder premiums are reduced by the factor $1 - \pi$. Hence we obtain, as a straightforward application of (10.3.1):

$$B(t + 1) = B(t) \cdot r_g(t) \cdot (1 - \pi)^{1 - \alpha} \qquad (10.3.3)$$

where r_g is the coefficient of normal growth (9.2.1). This kind of formula may give a basis for evaluating the profitability of sales strategies. A reduction in premiums can be seen as a reduction in the safety loading λ and will reduce the profit in year $t + 1$ per unit of premium income. On the other hand the new business will increase the volume of premium income.

When the market effects are to be modelled, as will be exemplified in section 14.3, separate modules may need to be created for the 'market' and for the insurer. Questions will need to be considered such as whether the decisions of the insurer may induce reactions among competing insurers.

(c) Risk of antiselection. It was concluded in section 10.1(g) that, from the point of view of the profitability and solvency of an insurance company, the total amount Λ of all of the safety loadings is relevant, rather than how this loading is divided between the individual policyholders. However, this latter aspect may have indirect long-term consequences which require attention. If the loading is not equitable, some policy groups may be overloaded in favour of others. For instance, if the standard deviation or the variance rule included in (10.2.6) is applied for the whole portfolio with uniform coefficients, large and dangerous risk units are likely to gain at the expense of small and homogeneous policies. This feature, which calls into question the practicability of many of the theoretical premium rules, was discussed by Borch (1962) and illustrated by an example in Beard *et al.* (1984), section 5.3. Depending on how price-conscious the market is, policyholders with policies which are more heavily loaded than elsewhere in the market can be expected to have a tendency to move to competing insurance companies, whilst those policyholders who can benefit from a lighter loading may move from elsewhere in the market. The result is an adverse selection of the risk properties of the portfolio, which worsens profitability. In section 6.6.6 this dilemma was illustrated by an example applying the ideas of multiplayer games.

Exercise 10.3.1 An insurer has experience which indicates that marketing policies to potential clients costs on average £10 per client and that every one in five of the contacts results in a new policy.

The average premium is £100, the cost of issuing a new policy exceeds the sum of expense and safety loadings by £30 in the first year and the policy is expected to give profit from the second year onward of £10 if it persists. Derive the coefficient α_m of (10.3.2) and evaluate after how many years the accumulated expected profit will cover the total initial expenses.

CHAPTER 11

Expenses, taxes and dividends

11.1 Expenses

Expenses, taxes and dividends will be dealt with briefly in this chapter. Reasonably accurate treatment may often be necessary in practical models, as the outcome can be quite sensitive to these issues. However, thorough consideration is beyond the scope of this book, so the treatment will be limited to such features as are relevant for our model-building. The construction of a detailed sub-model is left to the user, who can consult the standard study books of business administration to find applications which are suitable for his local environment.

The word **expenses** is used here as an umbrella term to cover all the different kinds of administrative and operational costs of an insurer.

In the foregoing chapters the expenses E of the insurance business were assumed bear a constant ratio e to the premium income B, that is

$$E(t) = e \cdot B(t). \qquad (11.1.1)$$

In fact, this might be a satisfactory first approximation, but for more in-depth analysis the changes in expense ratios may be significant. A weakness of this formula is that in long-term projections the premium income $B(t)$ is often subject to various changes which need not affect the expenses, at least not proportionally as (11.1.1) assumes. Tools for modelling expenses should, therefore, be available to risk analysts.

(a) Empirical data. Expense ratios depend first and foremost on the line of business. If a portfolio consists mainly of large insured units such as industrial plants, ships, etc., the expense ratio need not be more than 10% whereas the costs of insuring small units, homeowner property, motor cars, etc., are likely to be relatively high, and may

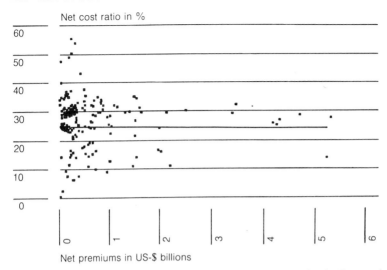

Net cost ratio in %

Net premiums in US-$ billions

Figure 11.1.1 *Net expenses as a ratio to premiums in a sample of US general insurance companies. Source: Sigma 4/91.*

reasonably require a loading of some 30–40%. An analysis of individual companies would reveal differences resulting from management actions, including the balance between different classes of business, the size of the company and the way in which it sells business. Substantial changes may take place as a result of economy drives, sales campaigns, etc. These are all of interest in modelling expenses at the company level. The significance of company variations is confirmed by Figure 11.1.1.

Further statistics on company expense ratios have been published, for example, by Malde *et al.* (1988) and Pentikäinen *et al.* (1989), section 3.5(c).

(b) Cost analyses. To understand the structure of expenses arising from administration, selling costs, claim settlement and all the other operations characteristic of an insurance company, it is useful to analyse how the expenses are made up as an aggregation of the various separate elements.

Let us assume, as an example, that the expenses are broken down into primary elements E_{hijk}, within the framework of a four-dimensional matrix, according to

- h, the line (class or group, for example, the sub-divisions used in rate-making);

- i, operation or function, e.g. issuing a new policy, collecting premiums, claim settlement, etc.;
- j, the part of the organization where the operation i concerning class h is performed; and
- k, the type of expenditure, such as salaries, material costs, rents, information technology, etc.

This kind of matrix permits the estimation of the cost of various combinations, for example, the total claim settlement costs for a class h, etc. This can be obtained by summing over the relevant indices:

$$E_{hi}(t) = \sum_{j,k} E_{hijk}(t) \qquad (11.1.2)$$

where h and i are assigned the relevant values for the line in question and for claim settlement respectively. Making use of these summations, various actual or planned operations can be evaluated, for example, the cost of a planned sales campaign.

Some of the costs cannot be allocated directly to any particular line, for example, general management. These are called **overhead expenses**, denoted by $E_0(t)$. The remaining expenses are attributable both to time t and to a specified line h. Hence the total expense can be expressed as

$$E(t) = E_0(t) + \sum_{h,i,j,k} E_{hijk}(t). \qquad (11.1.3)$$

A particular problem is how to divide the overhead expenses between the lines of business. No unique answer exists, and there are various ways of tackling the problem, which has much the same structure as the problem of allocating the safety loading to different sorts of policies. The problem is analogous to that considered in section 6.6.6. The papers of Borch (1962) and Lemaire (1984) are relevant. A possible solution, for example, would be to allocate the overhead expenses in proportion to the premium income, but it may be necessary to take into account the fact that certain lines of business require more management time. The total expenses to be allocated to a line h are given by

$$E_h(t) = E_{0h}(t) + \sum_{i,j,k} E_{hijk}(t). \qquad (11.1.4)$$

(c) Cost formulae. Once the costs by line have been obtained, the expenses entry in the comprehensive equations (1.1.1), (1.2.5) and (11.1.1) can be derived in terms of the ratios $e_h(t) = E_h(t)/B_h(t)$ for each line, and the total cost for (11.1.1) therefrom by a summation similar to (10.2.3).

A drawback in using formulae which relate all the expenses to premium income is that the absolute amount of projected expenses is affected by any change in the projected premium income $B(t)$, even when this is amended to correspond to the accruing claim experience, or is adjusted because of the competitive and cyclical market. Inappropriate changes may sometimes be brought about in the expenses if a formula of type (11.1.1) is applied indiscriminately. A more sophisticated approach would be to relate the expenses to the consolidated premium income to be introduced in section 12.3(c). It might be better still to establish an independent development formula for the expenses, for example as follows

$$E(t) = E(0) \cdot \prod_{\tau=1}^{t} r_g(\tau) \cdot (1 + i_E(\tau)) + E_m(t) \qquad (11.1.5)$$

where $r_g(t)$ is the growth factor of the portfolio (9.2.1), $i_E(t)$ the rate of cost inflation and $E_m(t)$ an auxiliary extraordinary cost item, for example in respect of a particular sales campaign, a major investment in information technology, or some other one-off expenditure. The cost inflation index should be an appropriate one; ideally a specific expense inflation index should be constructed (section 7.2(a)).

Although a major part of the expenses can be expected to increase broadly in proportion to the size of the portfolio, other expenses might be more or less independent of the volume of business, increasing only in line with an expense inflation index. These latter might be described as **fixed expenses**, in contrast to **variable expenses** which are proportional to business volume. In their management model Daykin and Hey (1990) also postulated **step increase expenses** which go up in discrete jumps, as certain business volume thresholds are passed. Any combination of these various types of expenses can be realized without special difficulty in a simulation model.

(d) Strategic impact variable. The term $E_m(t)$ in (11.1.5) allows some flexibility. For example, the company might plan a sales campaign and set aside money for it over and above the normal level of sales expenses accounted for by the first term of the right-hand side of (11.1.5). The term $E_m(t)$ can be used to account for such extra costs. It can be called a **strategic impact variable**, or **decision variable** in the terminology of dynamic programming. The variable needs as a counterpart an entry in the model to generate the anticipated response. In the sales campaign example this would be the expected increase in the volume of business, preferably measured as an increment ΔB of the premium income. Text-books on business economy propose several formulae for this sales response (Kotler, 1975). The simplest one is linear

$$\Delta B(t + \tau) = \beta \cdot E_m(t) \quad (\tau = 0, 1, 2, \ldots). \qquad (11.1.6)$$

The coefficient β is determined empirically. If the sales campaign also involves reductions in premium rates, then this formula can be used in parallel with formulae (10.3.1) and (10.3.2).

It is worth noticing that the possibility of inserting varying sales efforts into the model gives it a further dynamic dimension. In a long-term evaluation, additional sales efforts can be programmed to be triggered by a situation where the solvency ratio is becoming large and to be discontinued if the solvency ratio turns downward or falls below a specified level. Further dynamics are introduced if the variable E_m is also allowed to take negative values. It can represent the case where the insurer has reached a precarious position and has to resort to a reduction in normal expenses as a remedial action.

The same variable $E_m(t)$ can also be used for modelling other types of cost movement. For example, if money is invested in new office buildings, information technology, etc., the term may initially be positive, but turn negative after a period, when the investment can be expected to be showing a return in terms of reduced costs elsewhere.

11.2 Taxes

(a) An approximate formula. The legislation and current practice concerning taxes differ so much from country to country that it is not possible to propose any universal formula for them. In some countries the amount actually paid in tax is in practice relatively stable, owing in part to the possibility of equalizing annual results between successive years. Then a satisfactory approximate formula could be quite simply a linear relationship to premium income.

$$E_{\text{tax}}(t) = e_{\text{tax}} \cdot B(t). \qquad (11.2.1)$$

This rule can be made less coarse by incorporating a trigger module which stops or reduces the tax payment in cases where the financial position of the insurer is precarious.

It may often be convenient to incorporate this tax term into the general formula for expenses.

(b) Special formulae. For more ambitious practical models a detailed tax formula may be necessary, based on the evaluation of the taxable income, capital gains, etc. Daykin and Hey (1990) present an example of how the British taxation system can be modelled. An example from the USA was given by Gragnola (1985).

11.3 Dividends

(a) Profit-related formula. In practice insurance companies try to keep dividend payouts to shareholders fairly stable, giving compensation for inflation and providing a real return on the investment. A straightforward approach in modelling this is to relate the dividends to the profit $Y(t)$:

$$D(t) = d_y(t) \cdot Y(t). \qquad (11.3.1)$$

The profit in this context should preferably be the book profit stated in the company's income statement, reflecting such smoothing and equalization as will be discussed in section 12.3(c). The coefficient $d_y(t)$ is a parameter to be defined according to the application.

(b) Capital-related formula. An alternative approach is to relate the dividends to the share capital $U_s(t)$:

$$D(t) = d_s(t) \cdot U_s(t). \qquad (11.3.2)$$

In practice the coefficient $d_s(t)$ might be expected to be fairly constant from year to year. The shareholders are compensated for inflation and get potential real growth in value through the increase in value of the capital $U_s(t)$, which should be properly programmed into the model.

As in the comment on (11.2.1), this formula can be made less coarse by incorporating a trigger term which stops or reduces the dividend payment in cases where the resources are inadequate.

(c) A consolidated formula. Daykin and Hey (1990) proposed the following formula which explicitly implements the requirements of stability and compensation for inflation as well as providing for real growth:

$$D(t) = D(t_0) \cdot (1 + g)^{t - t_0} \cdot I(t) \tag{11.3.3}$$

where t_0 is the year before the date of assessment, g is the real rate of growth of dividends, and $I(t)$ is the retail (consumer) price index, taking $I(t_0) = 1$.

REMARK The selection of the dividend formula depends on the philosophy which is adopted for the ultimate purpose of the management. De Finetti (1957) and later Borch in numerous papers and many other authors have posed the problem of what the business strategy should be that gives a maximal profit to the shareholders, for example, in terms of the sum of discounted future dividends. In this kind of problem the dividend term has a central role, suggesting the profit-related formula (11.3.1). Business strategies, in theory and practice, will be discussed later in section 14.5, concluding that a balanced appreciation of shareholders', policyholders' and public interest leads to a formula of type (11.3.3).

CHAPTER 12

The insurance process

12.1 Basic equation

(a) Equation. We are now in a position to bring together the models which were developed for claims X, premiums B, expenses E and the return on investment J in previous chapters. For this purpose we will make use of an equation which combines these quantities with the solvency margin (risk reserve) U as follows

$$U(t) = U(t-1) + B(t) + J(t) - X(t) - E(t). \qquad (12.1.1)$$

This is, in fact, a reduced variant of the basic accounting equation (1.2.5).

Following the idea of introducing the features of the model step by step, some of the elements of the final comprehensive model, referred to in Chapter 1, will still be left to later chapters. Some empirical observations about the behaviour of relevant quantities are presented in section 12.2 and these are analysed in subsequent sections, in preparation for the modelling aspects in section 12.4 and for applications to be introduced in Chapter 13.

(b) Concepts and notations. **Underwriting** is the process whereby an insurer selects which risks to insure and the terms (policy conditions and premium rate, coinsurance, deductibles, limits, etc.) on which the business will be accepted (section 1.3(c)).

The **underwriting result** in general insurance is traditionally defined as the excess of earned premiums in the year over incurred claims and expenses.

$$Y_u(t) = B(t) - X(t) - E(t). \qquad (12.1.2)$$

The underwriting result (12.1.2) does not take into account the return on investments, although this is an integral part of the insurance operation, since the success of underwriting has historically always been measured without investment income, the latter being carried directly to the profit and loss account as an additional source of profit.

We define the **insurance result** to be the underwriting result, together with the associated investment return on assets backing the technical reserves. This is sometimes also referred to as the **trading result** or the **overall result**.

$$Y(t) = B(t) + J(t) - X(t) - E(t). \qquad (12.1.3)$$

Note that the $J(t)$ in (12.1.3) need not be strictly the same as in (12.1.1), because in this context it may be restricted to the yield earned on the technical reserves whereas in (12.1.1) the yield on the company's solvency margin U may also be included.

Sometimes the **operating result** is separately defined as the insurance result plus other investment income (for example, on shareholders' funds) plus other sources of profit (such as earnings from associated companies).

These results are often expressed in ratio form, usually as a percentage of earned premiums. We define a hierarchy of three such ratios, the **claim ratio**, the **combined ratio**, incorporating both claims and expenses, and the **trading ratio**, in which investment income on assets backing the technical reserves is offset.

$$\text{Claim ratio} = x(t) = \frac{X(t)}{B(t)}, \qquad (12.1.4)$$

$$\text{Combined ratio} = x_c(t) = \frac{X(t) + E(t)}{B(t)} \qquad (12.1.5)$$

$$\text{Trading ratio} = y(t) = \frac{X(t) + E(t) - J(t)}{B(t)}. \qquad (12.1.6)$$

REMARK   The combined ratio is, according to (12.1.5), the sum of the claim ratio and the expense ratio (e in (10.2.2) and (11.1.1)). Some authors use the written premiums B' in the denominator of the expense ratio instead of the earned premiums B, having regard to the fact that a major part of expenses, in particular commissions, falls due at the beginning of the premium term. This definition blurs the additivity of the two ratios since they have different denominators.

12.2 Empirical observations

(a) Empirical results. It is useful for understanding the insurance process and as a guide for model-building to consider some empirical observations of actual behaviour of general insurance in various countries. Some examples from the UK and USA are exhibited in Figure 12.2.1.

Corresponding data from Finland are given in Figure 12.2.2 where the graphs for the whole market are supplemented by information concerning individual insurers.

(b) Observations. A striking feature is the occurrence of more or less irregular cycles several years in length. This confirms the remarks made in section 9.2, in that context in relation to claim expenditure, but clearly capable of extension to the whole insurance process: each process is a superimposition of trends, cycles and short-term fluctuations.

It is clear that cycles play a significant role, but are they always present? In the statistics of various countries there are cases where the cycles can be seen clearly, as in the diagrams displayed overleaf; there are also cases where no cycles can be seen. This may be because they do not exist for the particular line of business, for example, in life insurance, or that they are hidden by the smoothing of published results, for example, through flexible use of margins in the technical reserves. Hewitt (1986) presented evidence to indicate that the strongly adverse cycle in the US market in the 1980s (Figure 12.2.1(a)) was offset by a considerable weakening of the reserves for outstanding claims (notwithstanding the fact that one might have expected strengthening to meet the increase in expected claims). The number of insolvencies is strongly correlated with the cycles; in many insolvent insurance companies it can be seen with hindsight that the reserves were significantly underestimated in the years before the eventual collapse (Report of the US Subcommittee on Investigations: Failed Promises, Dingell, 1990).

It seems clear that business cycles are so common in general insurance, and their impact so profound, that any risk theory model which claims to describe real-life situations must permit the user to evaluate the impact of any cycles which may be present. Section 12.3 will examine the causes and the mechanisms generating cycles from a mathematical point of view.

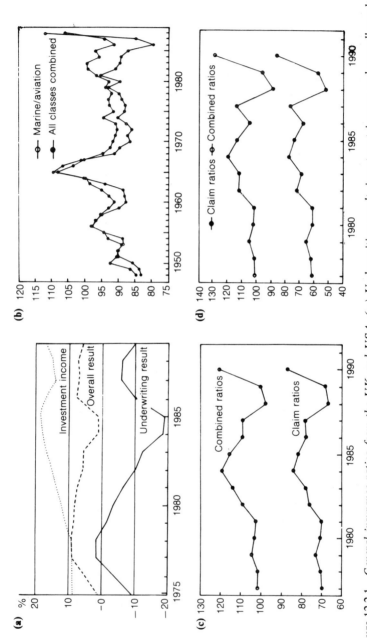

Figure 12.2.1 General insurance ratios from the UK and USA. (a) Underwriting result, investment income and overall result, USA, 1975–89, (b) Lloyd's trading ratios, 1948–88, (c) Total UK business, 1977–90 and (d) UK property damage business, 1977–90. Source for the US diagram: Sigma 6/91; UK: Lloyd's and Association of British Insurers.

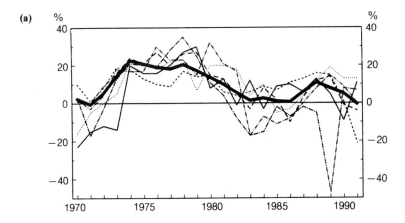

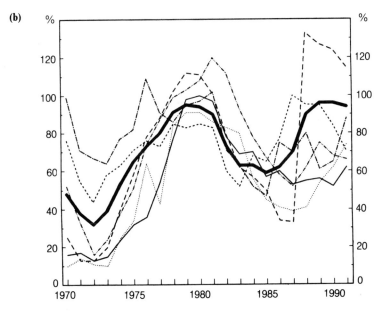

Figure 12.2.2   *The trading ratio (a) and the solvency ratio (b) (u = U/B)*
of the six largest Finnish non-life insurers, and the joint ratios for all insurers.
Source: Research Department of the Federation of Finnish Insurance
Companies.

REMARK 1   It has sometimes been argued that the term 'cycle' may be misleading. It could be misunderstood to indicate a regular fluctuation such as that implied by the sine curve (9.2.4) or by a stationary time series (section 9.2(c, d) and Appendix G, section G.3(a)), whilst the observed variations are often in fact rather irregular. However, in the absence of any more descriptive term we will refer to **trading cycle** and **underwriting cycle**, or more briefly **cycle**, following the current insurance terminology.

Figure 12.2.2 warrants some further observations. The curves for individual companies show a similar pattern, with only minor deviations, reflecting short-term random fluctuations. The cycle is effectively the same for each of the six insurers, so that we can speak about a **market cycle**. It is plotted in bold. This phenomenon seems to be common in industrialized countries.

The common scale for the two graphs of Figure 12.2.2 makes it possible to see the annual results reflected in the cumulative financial position, both qualitatively and quantitatively. The amplitude of the solvency ratio fluctuation is about three times that of the trading ratio curves. This is because the positive (and similarly the negative) results tend to occur consecutively in such a way as to reinforce their effect on the solvency ratio. The length and the phase of the waves are the same in both figures.

Note that the solvency ratio turned downwards from 1980 onwards, even though the insurance result was still positive. This was caused by the premium income $B(t)$, the denominator of the solvency ratio $U(t)/B(t)$, growing faster than the solvency margin $U(t)$. This demonstrates that the relative financial positive of an insurer may worsen even during periods when the insurance result is positive. In order to prevent deterioration of the solvency position, the solvency margin must increase (approximately) at least in proportion to the growth of the business volume, including both the effect of inflation and the real growth of the portfolio.

REMARK 2   Because the solvency margin in Figure 12.2.2 includes also the company's own equity capital, the issue of new equity affects the flow of the curves. This explains the jumps in two of the curves in the years 1987 and 1988.

(c) Summarizing, the following features may be observed

(1) The **amplitude of the cycles of combined ratios** and trading ratios is commonly of the order of 10–20%, in some cases even more (Figure 12.2.1).

(2) The **amplitude of the solvency ratio cycle** is, as a rule, greater than that relating to annual results, since several adverse or profitable years may occur in succession. Put simply, if the length of an adverse period is T and the loss on average is y, then the total accumulated loss is $T \cdot y$. In Figure 12.2.2 the amplitude of the solvency ratio cycle is about three times the amplitude of the trading ratio cycle.

(3) The **length of a cycle** varies from 3 or 4 years up to 8 or even 12 years.

(4) The appearance, amplitude and length of cycles vary a lot according to the class of business. In some cases they appear not to be present at all.

(5) Successive cycles are not necessarily similar in shape, size or length. This suggests that there are many different causes of cycles. One cycle may be caused by one factor and the next cycle by another, or perhaps by a combination of several factors.

(6) There is a significant degree of synchronization between the cycles in different countries.

12.3 Business cycles, analysis of causes and mechanisms

It is clear that underwriting cycles are one of the most important factors affecting the business results and, ultimately, the financial strength of insurers. Because of their significance it is useful to analyse the causes of cycles and to explore some of their features. The empirical observations presented in section 12.2 indicate that there are many potential background factors, which vary from period to period. Several factors may in fact operate simultaneously, with complicated interrelationships. Some of the most important factors will be discussed in this section. Before that it is useful to demonstrate by means of our model equation how some simple impulses affect the underwriting process. Real-life processes can be seen as superimpositions of numerous successive random impacts of this kind.

(a) Equations in ratio form. The formulation (12.1.1) is not always convenient when the analysis extends over several years. The relevant variables are continually subject to changes resulting from inflation and the real growth of the portfolio. The values assumed in consecutive years are not, therefore, directly comparable unless they are transformed to correspond to the same monetary values and portfolio size. For this purpose it is often useful to manipulate the formula into a form which deals with ratios rather than absolute variables. This can be done by dividing the variables by a volume measure $M(t)$, the choice of which should depend on the intended application. Possible candidates are the sum of insured amounts (property insurance), the number of policies multiplied by a suitable money-dimensioned factor, the total payroll (in workers' compensation), or, in the case of multiline general insurance, the premium income $B(t)$, the assets $A(t)$ or an estimate of the expected value of incurred claims. The premium income $B(t)$, or its net of expenses variant $P_\lambda(t)$, are the most common volume measures. $B(t)$ was used in section 12.1(b), where combined and other ratios were considered. However, premium income needs some modification if it is to be a suitable volume measure for some applications. This will be considered in section 12.3(c) below.

Instead of the key variables $U(t)$, $A(t)$, $X(t)$, etc., it is convenient to use the corresponding ratios obtained by dividing each one by the volume measure $M(t)$. If premium income is used as the measure, then the solvency ratio, asset ratio, loss ratio etc. result. These ratios will be denoted by the lower-case symbols $u(t)$, $a(t)$, $x(t)$, etc. corresponding to the upper-case symbols for the absolute quantities. The basic equation (12.1.1) is transformed by dividing it by $M(t)$

$$u(t) = r_M(t)^{-1} \cdot u(t-1) + b(t) + j(t) - x(t) - e(t) \quad (12.3.1)$$

where

$$r_M(t) = M(t)/M(t-1) = r_g(t) \cdot (1 + i(t)) \quad (12.3.2)$$

is the **incremental factor** of the volume measure. It can be decomposed into a factor r_g relating to real growth and a factor $(1 + i(t))$ for inflationary growth. The former includes, for example, increases in the number of risk units and changes in risk propensity. In our model-building r_g is specified as the growth of the expected number of claims n (9.2.1). The latter factor in (12.3.2) should be taken as the appropriate inflation for the particular application, e.g. claim inflation (section 7.2).

(b) Composite growth factor. For some applications it is useful to break down the return on investments into components relating to assets representing the solvency margin $U(t-1)$ on the one hand and the technical reserves and other liabilities $L(t-1)$ (1.2.3) on the other.

$$J(t) = j_u(t) \cdot U(t-1) + j_L(t) \cdot L(t-1) \quad (12.3.3)$$

where $j_L(t)$ is the rate of return allocated to the technical reserves and $j_u(t)$ is the rate of return on the solvency margin arising from the balance of the investment return. Then (12.3.1) can be transformed into an alternative form

$$u(t) = r(t) \cdot u(t-1) + b(t) + j_L(t) \cdot r_M(t)^{-1} \cdot l(t-1) - x(t) - e(t), \quad (12.3.4)$$

where

$$r(t) = (1 + j_u(t))/r_M(t) \quad (12.3.5)$$

is a composite factor introducing the effect of the rate of return on investments, inflation and the real growth of the portfolio. The numerical value of $r(t)$ is normally close to unity.

REMARK   It is a matter of judgement whether j_u and j_L are assumed to be different or are taken to be equal, corresponding to the average yield on all the assets of the insurer.

For some applications it is appropriate as a first approximation to amalgamate the third and fourth terms of the right-hand side of (12.3.4) by the use of discounted reserves ((9.5.2) and (9.5.16)).

The gross premium $b(t)$ can be replaced by $p(t) + e(t)$, where $p(t)$ is the premium income net of expenses (the same as p_λ in (10.2.5) with the subscript λ being dropped for brevity). Hence, we arrive at a simplified version of the algorithm

$$u(t) = r(t) \cdot u(t-1) + p(t) - x(t). \quad (12.3.6)$$

This version will be useful for analysing some of the fundamental characteristic properties of the insurance process.

(c) Consolidated premium income.   The premium income $B(t)$ was proposed above as a suitable volume measure when the ratio version of the equation (12.3.1) is required, particularly for multiline general insurance. In fact, conventional indicators such as the loss ratio, combined ratio and solvency ratio have $B(t)$ as the denominator. This choice of measure, however, is not ideal for all purposes. For instance, the variable $b(t)$ is always equal to unity in this formulation and, as a result, the effects of changes in premium rates are obscured. Furthermore, if the insurer increases premium rates in order to enhance profitability and solvency, the solvency ratio $u = U/B$ is decreased, giving a spurious impression of weakening solvency, though the situation is in fact quite the reverse. To overcome these drawbacks it is advisable to define the volume measure as some derivative of the premium income $B(t)$, e.g. by fixing the initial level at some average (estimated or assumed) level $B(t_0)$ and letting it move in line with volume, as defined by the incremental factor r_M (12.3.2).

$$M(t) = \bar{B}(t) = B(t_0) \cdot \prod_{\tau=1}^{t-t_0} r_M(t_0 + \tau). \qquad (12.3.7)$$

The term $b(t) = B(t)/\bar{B}(t)$ in (12.3.1) and (12.3.4) then indicates the relative deviation of $B(t)$ from its normalized or consolidated flow (12.3.7) and the effect of changes in tariff rates can be seen in the flow of $b(t)$, or similarly in $p(t)$ in (12.3.6).

(d) Mechanisms generating cycles. In order to provide some insights into the significant features which may be behind the cyclical variation of business outcomes, and the mechanisms which might induce them, it is useful to present some simple examples, using the basic equation in its reduced form (12.3.6). The presentation follows Pentikäinen et al. (1989).

(e) A test model. Let us assume that the safety-loaded risk premium p is controlled by adjusting its level according to the state of the solvency ratio u, with a two-year time delay. In fact, it can be assumed that $u(t-1)$ is known in year t, but the necessary amendments to premium rates cannot be effected until year $t+1$. Hence in practice the total delay is (at least) two years. If $u(t-2)$ is below a certain target level u_0, then the premium rates are to be increased from the basic level p_0 in proportion to the shortfall. Similarly, if $u(t-2)$ exceeds the target, the rates are lowered accordingly:

$$p(t) = p_0 + a \cdot [u_0 - u(t-2)], \quad (0 \leqslant a \leqslant 1) \qquad (12.3.8)$$

(adopting here the bold type-face notation to indicate the stochasticity of the testing algorithms; see the convention of Nomenclature). The coefficient a controls the smoothing effect of the system. The argument $t-2$ of u introduces the time lag into the model and $p_0 = E(x) - (r-1) \cdot u_0$.

The solvency ratio u is generated from the basic equation (12.3.6). To simplify further, the coefficient $r(t)$ (12.3.5) is assumed to be a constant r:

$$u(t) = r \cdot u(t-1) + p(t) - x(t). \qquad (12.3.9)$$

Eliminating $p(t)$ from (12.3.8) and (12.3.9), after some simple manipulations we obtain:

$$u(t) - u_0 = r \cdot [u(t-1) - u_0] - a \cdot [u(t-2) - u_0]$$
$$+ p_0 + (r-1) \cdot u_0 - x(t). \qquad (12.3.10)$$

Thus $u(t)$ is an autoregressive process of order two (Appendix G, equation (G.3.1)), where the last three terms together can be interpreted as noise which approximately fulfils the standard assumption for the theory, having regard to the fact that its expected mean is zero, owing to the definition of p_0.

The deviation $(p(t) - p_0)$ of the relative premium from its target level obeys, according to (12.3.8), the same law as $u(t)$ with a two-year delay and damped by the factor a.

REMARK Taking differences on both sides of (12.3.8), and having regard to (12.3.9), we obtain

$$\Delta p(t) = a \cdot [x(t-2) - p(t-2)] + a \cdot (1-r) \cdot u(t-3). \qquad (12.3.11)$$

If r is close to unity and a is not very large, the last term is insignificant and the premium change is approximately proportional to the difference between claims and premiums in the previous year. Thus rule (12.3.8) might also be interpreted as having a forecasting (experience-rating) feature because, according to the above equation, a prediction $p(t)$ for the future required relative premium rate is obtained by adjusting the previous premiums on the basis of the latest information regarding observed claims. Note that the hypothetical target rate p_0 does not appear in this formula.

We will now demonstrate the mechanisms implied by the above simple model by means of some examples shown in Figure 12.3.1. They may help to understand the complex phenomena of the real world, where the effects of innumerable minor impulses and changes of these simple types are superimposed.

The claims $x(t)$ are the primary driving process, and the premiums $p(t)$ and the solvency ratio $u(t)$ are determined by equations (12.3.8) and (12.3.10) applying the time series technique of Appendix G. $x(t)$ is initially assumed to be a steady process, i.e. constant and equal to x_0 for all values of t until an impulse or change emerges.

(1) A single impulse is given in year t_0 in the form of extra large claims. The response is a damped sine wave in both $p(t)$ and $u(t)$ if ((G.3.2) of Appendix G)

$$1 > a > \tfrac{1}{4} r^2. \qquad (12.3.12)$$

The wavelength T of the impulse response is given by ((G.3.5)

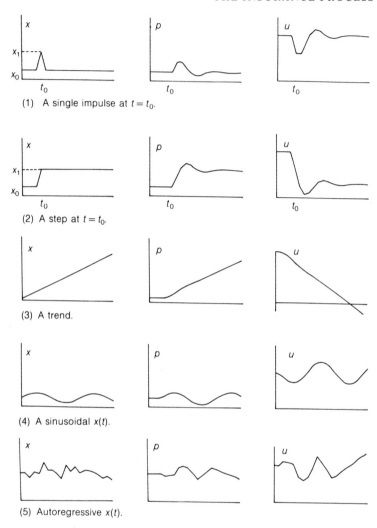

Figure 12.3.1   *System responses for alternative impulses.*

of Appendix G)

$$T = 2\pi/\arctan\sqrt{4a/r^2 - 1}. \qquad (12.3.13)$$

(2) $x(t)$ is assumed to be constant, but at time $t = t_0$ it steps up abruptly from level x_0 to level x_1, for example, as a result of a change in policy conditions or a change in claim settlement

practice. A damped sine wave again results. Notice that $u(t)$ stabilizes, but does not recover to its original level.

(3) Shows a case where $x(t)$ grows linearly as a result of a trend in risk-propensity. In this case $u(t)$ does not stabilize and is rapidly exhausted.

(4) The primary process $x(t)$ is assumed to be a deterministic sinusoidal wave

$$x(t) = \alpha \cdot \sin(\omega t + \phi). \qquad (12.3.14)$$

Then $u(t)$ and $p(t)$ are also sinusoidal waves, with the same wavelength. The amplitude of $u(t)$ is magnified, and can be evaluated by means of the frequency response functions of control theory (Pentikäinen et al., 1989, section 3.3.5).

(5) Finally, we assume that the claim process is a compound mixed Poisson process and that its cycle variable follows the second order autoregressive time series (9.2.7).

REMARK 1   As can be seen from Figure 12.3.1, the system, subject to the above conditions, has a tendency to overreact to adjustments to the premium rate, i.e. $u(t)$ may start to oscillate. This feature is a consequence of the two-year time-lag in (12.3.8). The correction $a(u_0 - u(t - 2))$ does not affect the premium rate $p(t - 1)$, nor does it affect $u(t - 1)$. Thus the inaccuracy in the premium in fact has an impact in two consecutive years, until it can be corrected. Furthermore, the new 'overcorrected' premium rate $p(t)$ again causes a deviation of $u(t)$ from the target level, which is then reflected in later premiums. If $a < 1$, however, the deviations will gradually be damped away, as can be seen from the impulse responses in Figure 12.3.1. If $a = 1$, the deviations will persist.

REMARK 2   If instead of (12.3.8) we had used the formula

$$p(t) = p(t - 1) + a \cdot [u_0 - u(t - 2)] \qquad (12.3.15)$$

the resulting process would not usually be stable. The condition for stability is now $0 < a < 1 - r$. This type of rule may seem to be quite natural and appealing, indeed, if a need to strengthen the financial position is revealed, why not correct the latest premiums! The resulting instability may, at first sight, be unexpected. The reason is shown by (12.3.15), according to which the corrections have a cumulative effect, whereas in (12.3.8) they have the same reference level p_0.

(f) Factors inducing cycles.

We can see that there are cases where the rating procedure itself produces cyclical variations in the solvency margins, premiums and loss ratios, in spite of the fact that there were no cycles in the original claim process. This results from what can be called the **time lag effect**, i.e. the premium control inevitably lags behind. Profitability and other relevant factors can only be ascertained after a certain delay and further time is required to implement corrective measures. If tariff bureaus and regulatory approval are involved, the process may take even longer. The total time delay is usually 1.5–2.5 years (Hart *et al.*, 1987; Beard *et al.*, 1984, section 7.6(e)).

On the other hand, measures to reduce the delays may introduce further inaccuracies, for example the claim data may be more incomplete and the ultimate outcome more uncertain. If claim variability is disguised by over- or under-reserving, the effective time lag may be even longer. Some lines, e.g. third-party liability and inwards reinsurance, may be seriously affected by late notification of claims.

Both the experience of actual behaviour of insurance processes and the various impact responses shown in Figure 12.3.1 suggest that there are several cycle-inducing background factors. Factors which may, directly or indirectly, be responsible for business cycles are considered briefly in the following paragraphs.

(1) **Claims** have their own trends, cycles and short-term variability, as was indicated in section 9.2. These give rise to fluctuation of the whole insurance process.
(2) **Premium cycles** can be induced by competitive strategies, where the theoretically necessary rate adjustments may be ignored, or by regulatory or political pressures, preventing implementation of commercially desirable increases.
(3) **Inflation**: steady inflation would not be expected to cause cyclical effects, since proper allowance can be made for inflation at future expected levels in determining premiums, reserves, solvency margins, etc. However, if the rate of inflation changes to an unexpected degree, there may be a tendency to provoke cycles in all the key quantities.
(4) **Fluctuations in interest rates and asset values** may also have a direct impact on underwriting margins. If the rate of return on the technical and other reserves is high, one may expect lower premiums to result, either because the expected cost of settling claims is adjusted downwards or because lower explicit profit margins are regarded as acceptable. High returns may lead to over-optimistic expectations of future investment returns and thus to operating losses when rates of return fall. Changes in asset values may have a significant impact on financial strength, inducing corrective action through premium rates.

(5) **Market-related causes**. The capacity available in the market may have a major impact on price in some classes, particularly reinsurance and commercial lines. Excessive capacity has a tendency to push down premiums and, conversely, lack of capacity forces prices up. On the other hand, the capital available depends partly on the past profits or losses of the industry, and partly on the anticipated profitability of the insurance business relative to other industries, particularly if there is free entry to the market. New entrants will be attracted to the market by the promise of high profits. It will also be easier to persuade existing shareholders to subscribe additional capital in such circumstances. Capacity to write additional premium income is determined by the capital and free reserves. These features provoke alternating expansion and contraction of capacity and consequent cyclical price variation.

The effects may also depend on sales practice. For example, Levin (1986) surmises that one reason for the deep downswing in the 1980s in the US market was the role of independent brokers in a deregulated market. According to Levin they have a tendency to identify lower-priced products in order to satisfy clients, whilst insurance companies have not for the most part devised any mechanism to monitor and control their products and premium rate adequacy in a deregulated environment. Levin claims that, as a consequence, even some of the biggest and most-respected insurance companies had no idea how many policies were on their books with deviations from the theoretically correct rate level, nor what was the total or average deviation.

So-called **innocent** or **uninformed capacity** may come into the market when profits have been high and delay the necessary premium corrections as profitability weakens, making the downswing phase deeper (as happened, for example, in the reinsurance crisis of the 1980s). Smoothing of underwriting results for published accounts may reinforce these effects, since losses may be disguised for a time. Furthermore, corrective action on premium rates may be hampered by the presence in the market of newcomers who have not been weakened by earlier losses and do not need to strengthen reserves in respect of past business, with less of a need as a result to increase premiums. In fact entry into the market may be particularly appealing when established insurers are being compelled to raise premiums after a period of heavy losses.

It is claimed that the insurance business has the basic attributes of a so-called **cycle-character industry** (Mooney, 1986). Insurance is a necessity, not a luxury, and demand for the product is relatively fixed in any time period, whilst supply is more variable. All that an existing insurance company needs in order to write more business is additional financial capital; it does not need to build new factories. The history of the insurance industry indicates that, when returns appear promising, the supply of capital can expand rapidly. Companies with surplus capital may seek to gain market share by cutting prices. For an individual company with unused capacity it may make sense to cut prices so that market share can be expanded or protected. However, for the industry as a whole price-cutting has a minimal effect on total demand. As a result the industry may engage in a period of fierce competition without it having much impact on total volume, but with a resulting fall in total premium income and profitability.

The reverse process can be observed in the recovery phase of the cycle. Owing to the relative insensitivity of demand to price, consumers will continue to buy the product in spite of significant price increases. Thus the recovery will be faster and more robust than in an industry where demand is more price-elastic. On the other hand, it may be argued that insurance demand is no longer as insensitive to price as formerly. As a result of the development of risk management, larger economic entities are inclined to use different kinds of self-insurance in order to minimize their insurance costs. For example, it has been estimated that the self-insured market in the US was about 18% of the whole market in 1983, 28% in 1986 and 33% in 1990.

(6) In a similar way, if there is access to reinsurance capacity at low prices, this may be reflected in prices in the direct insurance market.

(7) The international links through reinsurance and many of the background factors, such as the economic booms and recessions which are common to many industrialized countries, have a tendency to synchronize the insurance market movements internationally.

(g) Concluding remarks.   Some of the causes and effects discussed above will induce cycles for the whole market and affect all, or at least many, insurers simultaneously. On the other hand, some causes are specific to individual insurers. So far as experience is available, it seems that the former predominate and that company-specific cycles are of relatively small amplitude and importance (Figure 12.2.2).

Market cycles can, furthermore, be expected to vary greatly between different countries and markets. If the market is small, then the conduct of each insurer, and in particular that of any company with a substantial share of the market, can make an impact that creates or influences a cycle, either exacerbating or helping to control it.

One of the relevant factors is the price sensitivity of policyholders. This obviously depends on the extent to which brokers are used and can be very different for commercial policies and personal lines policies. A theory which satisfactorily explains the cycle phenomenon in one country and in one market, therefore, will not necessarily be valid in other countries and in other markets. Even successive cycles in the same market may have quite different causes.

(h) References.   In recent years several authors have studied control theory applications, cycling, etc., for example, Balzer and Benjamin, 1980; Smith, 1981; Pentikäinen and Rantala, 1982; Rantala, 1984, 1988; Venezian, 1985, and Taylor, 1988. A brief summary of control theory ideas is given in Pentikäinen et al., 1989.

Exercise 12.3.1   Let X_u denote the aggregate claim amount for a group of risks in the year u, and assume the variables X_u to be positive, independent and identically distributed. The risk premium P_t for the year t is given by the formula

$$P_t = a_1 \cdot X_{t-1} + a_2 \cdot X_{t-2} + \cdots,$$

where $\sum a_i = 1$. Then obviously $E(P_t) = E(X_t)$. Show that

$$E\left(\frac{X_t}{P_t}\right) > 1.$$

Comment on this slightly paradoxical result! (Hint for solution: Jensen's inequality.)

12.4   Simulation of the insurance process

As was noted in section 9.4, conventional analytical methods are not adequate for long-term considerations. This suggests that simulation should be the principal working tool. We now show how the insurance process (12.1.1) can be simulated. The necessary components have, in fact, already been derived in previous chapters and will be brought together in this section. The presentation is an extension of the basic model (5.5.1), with the return on investments, decomposition into lines of business and some other features now also included. On the other hand, we defer to Chapter 14 the treatment of how simulation of the insurance process could assist an insurer's strategic planning and how, among other things, interactions with the insurance market might be taken into account. A number of further extensions will be dealt with in Chapter 13 in discussing some applications.

(a) Basic equations. As was outlined in section 5.5, the simulation proceeds step by step from the initial accounting year t_0 to the succeeding years $t_0 + 1$, $t_0 + 2$,.... The transition from one year to the next will be carried out using a transition algorithm which treats the cash flow equation (1.1.1) and the accounting equation (1.2.5) in parallel. The former results in a sequence of asset amounts $A(t_0)$, $A(t_0 + 1)$,... which are needed for the simulation of asset movements, as well as for various cash flow considerations, in particular, when the break-up principle is applied. The accounting algorithm gives the corresponding sequence for the solvency margin U and for other variables which are required in conventional income and balance sheet statements and for many going-concern considerations. These two equations will now be employed in a simplified and modified form as follows

$$A(t) = A(t - 1) + B'(t) + J'(t) + \Delta^* A(t) - X'(t) - E(t) \qquad (12.4.1)$$

and

$$U(t) = U(t - 1) + B(t) + J(t) + \Delta^* A(t) - X(t) - E(t) \qquad (12.4.2)$$

where the primed quantities indicate the amounts paid in year t, as distinct from the unprimed quantities, which represent the earned or incurred amounts. The links between them were given in (1.2.1) and (1.2.2). J' is the cash income earned on the assets and $\Delta^* A$ the change in value of the assets, including the difference of values arising from new investments and from sold or matured assets ((8.5.1). Note: $J' + \Delta^* A = J$ in (1.1.1) and (1.2.5)).

The simulation is carried out by extending the procedure described in section 5.5. The primary problem is to find rules and sub-modules for the various entries in the above equations.

(b) Initial choices and definitions. The relevant model formulation, self-evidently, depends largely on the intended application. If it is aimed at supporting the analysis of the stochastic behaviour of a **particular insurer**, then the model should be fitted to the actual and anticipated structures of this business, having regard to whether the analysis is intended primarily for rating, reserving, investments, reinsurance, sales or to give a general account of the financial strength of the company. These special requirements may determine how detailed the model assumptions and parametrizations should be in the different sub-modules of the model.

Another application would be to obtain an idea of the general behaviour of a typical insurer, investigating the resilience to certain adverse events such as inflation, movements in capital markets, economic cycles, potential catastrophes, etc. Analysis of this kind assists understanding of the properties of the insurance process, may be useful both for management of a company and for a supervisory authority and is an efficient tool for educational and research purposes.

Irrespective of the ultimate application of the model, it is advisable to establish an initial set of assumptions to represent a notional model office scenario. Understanding of the simulated outcomes can then be deepened by making auxiliary simulations for suitably changed values of the initial parameters, i.e. by performing sensitivity analyses, for instance, to test how the behaviour of the process reacts to increased inflation, changes in the size of the portfolio, new combinations of asset types, etc. The outcomes of the changed processes can be compared with the original ones. The initial parameter set might be described as a **reference insurer** (or a **standard** or **model insurer**). The outcomes based on it serve as yardsticks for comparing the results of the variants. Examples of a reference insurer can be found in Beard *et al.*, 1984; Pentikäinen *et al.*, 1989; and Daykin and Hey, 1990.

A fundamental problem facing the modeller is to decide how many details and special features to include in the model. Pursuit of great accuracy and reliable representation of a large number of features may lead to a model of enormous size and complexity. This may create serious programming complications and greatly increase the time which is required to run the simulation. Furthermore, if the number of parameters becomes too great, setting initial values for all of them can be laborious. If no accurate data are available, shortcomings in parameter estimation may detract from the targeted reliability. Hence, compromises between accuracy and simplicity are essential. In what follows, alternative detailed and straightforward approaches are suggested in parallel, in order to provide choices for model-builders.

One of the initial choices is to decide whether, and if so how, the portfolio should be divided into parts which require special handling, for example, according to **class of insurance**, and possibly with further sub-division according to type within each class. If the number of classes or sub-divisions is large, it may markedly increase the complexity of the programming and the running of the model and may make interpretation of the results more difficult. It may be advisable, therefore, to combine classes which have similar properties into a few, internally fairly homogeneous classes. For example, classes where the risk sums are small, i.e. conventional personal lines insurance such as family policies, travel, motor car, etc., might constitute one of the combined classes. Another class could be made up of those policies which involve large-risk sums, for example, relating to industrial plants, ships and aircraft. A third class might consist of those risks which are sensitive to significant mixing variation, for example, credit insurance and classes that are particularly vulnerable to natural catastrophes. If the mixing variation and claim inflation can be expected to be approximately the same for several classes, then the technique suggested in section 3.2(d) can be used to combine these classes and to derive the necessary characteristics for the combination.

Some particular comments follow on each of the entries in the basic equations.

(c) Premiums *B*. Some simple premium formulae were presented in section 10.1 and equations (10.2.1) to (10.2.7). They require the expected value of claims $E(X)$, which may be obtained from the initial data for the simulation of claims (section 12.4(e) below), ideally using actual data from recent years.

If past data are not available, for example, if a fictitious or illustrative insurer is being investigated, the 'history' of claim expenditure can be simulated backwards for a suitable period. The premiums can be derived from these data by mimicking an actuary who calculates the rates from actual claim statistics. Note that the inaccuracies inherent from parameter error (section 1.3(e)) will influence the simulation process in the same way as they do in all practical rating exercises.

Other alternatives would be to apply experience-rating (section 6.5, and equation (10.2.8)), control theory (sections 12.3(e)) or to construct straightforward rules which fit the strategies of the company in question (the simple example in section 5.5(d)).

(d) Return on investments J' and $\Delta*A$. The construction of the model depends to a large extent on the intended application. If assets are required to match the liabilities not only by amount but also by term, they will need to be specified in classes and, at least in certain groups, be earmarked to belong to specific liabilities, in particular, to the outstanding claims and other technical provisions. A condition for an acceptable investment strategy is that the **break-up analysis** (section 9.5.4(b)) should give a positive outcome with a high confidence probability at the end of each year. This lays down certain constraints for the selection of investments. Subject to these constraints, the income and changes in value can be simulated as was presented, for example, in sections 8.4 to 8.6. Daykin *et al.* (1987, 1987a) considered the impact on run-off solvency probabilities of a variety of different asset distributions, showing that this aspect contributed materially to the security level for any given solvency margin.

In the going-concern case the procedure can be simpler. It is sufficient to specify the targeted relative distribution (w_k) (section 8.6(d)) of the investment types. The total amount of assets $A(t)$ is simulated from (12.4.1). For the simulation of changes in value it is assumed that, at the beginning of each year, the distribution of assets corresponds strictly to the targeted distribution (w_k), i.e. $A_k(t-1) = w_k \cdot A(t-1)$. If this is not the case, owing to the performance of the investments, disinvestments and changes in value, it is assumed that sales and purchases will take place until this distribution has been restored. Applying some of the methods of sections 8.4 and 8.5, the required increments $\Delta A_k(t)$, as well as the income rates $j_k(t)$, can be obtained. Finally, $J'(t)$ is the sum of its components.

By experimenting with different distributions (w_k), different investment strategies can be explored.

In the management model developed by Daykin and Hey (1990), a different asset distribution can be specified for the investment of technical reserves for each separate class of liabilities, and again for the free assets or solvency margin. The asset distribution is assumed to be rebalanced each year, once the technical reserves have been calculated, with appropriate purchases or sales to re-establish the specified distribution.

A more sophisticated version would be to permit the distribution (w_k) to change during the simulation period according to some rule which, for example, takes into account inflation or the simulated outcome for particular investment categories. The idea is akin to that mentioned in the third paragraph of section 8.2(b) concerning purchasing and selling equities according to the phase of the market cycle. Unfortunately, however, whilst it is easy to programme the model to buy equities at the bottom of the market and to sell them at the top of the market, it is not usually possible to be as successful as this in practice.

Daykin and Hey (1990) provide for the investment distribution to be changed after a specified period, or when a particular trigger point is reached by the solvency margin. However, this falls short of a fully dynamic investment policy.

For cases where the investment strategy is not changing and where parameters for the whole asset portfolio are available or can be derived, a simple approach would be to handle the assets together as a group, without dividing them into categories. The total amount $A(t)$ is again simulated from (12.4.1) and the total rate of the return, including both cash income and changes in value, for all assets together, from an algorithm of the type defined by equation (8.5.12). The parameters should be calibrated to match the overall variability of the assets. Periodical changes and potential crashes, as proposed in section 8.5(i), might be introduced to enhance the model. This simplified model could introduce the impact of asset variability into the total business flow in a satisfactory way in applications where the main focus is on aspects other than investments. This approach is not suitable, however, if different investment strategies are to be investigated.

(e) Claims X.   If the effect of run-off inaccuracy is to be taken into account, then the simulation should be performed as described in section 9.5.3. The paid claims are obtained for (12.4.1) from (9.5.1) and the incurred claims for (12.4.2) from (9.5.6).

Having regard to the relatively modest effect of run-off inaccuracy in many going-concern simulations, the procedure can be shortened significantly by ignoring run-off errors and assuming that incurred claims can be accurately estimated at the end of the year in which the premium is earned. Then X can be simulated in the way described in section 9.4(a) for the cross-sectional variable.

(f) Expenses E in (12.4.1) and (12.4.2) can be assumed, as a first approximation, to have a constant ratio $e(t)$ to the premium income ((10.2.2) and (11.1.1)). This assumption will be relaxed in Chapter 14. Various more accurate approaches were outlined in section 11.1.

(g) Organization of the total simulation. As can be seen from the foregoing sections, there are, according to the intended application, numerous alternatives and variants in the construction of a fully comprehensive model, the more so because it should be adapted to the user's computer facilities, software packages and programming culture. It is not felt, therefore, that a detailed flow chart would be helpful, other than to give examples in Appendix F; instead a number of comments and hints are given below. It might be helpful for the reader to look at the following text and section F.5 of Appendix F in parallel.

(1) The programme should initially assign numerical values to the parameters which control the sub-modules for inflation, investments, claims, premiums, etc. (a ready programmed data matrix or manual input). Lists of relevant parameters are given in the sections above or they can be picked up from the equations where they are included as coefficients and parameters (examples in section F.5(b) of Appendix F).

A major preparatory task is to find these numerical values. They can be derived from the insurer's claim statistics, past investment performance, etc., or be separately evaluated. It may not be possible to derive some of the parameters from the company's own files. Then it will be necessary to resort to industry experience or to an evaluation based on common sense. For example, the properties of mixing variables are of this type. Moreover, it is important to emphasize that the structures of many of the relevant time series may be changing rapidly, possibly being significantly different from period to period. An uncritical use of conventional methods of mathematical statistics might lead to erroneous or at least inefficient data. The potential need to split the observation and target periods into homogeneous sub-periods was discussed in sections 7.3(g) and 8.5(i).

In cases where great uncertainty remains about the parameter values, one approach would be to choose several sets of values, for example, pessimistic, optimistic and most likely, and to run the process using them in parallel. This is called **scenario technique**. It does not, of course, remove the uncertainty but helps to evaluate its impact on the relevant outcomes.

(2) One of the first requirements is to fix the time intervals, both for the historical data to be used and the future to be explored. Claims are simulated by filling in a matrix similar to that which was constructed in Figure 9.5.1 for considering the run-off pattern. In cases where the run-off is ignored (section 12.4(e), second paragraph), the matrix degenerates into a single dimension.

(3) **Initializing simulations.** Some of the variables do not depend on the outcome of the company's business year by year. They should be generated and stored in advance, before carrying out the main process, which will be described as the next step. Inflation is one such variable, as are the yields and price indices for equities, property and other asset types (examples in box 3 in section F.5(d) of Appendix F).

(4) **Year-to-year progression of the simulation.** Each of the entries in equations (12.4.1) and (12.4.2) requires its own sub-module, which should be run first. The outcomes can then be gathered up in a master module which simulates the asset amount $A(t)$, the solvency margin (surplus) $U(t)$ and other key variables. Thus the state of the business can be obtained successively for $t = t_0 + 1, t_0 + 2,...,$ $t_0 + T$.

(5) **Presentation of outcomes** requires special care and skills (and is often given insufficient attention). It is advisable to present the simulation outputs in both tabular and graphical form. There is usually a large amount of information and there are major difficulties in drawing out the most important features from the mass of data. A graphical presentation of the key variables may be the best, and possibly the only way, to maintain control. One of its major merits is that it can be made more accessible to people who are not familiar with simulation or other technicalities. An example of an alternative idea for graphical presentation will be given in section 12.4(i).

(h) Examples. In order to illustrate the simulation approach, a few examples from Pentikäinen *et al.* (1989) are displayed in Figures 12.4.1 to 12.4.5.

A couple of individual simulated outcomes are presented in Figure 12.4.1 in respect of the key variables: claims X, premiums B and solvency margin U. The differences between the parallel sets of

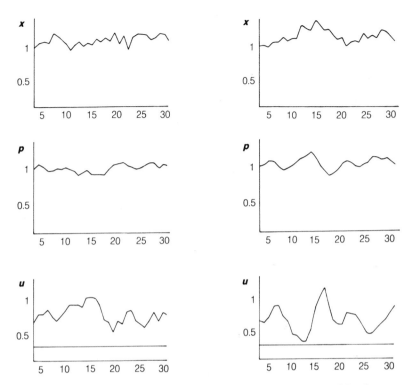

Figure 12.4.1   *Two individual realizations with outcomes presented for claims, premiums and solvency ratio.*

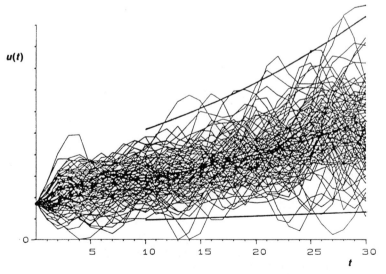

Figure 12.4.2   *A simulated **u**(t) process.*

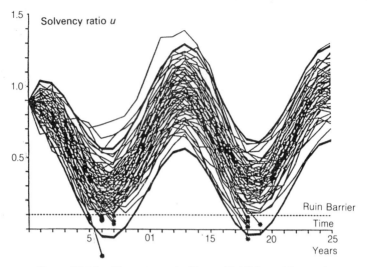

Figure 12.4.3   *The same as in Figure 12.4.2 but with a cycle.*

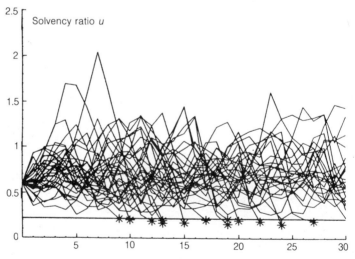

Figure 12.4.4   *Simulation of the solvency ratio of the reference insurer, with an equity-orientated investment strategy.*

diagrams are due entirely to the randomness of the simulation process.

The ratio form $u(t)$ of the solvency margin, and of other key variables, is appropriate for diagrams showing the long-term development (see section 12.3(a)), notwithstanding the fact that the calculations can be performed more easily using the absolute values $U(t)$, etc., transforming the results into ratio form only for the purposes of presentation.

Figure 12.4.2 shows the same process but with the simulation repeated numerous times, according to the Monte Carlo method (section 5.5(b)). This diagram reveals many of the properties of the process, for example its average level of flow and the range of variation, which indicates the risk of adverse random outcomes.

Figure 12.4.3 presents a variant where the process is, via the mixing variable, subject to a deterministic cycle. The diagram is shown with a **ruin barrier**. This could be a regulatory limit laid down by law or, in other applications, a limit determined by the management to assist risk control by reference to the capital at risk (sections 6.2 and 13.2). The asterisks indicate a **ruin**, i.e. a case where the solvency margin fell below the barrier and the process was terminated.

Figure 12.4.4 illustrates the use of sensitivity analysis. The insurance process shown in Figure 12.4.2 had the investment mixture (section 8.6(d)) of cash, bonds, property and equities in the proportions $10:30:30:30$. This was changed by increasing the share of equities so that the new proportions were $10:40:0:50$. As might be expected, the risk of ruin is appreciably increased.

(i) B, U-diagrams. A major difficulty in the use of models like that described above is that the number of variables, distributions and user choices can be very great. It is often quite difficult for the user to obtain an insight into the structure of the model, which is, in fact, a more or less complicated configuration in a multi-dimensional space. It is still more difficult to explain successfully the properties and results of the model to outsiders. The graphical presentation may then be an indispensable approach. It is a problem to reveal only the essential features. Figures 12.4.1 to 12.4.4 show examples of the graphical presentation of simulation outcomes. In this section an alternative approach is illustrated.

In Figure 12.4.1 the key variables x, p and u were placed in different diagrams. It may often be useful to study the relationship between two of them, for example, p and u. Our example shows a way in which it can be done.

The variable **U**, the solvency margin, is chosen as one of the key variables. It represents the **wealth** of an insurer. Another important feature to be followed and analysed is the **volume** of business. This can conveniently be described by the gross premium income on the insurer's net retention **B**, and this will be adopted as the second key variable. A survey of the overall situation can now be given by

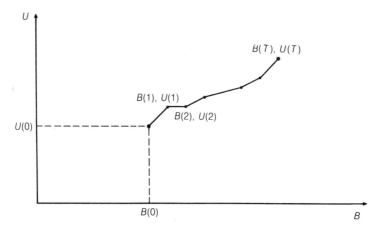

Figure 12.4.5 *An outcome of the underwriting process presented in the B,U-plane.*

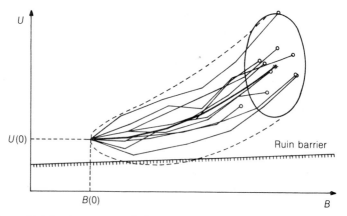

Figure 12.4.6 *An outcome of a business strategy in the B,U-plane. Stochastic bundle and its confidence region.*

plotting the variables **B** and **U** in the same plane. This is done in Figure 12.4.5, where the premium volume **B** is displayed on the x-axis instead of time, which was the argument in the previous figures. Each simulation step generates a new outcome for the variable pair **B**, **U**, representing the state of business at time t, the next step at time t + 1, etc. A sample path in the **B**, **U**-plane is a broken line, a 'random walk'. Increase in volume is indicated by movement to the right and increase in wealth by a movement upwards, reflecting the profitability of the business. In fact, this diagram includes the same information as the p and u diagrams of Figure 12.4.1 taken together.

Repetition of the simulation generates a bundle of sample paths. The final points represent a sample of the situation at the end of the planning horizon T. In Figure 12.4.6 it is assumed that not only solvency margin **U** but also the premium income **B** is stochastic. Then at a given confidence level the final points are two-dimensionally distributed over a region which is approximately elliptical in shape.

Graphical configurations such as Figure 12.4.6 give an idea of the assumed flow of business and the properties of chosen strategy options. The more the distribution ellipse at time T is situated to the right, the more expansive is the strategy option, i.e. the more business

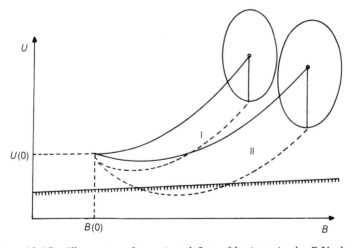

Figure 12.4.7   *Illustration of a projected flow of business in the B,U-plane according to two scenarios I and II. The developing financial position, in terms of premium income B and solvency margin U, is indicated by the lines which join up the points for the projected financial years t = 0, 1, 2, ... T. The ovals describe the confidence regions at the end of the year T and the dotted lines the paths of the lowest points of the successive ovals for the years up to T.*

is written. The vertical position of the ellipse indicates the solvency situation. If the ellipse or the bundle intersects the ruin barrier, it is an indication of a risky situation.

A B, U-diagram can be used to compare the future prospects on different scenarios. Figure 12.4.7 illustrates an example. Two flows of business are depicted, corresponding to two potential strategies. The development of both the growth of business volume and the financial strength can be seen at a glance. The ruin barrier indicates the level below which the solvency position is no longer satisfactory. The ovals represent confidence regions and the dotted lines the loci of the lowest points. Strategies should be chosen so that the risk of this lower line intersecting the ruin barrier is small.

CHAPTER 13

Applications to
long-term processes

13.1 General features

This section begins by studying the properties of the model of section
12.4. It is advisable to begin with a simple structure and then proceed
with more advanced variants.

(a) Some characteristic properties.   Some fundamental features of
the insurance model (12.1.1) can be seen from the simplified ratio
form (12.3.6). Using the safety loaded premium, it can be written

$$u(t) = r(t) \cdot u(t-1) + p(t) \cdot (1 + \lambda(t)) - x(t). \qquad (13.1.1)$$

where now

$$x(t) = E(x(t)) + \varepsilon_x \qquad (13.1.2)$$

and

$$p(t) = E(x(t)) + \varepsilon_p. \qquad (13.1.3)$$

These equations will be used in illustrative simulations. In order
to get an approximate idea of the behaviour of the process we assume
that

$$E(x(t)) = \mu_x' = \text{constant.} \qquad (13.1.4)$$

Replacing $r(t)$ and $\lambda(t)$ by their mean values r and λ, we can eliminate
$p(t)$ and $x(t)$ from (13.1.1) and manipulate it into the form

$$u(t) - \bar{u} = r \cdot [u(t-1) - \bar{u}] + \varepsilon_u \qquad (13.1.5)$$

where

$$\bar{u} = \frac{\lambda \cdot \mu_x}{1 - r}$$

$$\varepsilon_u = (1 + \lambda) \cdot \varepsilon_p - \varepsilon_x. \qquad (13.1.6)$$

This is a first order autoregressive time series (equation (G.2.4) of Appendix G). To describe a real-life insurance process stationarity is necessary, i.e. the flow of $u(t)$ must have finite bounds and must not behave irregularly. According to section G.2(b) of Appendix G, the condition for stationarity (see (12.3.2) and 12.3.5)) is

$$r = \frac{1+j}{r_g \cdot (1 + \bar{i}_x)} < 1. \tag{13.1.7}$$

If this condition is satisfied, the process tends to revert towards an equilibrium state with mean level given by (13.1.6). Figure 13.1.1 illustrates a process which satisfies this condition, balancing four forces: inflation, real premium growth, the yield on investments and the safety loading.

Condition (13.1.7) is not always satisfied. Moreover, even if it were, other factors such as dynamic control of premiums, return on investment, etc. would also be relevant and require attention. The dynamic control exerted by management is particularly important. In the event of adverse developments, it can be expected that remedial action will be taken, for example by amending premium rates. Pentikäinen et al. (1989) illustrated this effect by assuming a simple premium

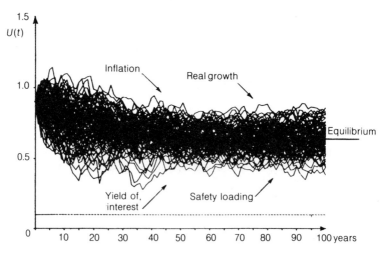

Figure 13.1.1 *Simulated realizations of a solvency ratio process with equilibrium in the presence of four forces.*

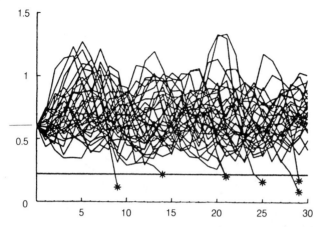

Figure 13.1.2 *A solvency ratio process controlled by the premium rule (13.1.8). The number of simulated realizations is 40. Paths which passed below a notional ruin barrier were terminated as insolvencies and are marked with an asterisk (Pentikäinen et al., 1989, p. 185).*

control of the form

$$p(t)/p_0 - 1 = \alpha_2 \cdot [u_{\text{target}} - u(t-2)] + \alpha_3 \cdot [u_{\text{target}} - u(t-3)] + \sigma_p \cdot \varepsilon(t), \tag{13.1.8}$$

i.e. if the solvency ratio deviates from a prefixed target level u_{target}, the premiums are changed so as to help to restore the level $p_0 = \mu_x(1 + \lambda)$. As was argued in section 10.2(g), the time lag cannot usually in practice be shorter than two years. When the coefficients are correctly specified, the control rule (13.1.8) keeps the process stationary, as is shown in Figure 13.1.2.

Figure 13.1.3 shows the effect of relaxing the control. The effect is dramatic. In order to prevent a very large number of realizations from 'ruin' (marked with asterisks), the safety loading was increased from 3% to 5%. Clearly, controls of this type should play an important part in practical model-building.

(b) Time horizon T. A number of important applications were dealt with in Chapter 6, including capital requirements, reinsurance, etc. The time horizon was limited to one year. This was convenient for the treatment of many problems, but the impact on the long-term development of the company could not be properly assessed. This

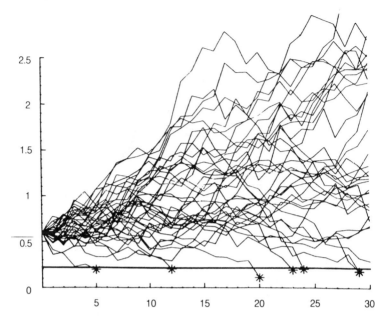

Figure 13.1.3   *The same process as in Figure 13.1.2 but with the premium control relaxed and the safety loading increased from 3% to 5%; 40 realizations.*

applies in particular to the effects of profits or losses, which may accumulate for several consecutive years as a result of business cycles. The one-year results may be a useful first approximation but it is desirable to check them and to amend them to ensure that the conditions of successful survival for a longer period are also met. The model should be constructed so as to permit the use of any time horizon T, with the value of T being selected according to the application.

If the risk theoretical evaluations are to support the company's conventional business planning, which as a rule will also include future projections, then T will naturally correspond to the time horizon of the planning process, say five years. However, the need to explore the effects of exogenous business cycles suggests a time horizon at least as long as one such cycle, incorporating both a peak and a trough. This might require 6 to 10 years. If the general behaviour of the insurer is to be tested, for example, the capacity to cope with adverse events such as high inflation, a crash in the value of equities or of other assets, competitive markets, etc. and possibly even the risk of several events occurring simultaneously, then it is advisable to consider quite a long period, say 30 years or even longer, in order to bring out potential equilibrium features. However, it should be remembered that, the further ahead the process is projected, the greater the uncertainty. It is essential to establish appropriate feedback mechanisms if sensible results are to be obtained.

In considering the problem of capital requirements a limit ε has to be set for the probability of ruin. Other things being equal, the probability of ruin increases the longer the horizon under consideration. With this in mind, it may be desirable to set different probability levels for different periods, for example $\varepsilon = 10^{-4}$ for $T = 1$ year, 10^{-3} for 10 years or 10^{-2} for infinite time (see next section).

(c) Risk of ruin. Evaluation of the probability of ruin, denoted by $\Psi_T(U)$ (section 1.3(c)), is one of the central tasks of traditional risk theory. We apply this concept in more general terms by speaking of capital at risk (section 6.2 and section 13.2 below). The simulation procedure of section 12.4 is very suitable for the purpose, as indicated by Figure 13.1.2. Once a ruin barrier has been defined, the simulated paths of the course of business which pass below the barrier are counted as ruins ($n_{\text{ruin}} = 6$ in the figure). Then the ratio of the number of ruins to the total number N of realizations in the simulation

$$\Psi_{30} \approx n_{\text{ruin}}/N = 6/40 = 0.15 \qquad (13.1.9)$$

gives an estimate of the probability of ruin. The example corresponds to a level of about $0.15/30 = 0.5\%$ insolvencies a year.

To make a more accurate assessment, more realizations are required. Confidence limits are discussed in Appendix F.

A visual inspection of a diagram of simulated paths can provide quite a good idea of the risk structure. Imagination can complete the bundle of the paths of the density configuration, which is illustrated in Figure 13.1.4 with a number of cross-sectional density distributions. The density of the simulated paths in different parts of diagrams like Figures 13.1.2 and 13.1.3 corresponds to the height of the density curve in the three-dimensional Figure 13.1.4.

(d) Infinite time ruin probability of classical risk theory. A common procedure in the risk theory literature is to allow the time horizon T to grow to infinity and to study the resulting ruin probability $\Psi_\infty(U) = \Psi(U)$ as a risk measure. In fact, it constitutes an upper limit

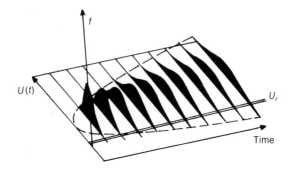

Figure 13.1.4 *A three-dimensional visualization of the random process $U(t)$. The function f is the density of the cross-sectional $U(t)$ for different t-values. U_r is a ruin barrier. The 1% confidence boundaries are plotted by dashed lines, which resemble a parabola.*

for finite-time ruin probabilities and leads to an elegant theory. The principal result is the so-called **Lundberg's formula**

$$\Psi(U) = C(U) \cdot e^{-RU}. \tag{13.1.10}$$

$C(U)$ is an auxiliary function with values in the interval $(0, 1)$. It is often replaced by unity as an upper limit estimate. R is known as Lundberg's coefficient or the adjustment coefficient (for further details see, for example, Beard *et al.*, 1984, section 9.2(b)).

Unfortunately, the different derivations of formula (13.1.10) have to a large extent ignored autoregressive inflation models, return on investments, business cycles, dynamic control, etc., which have been introduced as extensions of the model for long-term processes in the foregoing chapters, though these features are crucial for the behaviour of the processes, as various examples – both practical observations and models – have shown.

Another serious shortcoming is that the usual premises for (13.1.10) provide non-zero survival probabilities only if the solvency margin tends to infinity, as illustrated in Figure 13.1.5. This is due to the **paradox of classical risk theory**: if an upper limit, however large, is imposed to constrain the growth of the solvency ratio $u(t)$, then the infinite-time ruin probability Ψ is equal to unity, i.e. the ruin of every insurer is certain sooner or later!

REMARK This result can be understood by heuristic reasoning as follows. The cross-sectional probability of ruin $\varepsilon(t)$ at any time t is positive, although

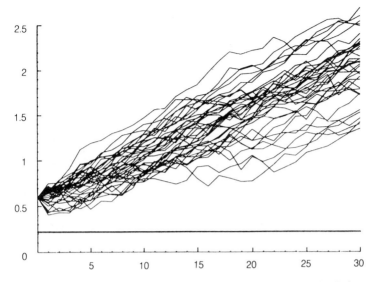

Figure 13.1.5 *The flow of the solvency ratio according to classical risk theory assumptions (Pentikäinen et al., 1989); 40 realizations.*

it might be very small. This follows from the Poisson (2.2.1) and the Pólya (2.5.5) formulae, which have a non-zero probability for arbitrarily large claims. The cross-sectional $\varepsilon(t)$ depends on the state $u(t-1)$ at the beginning of year t, being smaller for larger $u(t-1)$. If an upper limit is assumed for this quantity (supposing that no insurer can become infinitely rich), then a lower limit for $\varepsilon(t)$ clearly emerges. By quite general assumptions on the stationarity of the process there is a time independent lower limit ε valid for all t's, i.e. $\varepsilon(t) \leqslant \varepsilon$. Then the survival probability has an upper limit

$$1 - \Psi_T \leqslant (1 - \varepsilon)^T,$$

but this tends to zero as T grows to infinity, and hence $\Psi \to 1$. A more rigorous discussion can be found, for example, in Beard *et al.*, 1984, section 9.3(c).

13.2 Capital requirements of an insurance company

(a) Straightforward evaluation of the capital at risk can be made directly from the bundle of simulated realizations of the solvency

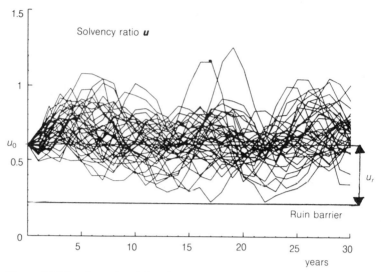

Figure 13.2.1 *A simulated flow bundle for a process which accumulates the underwriting profits and losses in a risk reserve (solvency margin).*

ratio process, such as was exemplified in Figures 13.1.2, 13.1.3 and 13.2.1. As in section 6.2 the consideration will be sufficiently general that the object of the investigation can be either the whole portfolio of an insurer or any part of it, for example, a line or a group of policies. Initial capital ratio u_0 is assigned to back the business. The simulated realizations spread over a confidence area, and, as a rule, both positive and negative values appear. The problem is to find a lower confidence boundary for them. The difference between the starting level u_0 and the lower boundary is denoted by u_r. It is the capital at risk, using the term introduced in section 6.2.

The capital at risk can be analysed by varying the assumptions for the various relevant background features in a way similar to that shown in section 6.2. Effects which have a greater impact on long-term considerations are, in particular, business cycles and the return on investments. The impact of a cycle was shown in Figure 12.4.3.

(b) The capital requirements of an insurer. Insurance management needs ways to cope with adverse time periods and to even out profits and losses between consecutive accounting years, i.e. to control the fluctuation of the solvency ratio $u(t)$. In broad terms the problem can be stated as follows.

In most countries there is a regulatory required minimum margin of solvency u_{req} (in ratio form), a wind-up barrier which every insurer must have in order to be permitted to continue writing business. For the survival of the company it is necessary to maintain the fluctuating solvency ratio $u(t)$ above this barrier. In this context the solvency ratio should be interpreted as consisting of both the book capital and the various hidden reserves, including the equalization reserve (if any), which will be discussed in the next section. If the solvency ratio $u(t)$ falls close to the barrier u_{req}, there is a considerable probability that normal random fluctuation could lead to a fatal drop in the margin below the barrier. It is, therefore, advisable to establish an early warning limit u_c, at a safe margin above u_{req}, for managerial use, and attempt to control the fluctuation of $u(t)$ so that the solvency ratio does not fall below it. Hence, the confidence area, or the fluctuation tube of the flow of the solvency ratio $u(t)$, should be positioned using suitable management control tools, so that its lower boundary is at u_c, as shown in Figure 13.2.2.

The area above the early warning barrier u_c in which the actual solvency ratio $u(t)$ should fluctuate can be called the **target zone**. It is of interest also to evaluate the upper limit u_{upper} for the zone. Management needs at least an approximate idea of the order of magnitude so as to keep the fluctuations of the solvency ratio inside the zone (see the discussion on equalization reserves in the next section). This upper limit can also be used as an indicator in deciding

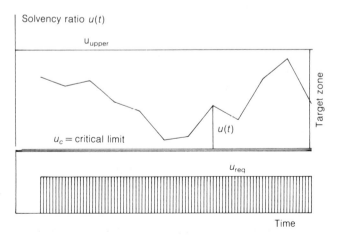

Figure 13.2.2   *Target zone for controlling the movement of the solvency ratio.*

how much of the profit can be distributed as dividends. A sound strategy is to retain sufficient resources to maintain the target zone.

It is useful to note that the level of the upper limit u_{upper} depends not only on stochastic variability, which is beyond the control of management, but also on the actions of management. Rating practice and outward reinsurance come into this category, for example. If the rates are changed quickly in the light of experience, the range of fluctuation of the solvency margin can be significantly reduced (this was demonstrated by Figure 13.1.3). Investment strategy may also have a major impact on the capital requirements measured by u_{upper}. If a high return is sought on the investments, the volatility of the return is likely to grow and there will be a possibility of capital losses. All of these contribute to the solvency margin fluctuation which influences the required size of the target zone.

(c) Equalization reserve. Notwithstanding that the range of fluctuation can be controlled to a certain degree by management action, such as reinsurance, rating, reserving, investing, etc., as discussed in the previous paragraph, it can never be totally removed. An insurance company, therefore, needs a capital reserve to cover any adverse annual results. On the other hand, it is natural that random profits should be stored in this same capital reserve to act as a buffer against the next adverse period. It depends on local conditions and tradition how the capital reserve, i.e. the solvency margin or its specified component, is constituted, for example by using book capital with hidden reserves consisting of overstatements of liabilities or understatements of assets. A solution which has been adopted in some countries is to establish a particular reserve for the purpose, usually called an equalization reserve.

A practical problem associated with the equalization reserve is its treatment for tax purposes. In fact, it will be operated as a special reserve only if transfers into the reserve can be made out of pre-tax profits, i.e. this reserve is understood as a technical reserve on a par with the premium and claim reserves. In order to obtain the consent of the taxation authorities the amount of the reserve should be limited so as to be reasonable for equalization purposes but not above that level. In fact, the problem is essentially the same as the assessment of a target zone with a well-defined upper limit (Figure 13.2.2).

Another condition is that transfers to and from the reserve should take into account genuine random fluctuation. This leads to rules which provide for that part of the underwriting profit which exceeds a long-term (trend-adjusted) average level to be transferred to the equalization reserve. Similarly, if the result falls below the average, the difference should (or may) be deducted from the reserve.

(d) References. Pentikäinen and Rantala (1982), EC credit insurance directive, Ajne and Sandström, 1991; Borregaard *et al.*, 1991; Morgan *et al.*, 1992; Hertig, 1992.

13.3 Evaluation of an insurer's net retention limits

It is useful to test the results of reinsurance considerations, for example the retention limit M obtained by the short-term methods of section 6.3, in the long-term environment. In particular, cycles can bring about accumulations of profits and losses which are not

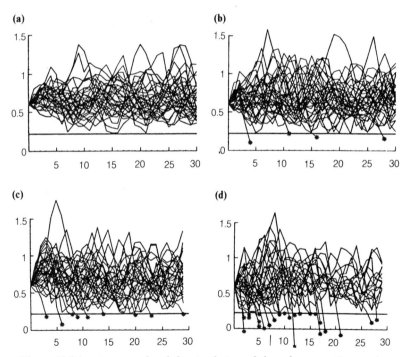

Figure 13.3.1 *An example of the simulation of the solvency ratio, varying the net retention: Case (a) $M = £0.1$ million, (b) $M = £3$ million, (c) $M = £4$ million and (d) $M = £10$ million; 40 realizations (source: Pentikäinen et al., 1989).*

fully revealed by short-term calculations. An increase in the net retention limit may, therefore, have a magnified effect.

The simulation introduced in section 12.4 is also useful for reinsurance control. Figure 13.3.1 shows an example. It is assumed that an initial amount of capital u_0 is provided to back the underwriting of a notional collective of risks (e.g. a line of business, a particular group of policies, or the whole portfolio of an insurer). Furthermore, the amount of the capital at risk (see section 6.3(b)) $u_r - u_0$ is given (in ratio form). The problem is to find a retention limit M which is as large as possible, subject to the constraint that the random loss must not exceed u_r by some predetermined confidence probability. When the type and details of the potential reinsurance treaty, as well as other relevant basic characteristics, are known, the simulation of the underwriting process can be carried out. The net retention limit is gradually increased, as stated in the caption to the figure. An appropriate order of magnitude for M can then be determined. More accurate evaluation is obtained by experimenting with values between the cases (a) and (b), and eventually running the simulation with a greater number of realizations.

CHAPTER 14

Managing uncertainty

14.1 Review of applications

(a) Development of an overall model of an insurer. In previous chapters a number of problems have been dealt with, mainly in isolation, such as the evaluation of capital at risk, reinsurance cessions, rating, reserves, run-off and investments. The potential linkages between these elements were not considered, nor the fact that an insurer is only a single player in the insurance and capital markets, so that the outcome depends to a large extent on the behaviour of the markets and not just on the decisions of the insurer itself. In this chapter the separate models will be brought together, outlining how an overall stochastic-dynamic model can be created and how it can be applied to support the management of uncertainty in a general insurance company.

First, the formulae components which were introduced step by step in the preceding chapters are summarized in section 14.2. Market aspects are dealt with in section 14.3 and ways of measuring the financial position will be considered in section 14.4. Then the necessary tools are in place to support managerial decision-making, which will be discussed in section 14.5. Finally, some aspects of regulation and public supervision will be addressed in section 14.6.

(b) Various uses of the models. The current practice of many insurers is to make statistical analyses of the flow of business and prepare budgets for future years as a basis for decision-making. Such projections may be quite modest in scope or there may be a comprehensive model structure for business planning, as described, for example, by Akhurst *et al.* (1988). As a rule, these models are deterministic, and they are not, therefore, able to cope fully with risks such as the fluctuation in claim expenditure, the return on investments, and many other sorts of uncertainty. A knowledge, both qualitative and quantitative, of these aspects is, however, essential for a proper evaluation of financial strength and viability. The idea of risk theory is to provide tools to complement traditional accounting and control systems, and to provide measures of the uncertainty. As will be seen in section 14.5, this can be done by making some key elements of the insurer's business planning model stochastic or, perhaps, by building a comprehensive overall stochastic-dynamic model to investigate important aspects of the business and to be operated in parallel and in coordination with traditional corporate analysis and planning systems.

Another area of application is **statutory solvency control** and regulation. In most countries a minimum level of capital and surplus (or solvency margin) is required to be maintained by insurance companies. Because the size of the margin should be related to the level of risk, risk theory has been applied in finding appropriate rules. For example, the solvency margins required by the European Community freedom of establishment directives are a rough approximation to the capital needs suggested by early risk-theoretical considerations (Campagne, 1957, 1961; De Wit and Kastelijn, 1980; Daykin, 1984). An early solution was implemented in Finland in 1953 (Pentikäinen, 1952) and further research is being carried out in Canada, the USA and the UK to develop more sophisticated risk-based capital requirements, taking into account a number of different risk factors (Butsic, 1992).

Solvency margin requirements may be supported by expert reports consisting of individual analyses on the state and prospects of the company. Another application of risk theory is in the construction of a framework for **equalization reserves**.

Whether or not a model is actually constructed, an understanding of the fundamental principles is very useful for actuaries and others seeking to manage uncertainty and to sell security. By linking together the various problem areas considered in previous chapters, this chapter aims to create a coherent view of insurance as a process in which external and internal administrative, economic and actuarial elements are all operating together. As such, it could also be a basic educational tool in the syllabuses of actuaries and for all involved in the technicalities of insurance and risk management activities.

REMARK Conventional corporate planning models are commonly aimed, among other goals, at forecasting the future flow of the business, usually for a short time horizon. These can be supported by stochastic elements such as are presented in this book, notwithstanding that our model is mainly intended to apply to longer time spans. Then the objective is no longer to make forecasts as such. The idea is to examine how the business in question will respond to various potential events, such as variation in the rate of inflation and in the return on investments, impulses caused by recessions or booms of the national economy, competitive pressures of the insurance market, abrupt decline in business volume, etc., and to find responses which offer adequate protection.

14.2 Basic equations

We will now summarize the models which have been introduced in earlier chapters. Their application is based mainly on the use of simulation technique, as was outlined in section 5.5 and further specified in section 12.4. The algorithms which control the flow of the business from year to year will, therefore, be the basic working tools.

(a) Cash flow equation. The comprehensive models are based on the cash flow equation (1.1.1) and on the accounting equation (1.2.5). The former is now recapitulated as follows

$$A(t) = A(t-1) + B'(t) + J'(t) + U_{new}(t) + W_{new}(t) \\ - X'(t) - E(t) - E_m(t) - E_{tax}(t) - D(t) \qquad (14.2.1)$$

and the accounting equation takes the form

$$U(t) = U(t-1) + B(t) + J(t) + U_{new}(t) \\ - X(t) - E(t) - E_m(t) - E_{tax}(t) - R(t) - D(t). \qquad (14.2.2)$$

The entries in these equations were defined in previous chapters. For the convenience of readers their meaning is reiterated and supplemented with references to the sections where their definitions can be found in more detail.

- A is the amount of assets (sections 1.1(b) and 8.6).
- B' and B are written and earned premiums (sections 1.1(b) and 1.2(b)). They may be modelled in the way outlined in Chapter 10. Examples of linkages are the simple market impulse rules ((10.3.1) and (10.3.2)) and the dynamics discussed in section 10.2(g) (see also example (13.1.8)).
- J' and J are the investment income and return on investments, the latter including both income from the assets and changes in their value (capital gains or losses) (section 1.1(b)). Modelling the return on investments was described in Chapter 8. The investment

strategy is incorporated into the overall model via these entries as an important source of dynamics and of risk.

- U_{new} is any new equity capital issued and W_{new} any new borrowings.
- X' and X are paid and incurred claims (section 1.2(c)). Their specification and simulation was considered in Chapter 9.
- E and E_m are general and special expenses (11.1.5). The latter will be activated only in the case of a special sales campaign or to cater for some other similar situation where the normal level of expenses presented by $E(t)$ is clearly exceeded. The possibility of using this particular entry contributes to the dynamics of the model. Evaluation of the return from a particular investment in administrative or selling costs will usually involve several entries in the basic equation (10.3.2).
- E_{tax} represents taxes (section 11.2) and
- D dividends (section 11.3).
- $R(t)$ gives the net cost of reinsurance cessions. It is useful for evaluating the cost of alternative reinsurance forms and retention levels. This approach presumes the use of gross of reinsurance values of variables throughout the model. Another approach is to use variables net of reinsurance. This is the common practice in risk theory. Then the R term disappears from (14.2.2) and the actual cost of reinsurance is hidden as unspecified reductions in other variables. Notwithstanding book-keeping practice, which is in any case moving towards more explicit disclosure of quantities both gross and net of reinsurance, the decision whether or not to reinsure, and to what extent, is largely within the control of management. It can be seen as one of the tuning devices in the overall dynamics of the insurance business, and should as far as possible be modelled separately (sections 6.3 and 13.3).

Inflation should be generated as a primary background variable which may be required in many of the other sub-models as discussed in Chapter 7.

A major problem is the interaction between the market and the insurer of concern. This will be discussed in section 14.3.

Another challenge is to find **dynamic control rules** which direct the flow of the solvency margin, investment strategy and other key variables in a way similar to that likely to be achieved by the actual management of the company in question. This leads to the treatment of business strategies in section 14.5.

REMARK   The treatment of the basic equations may depend on whether the technical reserves are discounted or not. In the former case particular corrective terms may be required, as was exemplified in equation (9.5.16), in order to achieve an exact balance of inflows and outflows.

14.3 The insurer and the market

In the preceding chapters insurers were considered in isolation, without any explicit regard for the interplay between different insurers operating in the same competitive market. Market effects have been introduced implicitly, since market dynamics are one of the most important causes of insurance cycles. This constraint will now be relaxed. It should be appreciated, however, that a complete and strict simultaneous treatment of several insurers taking decisions independently adds considerably to the dimensions of the problem, making it largely intractable, even with the use of simulation techniques. In fact, stochastic studies of insurance markets are still at a very preliminary stage and awaiting substantial further research efforts. Market effects are, however, too important to leave out and it is worthwhile adopting even very approximate approaches to explore the possible impact.

(a) One-unit formula. This very simple approach assumes that the insurer is able to ascertain satisfactorily the 'correct' premium level $b_0(t)$ (sections 10.2 and 10.3). The assumption is now made that, if the actual premium level $b(t)$ clearly undercuts this 'correct' level, it will attract new policies from the market, or, if $b(t)$ exceeds $b_0(t)$, policies will lapse or not be renewed. A rationale for this is that market prices, at least on average in the long run, have a tendency to adjust to the correct level. The elasticity of supply and demand would also suggest this sort of mechanism. It is worth recalling (remark of section 14.1(b)) that simulation models are often used to test the general resilience of an insurer to various types of events, rather than to make any specific forecasts of future scenarios.

The above reasoning can be represented by a formula which evaluates the relative change d_{vol} of the relevant business volume as a function of the relative deviation in the premium levels,

$$d_b(t) = \frac{b(t) - b_0(t)}{b_0(t)}. \qquad (14.3.1)$$

Pentikäinen *et al.* (1989) used the following illustrative rule

$$d_{\mathrm{vol}} = \begin{cases} 0 & \text{for } |d_b| \leqslant a \\ -g \cdot (d_b - a) & \text{for } d_b > a \\ -g \cdot (d_b + a) & \text{for } d_b < -a \end{cases} \qquad (14.3.2)$$

where a is a threshold parameter, which was assigned the value 0.1, and g an elasticity (gearing) factor, given the value 2.

This formula is a straightforward modification of the general rule (10.3.1).

(b) Two-unit market model. A more advanced approach can be developed to reflect the situation where the market is not affected to any noticeable extent by the pricing of the insurer in question. This could be the case if the market is very diversified or if the company is small compared to the whole market. The idea is to simulate both the market and the insurer in parallel (Figure 14.3.1).

The market gross premium level $b_m(t)$ (for each liability type), cycles, profit and other relevant features are first generated as if the market were a large insurance company. The methods outlined in section 14.2 apply. Next, the behaviour of the particular insurer is simulated. The idea introduced in the previous section can be employed, but now making the linkage with the market situation. If the premium level $b_i(t)$ of the insurer for a particular risk deviates significantly from the market premium level, the insurer can be expected to gain or lose business. (10.3.1) is an example of a simple formula for evaluating the relative change in the volume of business as a function of the difference in premium rates. As stated at the end of the previous section, rule (14.3.1) is a modified form of this and can be applied to this situation by replacing the deviation in premium levels from the 'correct' premium level by the deviation of the insurer's premium level from the market.

$$d_b = \frac{b_i(t) - b_m(t)}{b_m(t)}. \qquad (14.3.3)$$

Then rule (14.3.2) can be applied. Daykin and Hey (1990) used values

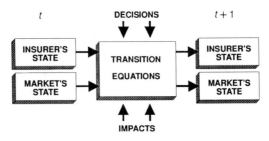

Figure 14.3.1 *Processing an insurer and the market simultaneously.*

$a = 0$; $g = 2$ or 1.5, whilst Pentikäinen *et al.* (1989) used $a = 0.05$ and $g = 2$. The g values may be different for the cases $d_b > a$ and $d_b < -a$.

Daykin and Hey (1990) have proposed further modifications to (14.3.3). First, it must be appreciated that it is often expectations rather than actual values which affect decision-making. This can be approximated by multiplying b_m by a factor $1 + g_s$, where g_s should be calibrated to describe the anticipated growth (or reduction) in market levels.

The movements in the level of market rates reflect the profitability of the market. If excess profits have been accumulated, it can be expected that they will be returned to policyholders, through reductions in premium rates, or smaller increases than might be justified by inflation and trends in claim costs. On the other hand, in the case of profits falling below target, any shortfall may be recouped from future premium increases. Daykin and Hey (1990) proposed adjusting the market premiums for deviations from the target level of profits in the previous year $t - 1$ (formula (23) in Appendix 1 of Daykin and Hey (1990)).

In order to make the model work, rules are required to show how the company premiums will be modified to react to market changes, i.e. what competitive strategy is adopted by the management. A simple approach is to assume that, in order to maintain its competitive position, the company follows the market. The procedure adopted can have a considerable impact on the stability of the business, as Figure 14.3.2 demonstrates. The degree of fluctuation, both of the annual results and, still more, the solvency ratio $u(t)$, has a tendency to increase significantly if the market is followed strictly, compared with the (fictitious) situation where the company operates in isolation without any market influences.

The assessment of company premium rates in general, and in particular in relation to the current levels in the market, is one of the key issues of any corporate model. The degree of flexibility depends, self-evidently, on the financial status of the company. A strong company has greater freedom for management to choose between various alternative actions in determining corporate strategy, as will be considered later in section 14.4(e). On the other hand, if the company is weak, problems may arise during low phases of market cycles and management's scope for action may be severely restricted. It is advisable, therefore, to allow plenty of flexibility in model-building, so as to permit users to adapt the procedure to any particular situation.

(a) 1.5

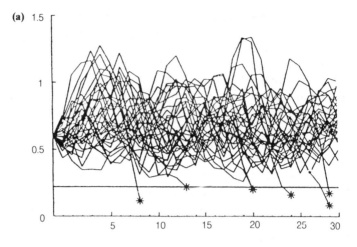

(b) 1.5

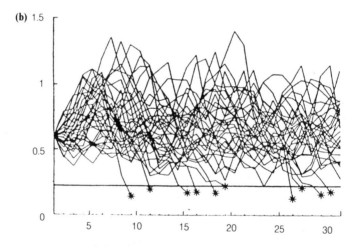

Figure 14.3.2 *Simulated solvency ratios (a) of a company operating in isolation, and (b) strictly following movements in market rates (an example given in Pentikäinen et al., 1989).*

One approach, proposed by Daykin and Hey (1990), is to define the company premium level relative to the market premium level by a link function $h(t)$, to be defined according to the current local circumstances:

$$b_i(t) = b_m(t) \cdot (1 + h(t)). \qquad (14.3.4)$$

For short-term simulations $h(t)$ can, perhaps, be taken to be a suitable constant. A simple formula for longer-term simulation is to relate it to the solvency ratio $u(t)$, for example by making it proportional to the deviation from the central level of the target zone presented in Figure 13.2.2

$$h(t) = c_h \cdot [\tfrac{1}{2}(u_{\text{upper}} + u_0) - u(t)], \qquad (14.3.5)$$

where c_h is a suitable constant coefficient.

More flexible variants can be constructed using the idea of dynamic programming which was outlined in section 5.5(d) and in Figure 5.5.3.

Another approach is first to simulate both the market and the company in isolation and then to take a mixture of the premiums

$$b(t) = c \cdot b_i(t) + (1 - c) \cdot b_m(t) \qquad (14.3.6)$$

where b_i is the insurer's premium level evaluated as in section 14.3(b). A value $c = 1$ for the distributor coefficient gives the pure company premium, whilst $c = 0$ gives the market premium; any mixture might be deemed suitable for intermediate values.

It is useful to recall the type specifications symbolized by (10.1.1). For corporate planning, the most general type is appropriate, incorporating as one control mechanism among numerous others in the implementation of the adopted corporate strategy. It is often difficult (and unnecessary) to express the ultimate solution in the form of a mathematical formula. Instead, it is advisable to experiment with various alternatives and, by trial and error, to look for an approach which fits both the market and company situation, as well as meeting the goals of the management.

In the case of multi-line insurers the above considerations should be applied separately for the different lines and the results then summed in order to arrive at the overall picture. However, lines with similar characteristics can be combined and low volume lines may be assimilated with other lines. Reference can be made to section 3.2(d), where the combination of different lines is discussed.

REMARK 1 The actual situation is usually much more complicated than the simple formulae (14.3.2) and (14.3.3) suggest. For example, competitive price reductions may relate only to one particular line. If a policyholder moves his policy of this type to the company offering the reduction, he may also move his other classes of insurance, irrespective of whether or not he can obtain any rate reduction for them.

The possibility should furthermore be kept in mind that competitors may follow suit in reducing premiums. Then the competitive advantage may be lost. These aspects open new avenues for considering the behaviour and interaction of insurers operating in the same market, as will be outlined briefly in section 14.3(c) below.

REMARK 2   Defining what a market is and what a market price is may also be a problem. This is because there may be frequent entries to and withdrawals from the market. Furthermore, a multinational company often gets only a small part of its premiums from a particular market. Such a market has only a marginal effect on its solvency and, vice versa, the effect of the company on the market may be negligible.

(c) Multi-unit models.   If the market is small, or if there are just a few leading companies, the action of any one of them can provoke reactions among the others. Then the validity of simple models is called into question. For example, increased sales efforts by any one of the insurers may very soon lead to reactions from the other insurers. If an insurer reduces premium rates for the purpose of acquiring additional new business, the outcome will differ greatly according to whether competitors follow suit or not. The elaboration of future scenarios requires more complex models than we have used up to now.

The idea can first be illustrated by a simple example, where three insurers I, II and III are present (Figure 14.3.3). The movements of portfolio volume for each company, measured by the premiums $B(t)$,

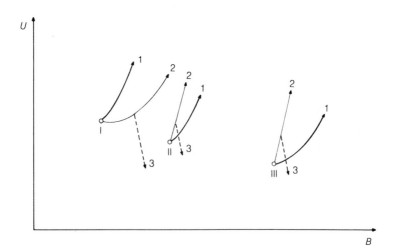

Figure 14.3.3   *Competitive market of three insurers I, II and III. The curves represent the expected flows of the insurers in the B,U-plane.*

and their respective financial strength, measured by the solvency margin $U(t)$, are both relevant, as well as their interrelationship, so it is helpful to use a B, U-diagram (section 12.4(i) and Figure 12.4.5). The initial positions of the three companies are marked by small circles.

First let us suppose that no special competitive activity is undertaken by any of the companies. Then the flow of business of the three companies may be as plotted by the thick curves marked 1. Each company has normal growth in both volume and solvency margin. Next, it is supposed that company I launches a sales campaign, making use of premium reductions, advertising, etc. If the other two companies do not react, the outcome might be as plotted by the curves marked 2. Company I benefits from a considerable growth in volume. The solvency margin may fall at first as a result of the campaign expenses, but the increased portfolio later leads to profits and a recovery. On the other hand, assuming that the market in question is relatively small and almost saturated, and that each insurer has a fairly sizeable share of it, the action of insurer I will be reflected in a reduction of the volume growth of its competitors.

Alternative 3, plotted by dashes, describes the possibility that, after a time lag, companies II and III will react with a similar sales effort, reducing rates and increasing advertising. It can be expected that the sales efforts of all three companies will neutralize each other and that equilibrium will again be reached, after some fluctuations. As a consequence, none of the companies will end up with any extra growth in volume. Moreover, the additional campaign expenses which they have incurred will leave each company weaker than before. Continuation of this process could soon create a critical situation in one or more of the companies and necessitate changes in business strategy to stabilize the position. Market mechanisms of this kind may be one of the major causes of cyclical variation.

Taylor (1986a) raised the question of whether it is wise for an insurer to follow the trend in market premiums during the low phase of the market cycle when the business can be expected to make a loss. Another strategy would be to keep the level of premiums at or above the break-even point, notwithstanding the possibility of losing business to competitors who are undercutting this level. When the market turns up, one can speculate that some of the lost business can be regained, since competitors may need to load their rates excessively in order to recoup earlier losses.

(d) Simulation of competition.   To make the multi-unit model workable, information is needed for each market unit and also formulae for the behaviour of the market shares. It is appropriate to adjust the response functions, as was illustrated by (14.3.2) and (14.3.3). The average level of premium rates of the insurers involved can be determined and used as the market premium $b_m(t)$.

Although models using general marketing theories are well developed and frequently applied to other industries, there is, surprisingly, a nearly complete absence of insurance applications, as Jewell (1980) stated in his model survey. Even though it is difficult, it may be possible to gain experience by trial and error, to get some idea of the very complicated interactions. One reason for the lack of literature is the tendency not to publish data of this kind. Annual company reports and general insurance statistics may possibly be of use as sources of information. For illustrative purposes the development of the market shares of five Finnish motor third-party liability companies is depicted in Figure 14.3.4.

In 1973 and 1974 company I applied rate reductions and gained market share until the rate reductions were matched by competitors. The market reactions depend, among many other things, on the degree of saturation of the market. An extreme case would be a

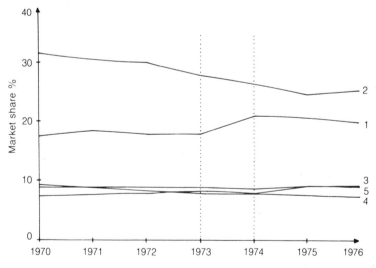

Figure 14.3.4   *Third-party motor insurance. Trend in market shares of the five companies as percentages of the whole market (Pentikäinen, 1979).*

mandatory insurance class such as motor third-party liability insurance. Because no level of market effort can increase the overall volume of business, growth of market share is possible only at the expense of other insurers. In the case of most voluntary classes of business, marketing efforts can be made to find potential customers who have not had insurance cover. In such a situation increased sales activity may, at least in part, increase the overall size of the market, as well as taking market share from competitors. Only experience can help in the construction of suitable response formulae.

(e) The analysis of multi-unit markets belongs to the extensive subject of **multiplayer games** and theories of economic behaviour. For example, there are theories of oligopolistic markets (Friedmann, 1977) where problems concerning collusion, equilibrium, etc. are examined. If the strategy changes are made at short time intervals the scenario is akin to systems which are treated by the theory of **differential games**, a survey of which has been edited by Grote (1975).

Both for business planning and for educational purposes so-called **business games** have been developed. For example the Staple Inn Actuarial Society in London has regularly organized a non-life insurance business game (Thompson, 1979). Teams play the role of the management of different enterprises and their decisions are simultaneously fed into a computer program to give the market reactions as output. These are then fed back to the teams for the next set of decisions to be made. The models outlined in this chapter could well be used for such purposes. If stochasticity is programmed into the model, it could give a new dimension to conventional game models, which are mostly deterministic.

The business game idea can also be applied for a company's internal use. The insurer can simultaneously 'play' his own company and all the relevant competitors. Then alternative strategies and counter-strategies can be analysed, as well as the consequences of potential outside events such as inflation, changes in capital markets, etc. which may affect different insurers in different ways. The example in Figure 14.3.3 illustrates the idea.

It is, of course, a major problem to program the behaviour of competing insurers. It is probably best to use quite simple models. A standard model could be constructed with a fairly large assortment of parameters to determine size, quality, strategy, etc.; values of these parameters can then be specified for each of the relevant insurers.

14.4 Measuring and managing financial strength

The importance of the financial strength, or the financial health, of
an insurer is self-evident, but it is not a trivial problem to find
appropriate ways of measuring the position and defining what is
satisfactory. In this section methods of measuring financial status
are considered first and then tools for monitoring and safeguarding
financial strength, as well as some related aspects. These are steps
in preparation for section 14.5, where they will be used as building
blocks in the management decision-making process for the company.

(a) Scenario technique. An approach which is in principle straightforward but in practice often rather laborious is to build a model
to describe the company of concern and project future outcomes.
The techniques outlined in previous sections can be applied for this.
By choosing various sets of assumptions (**scenarios**) the outcomes
can give an idea of future prospects in a variety of different circumstances, test the resilience of the company to adverse situations and
explore the properties of different strategies, as will be described in
a little more detail in section 14.5.

In Canada, for example, the appointed actuary of a life or non-life
insurance company is required to carry out a dynamic solvency test
each year in which a number of specified scenarios must be examined.
The actuary is also required to elaborate additional scenarios which
might be particularly challenging for the company (Brender, 1988;
Canadian Institute of Actuaries, 1992). Similar ideas of scenario-
testing are being developed by actuaries in the USA, usually referred
to there as **cash-flow testing**.

Even though the scenario technique, if performed well, can give
quite good information about the prospects of the company, it is
useful to be able to summarize the key features in a concise form as
indicators, for example as will be proposed in the following sections.

(b) Static measures. We will use the solvency ratio $u(t)$ (section
12.3(a)) as the main indicator of the financial strength. However, it
is more important to know how adequate it is for the maintenance
of solvency, i.e. to meet the various risks which could jeopardize the
company's position, than to know its absolute level. The target zone
introduced in section 13.2(b) may be useful for this purpose. The
position of $u(t)$ within the zone is a natural measure. A position near
the upper limit u_{upper} of the target zone indicates a good financial
status, and a position close to the lower boundary u_c a poor one.

This reasoning suggests the following measure as a 'yardstick'

$$y(t) = \frac{u(t) - u_c(t)}{u_{\text{upper}}(t) - u_c(t)} \cdot 100\%. \qquad (14.4.1)$$

The merit of this measure is that it is related to the individual risk structure of the company. To illustrate this by a numerical example, let the solvency ratio $u(t)$ (i.e. the solvency margin expressed as a percentage of the premium income) be, say, 50% for a company insuring a large portfolio of small homogeneous risks which are not very sensitive to business cycles, such as house-owners' or motor-car risks. For such a company the yardstick $y(t)$ corresponding to this level of solvency margin may commonly be of the order of 40% to 60%. On the other hand, the same solvency ratio value of around 50% would probably be inadequate for a company insuring big industrial plants, marine and aviation risks, etc. with a high net retention level. With the same solvency ratio the measure $y(t)$ in this example might be, say, only 10%. This indicates how misleading it would be to use the ratio $u(t)$ as the sole indicator.

(c) Dynamic measures. In those classes of insurance where market-related business cycles are strong, the indicator problem presents difficulties. As Figure 12.2.2 clearly shows, for example, insurers commonly have high solvency ratios during the high phase of the cycle and much lower ones in the trough. This feature is illustrated in Figure 14.4.1, by placing the fictitious insurers A and C in different positions and using the yardstick (14.4.1) as the scale of the vertical axis.

The measure $y(t)$ for the insurer A at position A_1 is about 80%, but, at the trough position A_2, it is no more than about 15%. Looking only at these numbers, one might be tempted to conclude that the insurer had suffered a serious adverse development. However, having regard to the market cycle, the situation might not be too alarming, because the position has not worsened in relation to major competitors and it is probable that the market cycle will soon turn up.

Another important feature can be demonstrated in Figure 14.4.1 by inserting a fictitious company C into the picture. It has a low solvency yardstick at the phase of the cycle when most competing insurers are much higher. When evaluating the prospects of this company at that time, it might be justifiable to regard the position as unsatisfactory. The level of market premiums will probably decrease. Competitors will be able to survive the anticipated poor results by

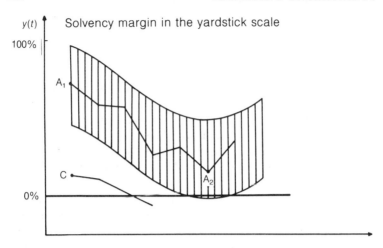

Figure 14.4.1   *The position of fictitious insurers A and C during a high and low phase of a cycle.*

allowing their solvency margin to fall. Insurer C has much less scope to do this.

The above example illustrates that a serious difficulty with a yardstick of the type (14.4.1) is finding a proper way to take cycles into account. One solution would be to extend the range $u_{\text{upper}} - u_0$ in the yardstick definition (14.4.1) so that the whole cycle, together with short-term random fluctuation, appears in the target zone. Figure 14.4.2 demonstrates this idea, with R_0 representing the static definition of the denominator of the yardstick, R_{cycle} the amplitude of the cycle in the solvency ratio and R_{tot} the expanded definition of the denominator of the yardstick.

However, the real situation would not be revealed very well in this way, unless guidance was given as to the expected range of the yardstick at any particular moment. Another approach would be to let the yardstick range $[u_{\text{upper}}, u_c]$ move downwards and upwards according to the cycles, as shown by the shading in Figure 14.4.2. This would result in what might be called a **dynamic yardstick**. Even though the construction of such a sliding scale might be difficult in practice, the philosophy behind it is useful to appreciate as a basis for an ad hoc assessment of the actual financial position of any insurer. In other words: it is not sufficient to look only at the data and indicators of a company in isolation; the phase of potential

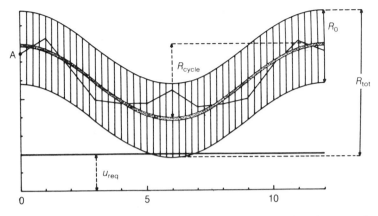

Figure 14.4.2* *The behaviour of the solvency ratio u(t) in broad terms. The bold line depicts a fictitious market cycle and the shaded tube area within which the solvency ratio of most insurers is likely to move. The weak broken line represents the flow of a particular insurer A.*

cycles and other aspects relating to the insurance and capital markets should be kept in mind.

What was said above regarding insurance markets also applies to investment markets. For example, Figure 8.5.3 shows an abrupt fall of 50% in equity prices. A number of insurance companies were probably 'technically insolvent' at that point. However, no action was taken to wind them up as it was expected that the market would recover; indeed the extent of the market recovery was already known before the accounts were published.

It is worth noting again in this context that it is essential to include the hidden reserves in the solvency ratio $u(t)$ in a reliable way when (14.4.1) or any other indicator scheme is applied. In fact it is likely to be an essential part of the solvency margin, and the part which can in practice most easily be used to equalize business fluctuations. This suggests that agreed actuarial standards should be applied in valuing both the assets and the liabilities in order to find a baseline by reference to which the hidden reserves can be measured. Without consistent application of standards, many indicators may be meaningless. Regardless of whether they are disclosed or not, they are indispensable for the insurer's own internal use (and for the supervisors).

REMARK It has frequently been observed in *a posteriori* investigations of bankrupt insurance companies that the liabilities had been significantly understated for a period before the formal discontinuation of activities. It is clear that correct disclosure of the 'negative' hidden reserves could have reduced the eventual losses, either by accelerating the discontinuation of business or, perhaps, by giving the signal to take remedial actions at a time when it could still have been effective.

(d) Early warning systems. Even though the correctly evaluated solvency ratio, and the more sophisticated versions such as (14.4.1), are good indicators, the overall financial situation of an insurance company is too complicated to be described fully by a single indicator. Insurers, rating bureaus and supervisors, have, therefore, developed sets of indicators which focus on different aspects of the business. Trends in these indicators can give some idea of the company's financial strength and serve as early warning signals for impending trouble. It has been stressed above that the insurer's position relative to others in the market is one of the relevant characteristics.

A well-known example of an early warning system is that applied by the state insurance supervisory offices in the US (a concise description by Hachemeister is given in Appendix C of Pentikäinen *et al.* (1989)). It comprises 11 such ratios, with **acceptability ranges** for each.

(e) Discussion on the importance of adequate financial strength. It may be useful, in anticipation of the description of managerial decision-making in section 14.5, to provide an overall view of some consequences of different levels of financial strength.

Adequate financial strength is a major advantage in insurance markets, allowing greater latitude in strategic planning, for example, when dealing with marketing issues. It enhances capacity to write business, both reinsurance and direct, and also to accept new types of business, for example high risk business which may nevertheless offer the prospect of excellent returns. It makes it easier to retain clients during a low phase of the market cycle, notwithstanding the fact that profitability may temporarily be poor.

Security is an important aspect which policyholders (and brokers) should take into account in choosing an insurer. Many market analysts and rating bureaus look at the security of insurers and publish ratings or solvency assessments. A strong insurer can expect to receive more business, and more profitable business, than weaker competitors. Security aspects have become particularly important in reinsurance markets owing to the scope for very large losses and recent unfortunate instances of reinsurance failure. A company which is financially strong can charge higher premiums.

Financial strength also makes it possible to maintain a relatively high net retention as regards outward reinsurance or retrocession and to accept larger risks by way of inward reinsurance. This increases the scope for profitability.

The share price of a proprietary insurance company depends on its financial strength. This feature has been explored by Reid (1984). A strong company has a better chance of acquiring new share capital. Moreover, owing to the minimal risk of losing the share capital, shareholders may be satisfied with somewhat lower dividends than would be required in cases where a potential loss of money may be perceived.

One should appreciate that maintaining a good level of solvency may involve a certain cost. The solvency margin, or at least a substantial part of it, must be readily available to cover fluctuations in cash flow. This liquidity requirement may worsen the yield obtained. On the other hand, those aspects dealt with in previous paragraphs may increase overall profitability. It is not possible without detailed study to conclude which of these opposing factors carries most weight (Pentikäinen *et al.* (1989), sections 5.2(i, j)).

It is useful also to look at the consequences of poor financial strength. They are very much mirror images of the benefits of a good position. The need to acquire new clients and to preserve the existing portfolio may lead to trade-offs in regard to the price and the quality of insured risks which would worsen profitability further. A weak insurer may have difficulty in engaging and retaining qualified staff. Reinsurers may shun such an insurer, market analysts may publish unfavourable indicators, and the possibilities of offering a good level of service are restricted. Moreover, the continued existence of the insurer is vulnerable at all times to adverse phases of the market cycle and to pure random fluctuation. All in all, poor financial strength creates difficulties which may have a cumulative effect and could in the worst case be fatal. It is important, therefore, to be aware of the problems well in advance and to know the company's actual financial position and its likely trend sufficiently early to be able to take remedial action before the business has fallen into a lethal spiral. Early warning indicators are not only necessary for supervisors; the management of the insurance company needs them even more!

(f) Tools for building and maintaining financial strength. Nearly all of the actions described in previous chapters have a direct or indirect

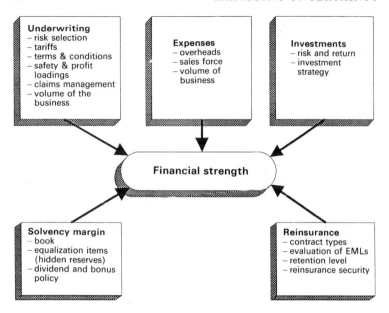

Figure 14.4.3   *The factors on which financial strength rests.*

impact on the total financial position of an insurance company, or as we have termed it, on its financial strength. This is illustrated in Figure 14.4.3 which expresses the basic equation (14.2.2) in diagrammatic form.

The various entries in the diagram are the tools and devices which insurance company management has available to control the operations and activities of the company.

Even though it might be possible to analyse the individual entries successfully, it is a major problem to contemplate them all together. There is quite complicated interplay between the entries. Moreover, background factors such as inflation, market movements, etc. simultaneously affect several different aspects. A high level of solvency margin permits higher net retentions and consequently cheaper reinsurance; a poor margin requires more conservative reinsurance arrangements. A good return on investments gives greater latitude in premium rating, as was discussed in section 14.4(e), etc.

All in all, we are dealing with a very complicated multi-variable system. The model which was outlined in section 14.2 provides a tool for managing it. Notwithstanding the fact that many particular problems can also be handled by conventional analytical methods, the overall model is tractable only by simulation.

(g) Unquantifiable risks. In previous chapters and sections the analysis of the risks which threaten to jeopardize an insurance company's financial soundness has been focused on the insurance process, including underwriting, investment performance and asset risks. In practice an insurance company also faces a number of other risks, which may be quite critical, often more so than the normal insurance risks discussed so far, and which are by their nature unquantifiable and unpredictable.

A frequent reason for an insurer getting into financial difficulties is management incompetence: more specifically, misfeasance or even malfeasance. It may be seen, for example, in overambitious efforts to conquer market share or to expand into new markets, speculative investments, deficient underwriting, such as the acceptance of risks with possible moral hazard, poor product design, inadequate rating and deficient reporting and control systems.

Other potential hazards within a company are, for example, inadequate computer security, accidental data loss and fraud perpetrated by employees or directors.

External hazards which may not be readily quantifiable are, for example, an unexpectedly high inflationary shock and sudden realignments in currency markets. Although partly considered in section 12.3 in relation to the modelling of cycles, a recession in the national economy, or a slump, may adversely affect an insurance company in several different ways; various classes of asset may simultaneously experience losses, premium income may shrink at the same time as claim outgo increases, etc.

Terrorism and criminal acts could present further problems and a sudden deterioration of financial condition could be brought about by the insolvency of a major reinsurer. Changes in judicial practice and interpretation have become a major problem, particularly in the USA for liability business, medical malpractice and environmental risks. Sudden changes in taxation may also create unexpected problems.

These kinds of risks are unquantifiable and are generally impossible to predict. The 'Failed Promises' report of the Dingell Committee in the USA found many of the major causes of insolvency in the companies investigated to involve management failures, ranging from poor management control through incompetence to fraud.

This being so, it is reasonable to question whether there is any point in making sophisticated calculations of the normal quantifiable risks when many of these highly significant and potentially dangerous factors cannot be handled in this way. One answer to this question with regard to unquantifiable risks is to organize adequate audit and internal control within insurance companies, supplemented by effective public supervision, including aspects such as whether the directors and managers are fit and proper persons to manage an insurance company. If internal and external supervision is lax, no amount of sophisticated modelling can guarantee the safety of the commitments. On the other hand, supervision cannot and must not take away from managers the responsibility for good underwriting and for controlling claims and for managing changes in premium rates to cope with inflation and with underwriting cycles. These factors necessitate good financial management of reserves and adequate solvency margins, including, perhaps, equalization reserves. Satisfactory control can only be achieved by a combination of technical skills, good management and firm supervision.

Of course, reserves and margins which may have been set aside primarily for underwriting risks can in an emergency be drawn on to cover the costs of unquantifiable risks. An effective combination of controls, as suggested above, should limit the extent of possible damage so that the probability of exhausting the reserves and the solvency margin remains slight. Further aspects of public regulation and national compensation fund schemes are discussed in section 14.6.

Many of the unquantifiable risks are not peculiar to the insurance industry; in fact, they are common to all business ventures. Managing quantifiable risks is, however, a special feature of insurance, being different from other industries.

14.5 Corporate planning

In previous sections we have considered how to build models to explore the prospects of an insurance company and to analyse the uncertainties involved. We have considered how to assess financial strength and the implications for managing the flow of future business. We have not discussed, however, how these tools should be used in the management of uncertainty. In this section we enter the realm of what is known as corporate planning. We point to ways in which risk theory methods may be used to support overall managerial decision-making. First, it is necessary to consider management goals and how they can be linked to the models which have been outlined above.

(a) Classical strategy approaches. In earlier literature the **maximization of profit** was often assumed to be the only, or at least the fundamental, business goal. When applied to the insurance environment, this might be the point of view of an investor who owns shares in an insurance company, expecting a profit that is competitive with other investment possibilities in the capital market. Some authors, for example, de Finetti (1957) and Borch (1963), have dealt with this kind of problem by focusing on the maximization of the discounted value of future profits or what in modern terminology might be described as the **appraisal value** of the company as a going concern. The simulation technique which has been outlined in previous sections lends itself to this type of analysis. Future profits should first be generated and then rules given about what proportion may be paid out as dividends $D(t)$. The discounted sum of the terms $D(t)$ $(t = 1, 2, ..., T)$ is a realization of the required appraisal value, and by the Monte Carlo method the d.f., mean value, standard deviation, etc., can be evaluated. This idea will be developed in more detail for life insurance in section 15.2 (15.2.9).

From this perspective, insolvency has only an indirect role to play, in so far as it ends the flow of profits. Limited liability of equity owners ensures that shareholders have no personal liability in the event of the insurer's insolvency; the value of their investment simply falls to zero. It may be worth noting, incidentally, that this is not the case for names at Lloyd's, who are personally responsible for losses down to their last penny. Shareholders cannot be expected to be particularly concerned for the policyholders or beneficiaries in the event of insolvency. An optimal shareholder strategy may be one which takes out high levels of profit from the insurance undertaking in the short-term, but fails to let the solvency margin grow sufficiently to guarantee longer-term security. In fact, a low level of capital may generate a higher rate of return for shareholders than a higher level of capital, since profit may not be proportionate to the capital supplied. If the investor has already had a good return in terms of dividends and has diversified his own investment porfolio well between numerous insurance, industrial, commercial, etc. equity shares, insolvency of an insurance company may not be of undue concern.

This kind of strategy would certainly be found wanting from the point of view of policyholders and the public interest. It is not really acceptable from the point of view of the insurance industry, because it would tarnish the reputation of the whole industry. A further consequence might be that the eventual losses would fall to be met by sound insurers and their policyholders, for example within the framework of a national guarantee system.

It is generally accepted nowadays, not only in the insurance industry but also more widely, that profit maximization is too narrow as an unqualified business goal.

Substantial private entrepreneurs are disappearing from the scene and most large enterprises are now run by professional managers. Galbraith (1973) describes in sarcastic terms how the interests of professional managers may not be so much in the maximization of dividends to a large and often diffuse group of shareholders, but rather in safeguarding the stability and continuity of the enterprise. A common status symbol is to achieve a strong rate of expansion in business volume. It should also be borne in mind that many insurance companies are mutuals, in some countries a substantial part of the market. The Galbraith view applies even more to them than to proprietary companies.

In the real world, however, the relationships are more complicated than the simplified Galbraith doctrine might suggest. For instance, the interests of shareholders and management can, at least to some degree, be aligned by giving the managers bonuses which are proportional to dividends, by giving them options to buy shares, or by setting remuneration according to some measure of added value. The prevalence of hostile take-overs can have similar effects. On the other hand, it can be argued that trends of this kind may lead to an unhealthy short-termism, with efforts focused on increasing the short-term returns at the expense of measures which would be desirable to safeguard long-term survival.

(b) Strategic triangle. There are alternative schools of thought regarding an optimal business strategy. The desired future outcome can be formulated as a set of goals, with a programme for achieving them. Financial strength is one of the goals. In fact, it is always present as a necessary, but not sufficient, prerequisite, because the other goals can only be achieved by assuming that the insurance company continues in existence!

A common approach is to condense the essential business goals in the form of a strategic triangle as set out in Figure 14.5.1.

The financial strength entry implies as a partial goal the **safety**

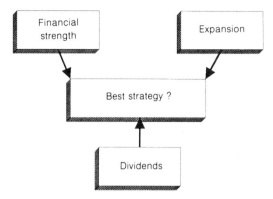

Figure 14.5.1 *The strategic triangle showing three principal aspects in the definition of business goals.*

first principle, i.e. in terms of section 13.1(c) the probability of ruin $\Psi_T(U_0)$ should be small

$$\Psi_T(U_0) \leqslant \varepsilon \qquad (14.5.1)$$

where the limit ε and the time span T are input parameters of the overall model. Any strategy which does not satisfy this condition should be rejected.

The triangle diagram formally describes the three principal goals as alternatives which may be mutually contradictory. In the long run this need not be the case, as was evident in the discussion on the consequences of financial strength or weakness in section 14.4(e). For example, the line of thinking, whether explicitly defined or subconscious, may be as illustrated in Figure 14.5.2 (Pentikäinen *et al.* (1989) section 6.1(d)). Good financial strength enables the business

Figure 14.5.2 *Strategic working schedule.*

to grow, and this can be expected to increase future profits. Then financial strength will be further reinforced, etc. Hence, the goals presented in Figure 14.5.1 are likely to support each other, providing, of course, that the newly acquired market share is composed of good risks and premiums are soundly rated, etc.

If the business goal of an insurer is a combination of partial goals, for example as was exemplified in Figure 14.5.1, the question arises as to whether it would be possible to formulate them as exact rules, for instance, by using mathematical formulae that could be incorporated into a model as dynamic control modules. Simple examples are given by the premium control formula (13.1.8) and the rules demonstrated in Figure 5.5.3. In particular, in long-term simulation models, at least one such control device seems to be necessary, as was demonstrated by the example of Figure 13.1.3. It might not be too difficult to express the business targets in terms of rules which operate as control signals at each stage of the simulation process, according to the then current position.

A conventional approach for handling business goals in the econometric literature is to introduce the concept of **utility** and to construct utility functions to measure the desirability or otherwise of business outcomes (see section 6.6.5(b)). These are usually simple one-variable functions, with the single variable commonly being taken as wealth, or in our case the solvency margin. This is in effect an extension of the profit maximization approach mentioned above. In fact, the latter is equivalent to the special case where the utility function is linear. However, as was stated above, this approach is too narrow in its scope.

The shortcomings of one-variable utility functions can be addressed by introducing multi-variable utility functions (Kotler, 1975). A different variable could be attributed, for example, to each of the entries in the strategic triangle of Figure 14.5.1.

Utility theory may be useful in clarifying the philosophy of management performance and for its educational value, but it does not seem to have given rise to practical applications in the insurance industry as a description of management behaviour.

(c) Corporate planning. As stated already in section 14.1(b), there are several ways of utilizing risk theory techniques to support managerial decision-making. One such is to assimilate stochastic elements into the existing statistical, accounting and corporate planning schemes which many insurers have in practice on a more or less comprehensive scale. The purpose is to help in evaluating the uncertainties involved in the various components of the business flow, for example to give an idea of the required capital at risk with various reinsurance arrangements or investment alternatives. For limited problems of this nature, models concentrating on particular aspects, as described in previous chapters, might be sufficient.

A comprehensive review on corporate planning was prepared by a working party of British actuaries (Akhurst *et al.*, 1988). This referred to the possibility of transforming elements of the conventional deterministic corporate model into stochastic form. In fact, this is a natural approach and can be done step by step by incorporating stochastic elements into the overall scheme. This may not be as laborious as building a new all-embracing model as a single project.

(d) Integration of the sub-models into an overall model. An ambitious alternative is to build a model to simulate the whole company along the lines described in preceding sections. It should be built around the basic equations (14.2.1) and (14.2.2), with separate modules for each of the entries in these equations. Moreover, regard should be had to the market, as was outlined in section 14.3.

Building such an overall stochastic-dynamic model is a major task. An example of the model's flow chart is given in section F.5 of Appendix F. There may be problems in finding suitable numerical values for all the many model parameters and in constructing relevant correlations between the entries. Note that dependencies between key variables arise, among other reasons, because inflation was chosen as the main driving force which affects both the assets and liabilities (section 8.2(c)).

It is necessary to have good cooperation between different parts of the organization in order to collect the information which has to be fed into the model. In a large company many people would probably be involved with this activity. They should be well informed and well motivated. It will only work successfully if senior managers are involved in using the output from the model and are committed to it (Mariathasan and Rains, 1993).

As is the case generally in corporate planning, the model-building process can itself be instructive, revealing many important features in the behaviour of the business and giving a deeper appreciation of their properties and associated uncertainty. If, as can be expected, the model, when operational, can provide an assessment of the current financial position of the company, give timely early warnings and evaluate the risks and rewards of planned business actions, this could well repay the efforts expended in building it. Furthermore, the model can initially be limited to the more important and financially significant elements, in order to economize on the work involved and limit the development time required. The model can be completed later on step by step, if desired.

The problem of finding values for parameters should not be exaggerated. As was stated in section 3.3.1(a), their derivation can be organized to link up with the company's other data processing systems. In some cases industry-wide experience can be utilized. In other cases precise absolute values may be of less interest than exploring the effect of changing the assumptions. In all cases there is a need for professional judgement, taking into account the model as a whole and the combined impact of different parameters.

Some ready-made software packages are available for carrying out an overall analysis of an insurance company. The Institute of Actuaries in London can supply such a model on disk, designed by Daykin and Hey (1990). This permits a great variety of features and alternatives to be chosen by the user.

Last but not least, it is important to present the results of the model in a form which allows for a clear review of the current state and future development of the company without undue effort on the part of users. The presentation should be accessible to those who are not familiar with the details of the model. It is desirable that the model should be supported by well-designed graphics.

In practice a comprehensive model will be successful only if the idea is understood, supported and also used by the senior management of the company.

(e) Other uses of risk theory models have already been referred to in section 14.1(b). For scientific and also practical research into the properties of insurance as a stochastic process it is useful to construct a reference (standard) insurer and to model it, either in its totality or with consideration limited to those features which are relevant for the particular aspect of interest. In particular, simulation permits the investigation of problems which are intractable by conventional analytical methods. In fact, the advent of the modern computer ushered in a new era in the development of practically-orientated risk theory.

Overall models may have a particular application for educational purposes, from simple models up to dynamic business games.

14.6 Public solvency control

The principal goal of public supervision and regulation of the insurance industry is to safeguard claimants, policyholders, etc. from the disastrous consequences of the insolvency of an insurer. This requires arrangements for preventing, as far as possible, insurers from becoming bankrupt. Ideally, ways and means should be found to stop a company before it gets into too precarious a position, if possible while the assets still exceed the liabilities, so that all the commitments entered into can be fulfilled.

Supervision is usually implemented by testing the financial position of each insurer at regular intervals, normally annually. If there appears to be a substantial risk of insolvency, urgent remedial measures will be required or the company's licence to write insurance business will be withdrawn. In the last resort the company may need to be wound up. The problem is to arrange the system of testing so that the probability of failure is minimal. In this section we discuss how risk theory can be utilized for this purpose. (Discussion of regulation in more general terms is beyond the scope of this book, but is given in Pentikäinen and Rantala (1982), Daykin et al. (1984 and 1987) and Pentikäinen et al. (1989).)

(a) **The problem** can be formulated as is illustrated in Figure 14.6.1. The solvency margin $U(t_1)$ at the end of an accounting year t_1 is evaluated by (or notified to) the supervisors. Having regard to the practical arrangements and the inevitable time delay, this can be done only at some later time point τ_1. The supervisor has to decide whether the financial position is sufficiently sound to give the company permission to continue normal activities up to the next testing point τ_2. Permission should not be granted if there is a significant probability ε that the company will be insolvent by the next accounting time point t_2 (the end of the accounting year) or will be in such a weak position that there is then a considerable risk of imminent insolvency. This would be the case if insolvency could occur so soon after time point t_2 that there would no longer be sufficient time for successful intervention, for instance withdrawal of the licence, before it occurred.

To formalize this requirement let us assume that (Figure 14.6.1) the supervisor will not have information about the situation at time t_2 until time τ_2 and that the earliest possible time that he could intervene, should it then prove necessary, would be at $t_1 + T$. This

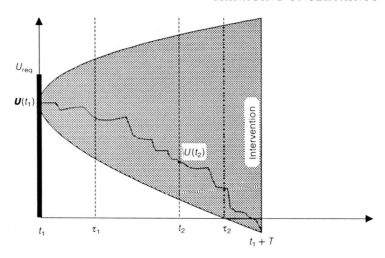

Figure 14.6.1 *Schedule for testing solvency. The confidence region is shaded and an illustrative flow of business plotted.*

results in the condition

$$\text{Prob}\{U(t_1 + T) < 0 \,|\, U(t_1) = U\} \leqslant \varepsilon \qquad (14.6.1)$$

where T is the total time lag from t_1 up to the hypothetical intervention.

The condition (14.6.1) sets a lower limit for U which will be called the **required solvency margin**, denoted by $U_{\text{req}}(t_1)$:

$$U(t_1) \geqslant U_{\text{req}}(t_1). \qquad (14.6.2)$$

i.e. $U_{\text{req}}(t_1)$ is the smallest value of U which satisfies (14.6.1).

The modelling techniques which have been developed in earlier chapters can be used to evaluate the limit (14.6.2) for any particular case where the necessary parameters and preconditions are given. However, the time lag T and the confidence (ruin) probability ε should be determined as a matter of policy or convention, as will be discussed briefly in sections 14.6(b) and (c) below. Furthermore, it needs to be clear what procedure is envisaged in the case of insolvency, along the lines set out in section 14.6(d).

In the special case where the risks related to assets and return on investments are slight, for example because of conservative investment policy or effective immunization (sections 8.3(a) and 8.6(h)), an approximate solution for (14.6.2) can be obtained by the short-term technique introduced in section 6.1, using (6.1.8) or (6.1.13).

Note that the problem is very much the same as in the evaluation of the capital requirements of an insurer in section 13.2(b). The difference is mainly in the perspective. In section 13.2 the survival of the company in the long term was regarded as a key management objective. Here only a relatively short-term perspective on solvency is required by the supervisor since there is always the possibility of withdrawing the insurer's licence. Solvency regulation is established to protect claimants and policyholders, not insurers.

(b) The test time span T has often been taken to be one year, e.g. in the deliberations leading to the EC directive. Both the Finnish Working Party (Pentikäinen and Rantala (1982)) and the British Solvency Working Party (Daykin *et al.*, 1984, 1987) argued that this period was too short to allow for timely intervention by the supervisor in cases where the business develops adversely. It always takes several months before information on the situation is available to the supervisors. Further time is bound to elapse after that before the supervisor can actually intervene. The delays will depend in practice on national legislation and judicial procedure. The British Solvency Working Party originally proposed $T = 18$ months but in later calculations used a two-year horizon for convenience and to allow a slight margin.

The choice of T has a marked influence on the level of the required solvency margin. If U_{req} is scaled so as to be 100 for $T = 12$ months, it can be estimated to be some 120–130 for $T = 18$ months and 140–160 for $T = 24$ months (Pentikäinen and Rantala, 1982, pp. 4.2–28 and Beard *et al.*, 1984, p. 280 where the going-concern basis was used); see also Exercise 14.6.1.

(c) The ruin probability ε has to be set by some consensus. Some idea of the importance of the choice can be obtained from rough estimates presented by Pentikäinen and Rantala, 1982, I, pp. 4.2–26 and 4.2–44, and Pentikäinen *et al.*, 1989, section 5.1. If the required solvency margin is scaled to be 100 for $\varepsilon = 0.01$, it may be some 70–80 for $\varepsilon = 0.05$ and some 120–140 for $\varepsilon = 0.001$ (Exercise 14.6.1).

(d) Procedure on insolvency. There are two different approaches to the problem of insolvency: the going-concern and the run-off philosophies. The choice between the two can fundamentally affect the choice of the appropriate solvency test.

In the **going-concern** alternative one tests only whether the best estimate of the assets exceeds the best estimate of the liabilities at the end of the test period T, for brevity scaling the initial time $t_1 = 0$ in Figure 14.6.2(a). This normally assumes that the insurer's usual activities will continue during this period, including the acceptance of new business. This philosophy postulates that at the end of the period the solvency margin $U(T)$ is still non-negative and that business continues to be written. Hence, it is reasonably likely that the portfolio, having a positive value, can be transferred to some other insurer who would take over both the liabilities and the assets and continue the business.

In the **run-off case** it is assumed that the insurer's licence is withdrawn at the intervention point T. No new business is written thereafter nor is any business renewed. The claims and other commitments in respect of premiums already received are settled as they fall due and paid from the assets. The assets remain invested, yielding interest and possibly capital gains (or losses) until they have been used up.

The solvency test is to evaluate whether the existing assets are sufficient to meet all the outstanding claims and other commitments. The risks of adverse fluctuation of asset values and outstanding liabilities during the notional run-off of the business should be taken into account, as well as the timing of claim settlement. Because the volume of business falls steadily during the notional run-off, it is increasingly vulnerable to risk fluctuations. Moreover, there may be special risks associated with forced realization of the assets when markets are depressed. Provision also has to be made to cover all the expenses which will arise during the run-off of the business.

The fundamental difference between the going-concern and run-off approaches is illustrated by Figure 14.6.2.

REMARK   The run-off approach is sometimes described as the **break-up** philosophy. The former term is preferable in order to distinguish it from the **winding-up** situation, which usually arises only if the assets are inadequate even to cover the liabilities. Here a winding-up petition is made to the court, the assets are realized as soon as possible and the available monies are distributed to creditors, including policyholders and claimants.

(e) Discussion.   The going-concern philosophy seems commonly to have been assumed when solvency requirements were established in various countries, including when the EC rules were proposed, even though the going-concern concept was not specifically referred to

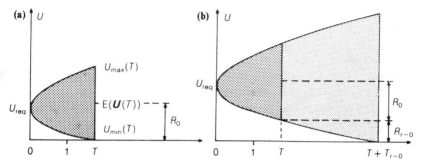

Figure 14.6.2 *The principles of establishing the required solvency margin* U_{req} *for (a) going-concern and (b) run-off cases. In the run-off case it is assumed that normal business dealings continue only up to time T, whereupon the issuing of new policies is stopped and the accrued liabilities are simply paid out when they fall due. The broadening of the confidence area results from the run-off uncertainties and is indicated by* R_{r-0}.

in the earlier literature. The background was that portfolios of weak insurers could usually be transferred to other companies. So the inconvenience and special risks involved with the winding-up or run-off options could be avoided.

The British Solvency Working Party (Daykin *et al.*, 1984, 1987) raised the question of the adequacy of the going-concern approach, referring to the fact that, as a result of the run-off uncertainties, claims could often not be paid in full, even though the balance sheet value of the assets might have exceeded the stated value of the liabilities when new business ceased. Therefore, in insurance markets where the risk of ruin is not negligible, and there cannot be any certainty of being able to transfer the portfolio to another insurer, the going-concern basis is insufficient. A prudent approach to supervision points to the run-off principle.

When considering the merits and demerits of the alternative procedures, one must bear in mind that the run-off condition markedly increases the required solvency margin (or other test requirements). The examples presented by the British Solvency Working Party (Daykin *et al.*, 1984, 1987), and by Pentikäinen and Rantala (1986) seem to indicate that adoption of the run-off basis increases the required margin by some 20 to 30 percentage points of premium income, and, in the case of predominantly long-tailed business, by still more.

Furthermore, the condition that the business should at any moment be able to satisfy a run-off test may impose restrictions on investment policy, since in principle not only the amounts, but also the term, of the assets should be matched to the pattern of liabilities.

A feature which may increase the possibility of getting the portfolio transferred is the fact that the accepting company may attribute some (possibly considerable) value to the portfolio based on the profit it might yield in future years. Of course, this may not be the case if the risk selection, rating or some other relevant feature has been thoroughly unsound, or if it is probable that a large proportion of the policyholders will lapse (which will depend on the practices and features of the market and the type of policies). The question of assessing the value of a portfolio was discussed at the International Congress of Actuaries in Zürich, 1980 (Transactions Vol. 3).

For the time being, and until further research results are available, it seems that this is a question which has more than one correct answer. The answer clearly depends on the character of the national market and the relevant legislation, as well as on the prevailing business practices (for example, whether the attitudes of management are strictly commercial or include a strong sense of social responsibility). Above all, it depends on the approach adopted by the supervisor. This can be a major factor in determining whether insurers which get into difficulties are treated as bankruptcies, closures to new business or simply transfers of business to another insurer.

If an insurer ceases to be viable, the best solution may be to transfer the portfolio to some other insurer which would then take over all of the responsibilities in relation to the existing insurance contracts. However, this will not always be in the interests of the other company and may create the wrong incentives in the market. Suitable provisions should be included in the insurance company legislation to facilitate the practical procedures of such transfers where they are appropriate.

Regardless of whether the going-concern basis or the run-off basis is assumed, it is still useful to refer to the possibility of supporting the solvency control system by introducing a national **compensation fund** to serve as the ultimate link in the security chain. In the unlikely event of an insurer not being able to meet its liabilities, the available assets can be topped up by a levy on all the other insurers, so that claims can be settled. Provision should be made for benefits to be reduced if the benefits promised were clearly excessive relative to the premiums charged. It is normally thought to be desirable for the ultimate risk to be co-insured with the policyholders (for example, only 90% of claims met from the compensation fund) to reduce the incentive for policyholders to choose a cheap, but inadequately capitalized, insurer.

(f) Required margins of solvency. Early insurance legislation prescribed minimum levels of share capital, or required margins of a flat-rate type, i.e. some fixed sum which every insurer should have as a condition for permission to operate. Both practical experience and theoretical considerations showed that this approach was inadequate. A fixed sum which might be sufficient for a small company would offer no protection for a large company. In recent decades the trend has been towards advocating risk-based capital requirements (section 6.2(c) and Pentikäinen, 1952; Pentikäinen and Rantala, 1982; Daykin *et al.*, 1984, 1987).

The EC freedom of establishment directives (for non-life insurance in 1973 and life insurance in 1979) used a simplified risk-based approach originally proposed by Campagne (1957, 1961). For non-life insurance a solvency margin of 18% of premium income up to a defined level and 16% of premium income above that level reflected the broad shape of the risk as it was related to the volume of business. However, no account was taken of the variation in risk between classes of business and asset risk was completely ignored. Moreover, this approach is counter-intuitive in requiring higher solvency margins when premiums are high than when they are inadequate.

In planning the regulatory solvency requirements the reasoning has been the same as was described at condition (14.6.1) above. The desire has been to lay down fairly simple rules in legislation and directives, such as was given in (6.2.3). With hindsight, the EC requirements can be seen more as offering a logical structure for intervention by supervisors than an attempt to offer any particular enhanced level of security to policyholders. A critical factor is how the assets and the liabilities themselves are valued. What may appear to be a weak solvency requirement may not be so if substantial margins are held in the technical reserves and in the asset values.

The final outcome in the EC was a compromise between conflicting requirements and politically could not have been allowed to have had the effect of making large numbers of existing companies in breach of the new requirement. The original proposal for a margin of 25% of premiums was, therefore, considerably watered down (Daykin, 1984).

Recent research works, including those referred to above, have shown that the actual capital requirement to achieve a reasonable level of security can be considerably higher than the current EC and other rules suggest. Furthermore, the required margin appropriate for the condition (14.6.2) depends to a large extent on the structure and size of portfolio as well as on the investments held, the reinsurance arrangements and other factors. Moreover, the capital requirement clearly depends on the stage of the market cycle, as was discussed in section 14.4(c).

The very approximate estimates given in the papers referred to above (also Pentikäinen et al., 1992) seem to show that in the least risky cases the solvency margin should be about 30%, but for portfolios involving higher risk, and in particular being subject to market cycles and investment risk, the need may be much higher, commonly 50% to 100% and even more, depending on the phase of the cycle. Hence the conclusion has been drawn that there is no simple rule which would be adequate for a universal use. In principle the requirement should be built up from a series of components relating to classes of business, investment risks, reinsurance risk, etc.

Any formula approach inevitably has its limitations. Ideally the capital requirement should be assessed individually for each insurer, perhaps by a designated professional person. An individual expert analysis made by an appointed actuary has been a tradition in the UK for life insurance for many years. The appointed actuary is responsible not only for an annual valuation of the liabilities but for the overall financial management of the business, taking into account both assets and liabilities and a long-term time horizon. He is required to monitor the position continually. The British Solvency Working Party (Daykin et al., 1987) and Daykin and Hey (1990) recommended an actuarial report on the financial condition of general insurers. The subject is discussed by Pentikäinen and Rantala (1982) and by Pentikäinen et al. (1989), such a system having been in action since 1953 in Finland. New insurance legislation in Canada in 1992 introduced a requirement for each general insurance company to have an appointed actuary, who will be required to prepare an annual financial condition report.

(g) A short-cut formula. If an individual expert analysis is not performed, it can perhaps be replaced by an approximate evaluation, as will be described in this section. By choosing conservative values for the coefficients, a safe estimate can be achieved. A convenient approach might be to make the short-cut test first. If and only if it gives a negative score would a more exact individual analysis be required. This was the earlier practice in Finland, but after some experience the individual analysis has been replaced by the short-cut, which has to be performed according to detailed directives of the supervisor (Pentikäinen *et al.*, 1989, p. 221).

The short-cut formula can have the following form (Pentikäinen and Rantala, 1982, item 5.3.3 and Pentikäinen *et al.*, 1989, p. 222), referring to the going-concern position at time T, see Figures 14.4.2 and 14.6.2

$$U_{req} = a \cdot B + b \cdot \sigma_X + c \cdot A + d \cdot \sigma_A + R_{skew} + R_{cat}, \quad (14.6.3)$$

where the coefficient a incorporates in the formula the risk associated with business cycles or adverse trends (or inadequate safety loadings in the premiums) and B is the retained premium income. The second term reflects the ordinary compound mixed Poisson fluctuation (with the standard deviation calculated specifically for each company (3.2.14)).

The third and fourth terms are concerned with the asset risks, A being the amount of the assets and σ_A the standard deviation associated with the variation of the return on investments.

The term R_{skew} is needed only for very small companies, e.g. captives, to rectify the excessive skewness of the claim expenditure which is typical for small risk collectives or collectives involving very large risks.

The term R_{cat} is needed in cases where there is an exceptional risk of catastrophes.

REMARK   The standard deviations needed for the formula (14.6.3) can be calculated case by case from the statistics of the insurer concerned. The calculation of σ_X can often be rationalized by manipulating the expression (3.2.14) as follows:

$$\sigma_X^2 = n \cdot a_2 + P^2 \sigma_q^2$$

$$= \frac{a_2}{a_1 M} \cdot n a_1 \cdot M + P^2 \sigma_q^2$$

$$= \beta \cdot P \cdot M + P^2 \sigma_q^2 \quad (14.6.4)$$

where

$$\beta = \frac{a_2}{a_1 M} \quad \text{and} \quad P = n \cdot a_1. \quad (14.6.5)$$

Here the a's are the moments of the claim size distribution and M is the maximum net retention per risk unit. A merit of the coefficient β is that it is approximately immune to the effects of inflation. It is not necessary, therefore, to recalculate it every year. The Finnish practice has been to derive these coefficients for the most important insurance classes from the insurers' combined statistics and present them in the form of a table for companies' common use. P can be approximated by the risk premium income of the company.

(h) Risk-based capital – Life Insurers. In the USA an analogous formula for risk-based capital for life insurance consists of four components.

C1 – Asset risk, including risk of default, concentration risk and risk of reinsurance non-recoverability.
C2 – Insurance risk, arising from high levels of claims.
C3 – Interest rate risk, arising from changes in levels of interest rates.
C4 – Business risk, including expense overruns and management incompetence.

These risks are combined using the following formula:

$$\sqrt{(C_1 + C_3)^2 + C_2^2} + C_4$$

(i) Risk-based capital – Property/Casualty. In 1993 the National Association of Insurance Commissioners (NAIC) adopted a risk-based capital formula for property/casualty insurers. This contained the following components.

● loss and loss adjustment expense reserve risk
● pricing risk
● credit risk
● investment risk

These risks are combined using the following formula:

$$RBC = RBC0 + \sqrt{(RBC1)^2 + (RBC2)^2 + \cdots + (RBC5)^2}$$

where the components are

RBC0 – Investment in insurance affiliates; off-balance sheet risks; contingent obligations.
RBC1 – Fixed interest investment.
RBC2 – Equities; real estate investments; and other asset risks.
RBC3 – 50% of credit risk (receivables and reinsurance recoverable risk).

RBC4 – Loss reserve risk; reserve growth risk; plus 50% of credit
   risk.
RBC5 – Written premium risk; plus premium growth risk.

Exercise 14.6.1 Evaluate the required solvency margin U_{req} using
(6.1.10) with parameters (see also (6.2.1)): $n = 10\,000$, $m = £0.004$ million,
$\lambda = 0.05$, $\sigma_q = 0.04$, $r_2 = 40$, $T = 1$ year and $\varepsilon = 0.01$. What is the
relative increment if (1) T is taken as 2 years, or (2) ε is taken as 0.001?

CHAPTER 15

Life insurance

The foregoing methods were developed to fit insurance classes in which premiums can be changed at the renewal dates, as is typically the case with general insurance business. They could also be applied to life insurance for short-term analysis or where life business is written, or premiums guaranteed, for periods of a year or less. However, for long-term investigations there is a crucial difference between general insurance business and most life insurance business. Life insurance premiums are usually fixed for the whole term of insurance, often for a decade or more. Whereas the non-life insurer can adjust the current premium rates fairly rapidly in the event of an adverse development of the business, the life insurer usually has no similar tool available in respect of policies in force. This presents difficulties for both rating and reserving. The life insurer has to be prepared to cope with potential fluctuations and changes in the return on investment, expenses, mortality, invalidity, etc. which might occur in the more or less remote future. The threats posed by inflation, for example, assume greater importance in this environment. Life insurance, therefore, requires it own techniques, which are the subject matter of this chapter.

15.1 Recapitulation of some basic formulae of life insurance mathematics

(a) Equivalence equation. Consider a cohort of people of the same age and sex, with the same mortality characteristics, and with identical policies issued at time $t = 0$. Let $l(0)$ be the initial number of persons in the cohort at the common entry age x_0 and let $l(t)$ be the number still alive at the end of year t. The expected cohort size obeys the algorithm

$$l(t) = l(t-1) \cdot (1 - q(t)) \qquad (15.1.1)$$

where q is the rate of mortality in year t, i.e. the proportion of those alive at $t - 1$ who are expected to die within the next year. Whilst we are considering a single homogeneous cohort we can restrict attention to the mortality rate q as a simple function of time. However, q will in practice vary according to the characteristics and attained age of the cohort.

We will specify a general model, so that the commonest types of life insurance can be considered as special cases. We define the benefits to be paid in years $t = 1, 2, \ldots$ either in the case of death, or in the event of survival, as follows

$S_d(t) =$ payment in the case of death in year t;
$S_e(t) =$ single payment if the cohort member survives for t years;
$S_p(t) =$ pension (annuity) paid for surviving cohort members in year t.

The following special cases can be identified:

(1) If $S_d(t) > 0$ for $t = 1, 2, \ldots, n$, $S_d(t) = 0$ for $t > n$ and if the two other benefit sums are equal to 0, we have conventional **term insurance**, where n is the term of policy.
(2) If $S_e(t)$ is non-zero only for $t = n$ and all the other benefit sums are equal to zero, we have a **pure endowment insurance** of term n.
(3) The combination of (1) and (2) is an **endowment insurance**.
(4) If $S_p(t)$ is > 0 for $t \geqslant w - x_0$, $S_p(t) = 0$ for $t < w - x_0$ and all the other benefit sums are equal to 0, a **pension annuity** is in question, with w the pensionable age.

The benefit sums S are assumed to be equal for all the cohort members. This assumption will be relaxed later.

With traditional non-participating policies the sums S are fixed in advance and remain constant, i.e. they are independent of time t. However, a more general definition is necessary to allow for changes resulting from bonus or indexation rules, or to allow for the sums to vary with t in order to provide a reducing or increasing level of protection, or higher protection at some specific age or time.

In detailed life insurance formulae great care is usually taken over the precise incidence of payments. For our purpose it will be sufficient to assume that all transactions, including payment of premiums, claims, etc., take place in the middle of the year, whereas status variables, for example, the size of the cohort, reserves etc. refer to the end of the relevant year. Furthermore, it is assumed that the cohort receives a total return on its funds of rate $j(t)$. This can be specified to differ from year to year, though it is treated as constant in most classical applications. For discounting, the following notations and definitions are introduced, along the lines adopted in the context of equations (9.5.4) and (9.5.5)

$$v(t) = 1/(1 + j(t)),$$
$$v(t, s) = v(t + 1) \cdot v(t + 2) \ldots v(s - 1) \cdot v(s)^{1/2} \quad \text{for } s \geqslant t + 2$$
$$v(t, t) = v(t)^{-1/2} = (1 + j(t))^{1/2}, \tag{15.1.2}$$
$$v(t, t + 1) = v(t + 1)^{1/2}.$$

The function $v(t, s)$ discounts from the middle of year s to the end of year t. In the special case where the rate of return j is constant, (15.1.2) reduces to

$$v(t, s) = v^{s - t - 1/2} \tag{15.1.3}$$

where $v = 1/(1 + j)$.

The fundamental equivalence principle of life insurance states that the premium reserve at the end of year t is equal to the difference between the discounted expected values of the future outflows and inflows. For a particular cohort of lives

$$l(t) \cdot V(t) = \sum_{s=t+1}^{\infty} l(s - 1) \cdot v(t, s) \cdot [S_e(s) + S_p(s) + q(s) \cdot S_d(s) - B(s) + E(s)] \tag{15.1.4}$$

where $V(t)$ is the premium reserve at the end of year t, $B(s)$ is the premium treated as payable in year s and $E(s)$ is the loading for expenses, these three quantities being per capita amounts corresponding to the assumed relevant benefit sums S. In the case of endowment policies and term policies the summation in (15.1.4) ends at the maturity date, i.e. at $s = n$, for both benefit payments and premium rates, i.e. $B(s) = 0$ for $s > n$.

Note that, for notational convenience, the variables, V, B and E are calibrated to correspond to the level of the benefit sums S, whereas the common practice in life insurance mathematics is to define them per unit benefit and then to multiply by the relevant sums assured or annuity amounts.

Equation (15.1.4) is handled deterministically in classical life insurance practice by specifying calculation bases for the relevant entries of the equation. If the pattern of premiums is defined, for example they are constant during a specified term of payment or there is only a single premium, then the condition $V(0) = 0$ uniquely determines the net premium $B(t)$, i.e. the premium calculated on the reserving basis without any allowance for expenses.

$$V(0) = 0 \Rightarrow B(t). \tag{15.1.5}$$

The value of $B(t)$ can be found by solving from (15.1.4) with the left-hand side set equal to zero. The net premium for reserving purposes may differ from the actual premium charged, since in some jurisdictions it is usual for the reserves to be established using more conservative assumptions than those employed in setting the premium rates. If the actual premiums are used in (15.1.4), it follows that $V(0)$ will in these circumstances be greater than zero. In this case $V(t)$ is described as a **gross premium reserve**.

(b) Profit algorithm. A key variable in the subsequent considerations will be the **profit function** $Y(t)$. It is defined as the difference between the left-hand side and the right-hand side of (15.1.4). Replacing t by $t-1$, having regard to (15.1.1), observing that $v(t-1, s) = v(t) \cdot v(t, s)$ and moving the reference point for discounting to the end of year t, the following expression is obtained

$$\begin{aligned} Y(t) = l(t-1) \cdot \{ & v(t)^{-1} \cdot V(t-1) - V(t) + v(t)^{-1/2} \cdot [B(t) \\ & - S_e(t) - S_p(t) - q(t) \cdot (S_d(t) - v(t)^{1/2} \cdot V(t)) - E(t)] \}. \end{aligned} \tag{15.1.6}$$

The terms inside the braces represent the inflows and outflows attributable to the cohort during year t, expressed in terms of the yield of interest and the mortality in year t, together with the development of the cohort reserve $l \cdot V$.

If the experience is strictly in accordance with the reserving basis for (15.1.4), then $Y(t)$ is identically equal to zero. However, if we then substitute the values of the interest yield, mortality and expenses actually experienced, represented by an asterisk, we obtain the profit (or loss) accumulated to the end of year t as

$$\begin{aligned} Y^*(t) = \ & l(t-1) \cdot V(t-1) \cdot [j^*(t) - j(t)] \\ & + l(t-1) \cdot [v^*(t)^{-1/2} - v(t)^{-1/2}] \cdot [B(t) - S_e(t) - S_p(t)] \\ & - l(t-1) \cdot [v^*(t)^{-1/2} \cdot E^*(t) - v(t)^{-1/2} \cdot E(t)] \\ & - l(t-1) \cdot \{ v^*(t)^{-1/2} \cdot q^*(t) \cdot [S_d(t) - v^*(t)^{1/2} \cdot V(t)] \\ & \qquad - v(t)^{-1/2} \cdot q(t) \cdot [S_d(t) - v(t)^{1/2} \cdot V(t)] \}. \end{aligned} \tag{15.1.7}$$

The four components of profit ($\pm$) can be recognized as

(1) Interest profit and capital gains on the invested reserves.
(2) A part year's interest profit on the net incoming cash flow during the year.

(3) Expense profit if actual expenses deviate from the assumption.
(4) Mortality profit.

To avoid unnecessary complications, lapses are ignored here, although they will be introduced in section 15.2(g). The interest yield is assumed to include changes in capital values as well as investment income.

Equation (15.1.6) is a discrete variant of the well-known Thiele's differential equation. A more general version, such as (15.1.7), can be obtained if we distinguish between the premium calculation basis and the reserving basis. In this case $V(0)$ and $Y(t)$ are non-zero. For non-participating business it would be usual for the reserving basis to contain greater margins for caution than the premium basis, in order for the actuary to be satisfied that the reserves are adequate to enable the future liabilities to be met with a high level of probability. The premium basis may represent a more commercial approach and be closer (although probably still on the safe side) to the anticipated actual outcome. As mentioned in section 15.1(a), this leads to a positive reserve immediately the business has been written; in other words capital has to be supplied, either by shareholders, participating policyholders or from the **estate** (non-attributable surplus) to enable the business to be written. In subsequent years $Y(t)$ can be expected to be positive, providing a return to the providers of capital.

For participating business the reserving basis will not necessarily be more cautious than the premium basis. Many countries use passive valuation systems in which the same bases are used for reserving as for setting premiums, and surplus is allowed to emerge according to the actual experience. Elsewhere, more conservative reserving bases will imply a need for capital, as for non-participating business. On the other hand, less conservative reserving bases will produce immediate release of profit when the business is written. This is not usually appropriate for the statutory valuation required by the supervisor, but it may be appropriate for internal management purposes, for example in planning bonus strategies, or for reporting the value of the business to shareholders.

Where the reserving basis differs from the premium basis, the actual profit emerging in year t will include a planned component arising from the difference between the bases, and a further component arising from the difference between the actual experience and the assumptions of the reserving basis. Sometimes a distinction is made between the **first order basis**, used for setting premiums (and in some countries also the reserves) and the **second order basis**, representing the expected outcome. We have avoided this terminology as it does not adequately distinguish between the various different possible sets of assumptions.

Equation (15.1.6) provides a useful starting point for considering a stochastic model of the life insurance process, as will be explored in the subsequent sections.

15.2 Stochastic cohort approach

Most of the quantities in (15.1.6) and (15.1.7) can be made stochastic by appropriately adapting the approaches which were derived in previous chapters for the general insurance process. A model can then be constructed to evaluate the uncertainties affecting the long-term development of the cohort. The ideas will be extended to the whole portfolio in section 15.3, using the cohorts as building blocks.

(a) Mortality. The deterministic (expected) number of deaths from the cohort of members in year t, $d(t) = l(t - 1) \cdot q(t)$, is replaced by a random variable $d(t)$. It can be generated using the technique which was introduced for the simulation of the number of claims in section 5.3. The deterministic $d(t)$ is used as the expected number $n(t)$ of events, assuming the Poisson law is postulated. Then (15.1.1) should be replaced by

$$l(t) = l(t - 1) - d(t). \tag{15.2.1}$$

In most applications realistic assumptions will be used for the evaluation of $d(t)$.

REMARK Some authors favour the binomial law in place of the Poisson. In fact, the former is a natural choice for small collectives and in particular when, as will be permitted later, the benefit sums differ between cohort members. Then the surviving cohort can be simulated year by year, perhaps person by person, with $q(s)$ being the binomial probability of exit. However, the difference in final outcomes between the Poisson and binomial models can be expected to be insignificant for large cohorts, and even for smaller cohorts in the context of large portfolios, where the inaccuracies for individual cohorts can be expected to cancel each other out. A binomial random number generator was presented by Pentikäinen and Pesonen (1988). (Comparisons of the binomial and Poisson approach are given in Appendix A, section A.1(c).)

(b) Inflation is particularly important in long-term applications. The rate of inflation $i(t)$ and the index

$$I(t) = I(0) \cdot \prod_{\tau=1}^{t} [1 + i(\tau)] \tag{15.2.2}$$

can be generated as in Chapter 7 (see (7.3.1) and (7.3.2)). The potential need to test for the consequences of an irregular pattern of inflation, for example the shocks which were discussed in section 7.3(g), should be kept in mind.

REMARK   In some applications the relevant variables may refer to the end of the year t and in some to its mid-point. Rigorous consideration could require an extra half-year of inflation corresponding to the factors provided by the exponent $\frac{1}{2}$ or $-\frac{1}{2}$ in formulae (15.1.2). For the sake of simplicity we omit this in the stochastic considerations where it is not very significant. It is left to the reader to complete the formulae in respect of such details in cases where it may be necessary.

(c) Yield of interest is another important potential source of uncertainty. Simulation of investment returns was considered in Chapter 8. Both the income yield and capital gains should be taken into account. The risk and return depend on the investment strategy adopted by the management.

Immunization was discussed in section 8.6(h) as one way of arranging protection against movements in the return on investment. A stochastic approach to immunization was considered in section 8.6(i). Immunization can generally be achieved in respect of fixed and guaranteed benefits by investment in fixed interest assets. The average duration of the assets and liabilities should match and the spread of the assets should be greater than the spread of the liabilities. The funds supporting non-guaranteed benefits (terminal bonuses and other future bonuses) may be able to be invested in equities and property in order to increase the expected return, albeit at the cost of greater volatility.

Statutory requirements for the valuation of assets and the determination of the level of the reserves when they are backed by equities and property may have an important influence on the balance sheet and on the emergence and distribution of profit from investments.

(d) Distributions of benefit amounts.   It was assumed in section 15.1(a) that the various benefits S were the same for each cohort member. For some purposes this may be appropriate, in particular when the volatility inherent in specified assumptions, for example mortality, yield of interest, inflation, etc. is being tested in isolation. For more general applications this assumption should be relaxed, permitting differing benefit amounts S for different cohort members. The techniques given in section 3.3 for the construction of the claim size d.f. can also be applied in this context. In subsequent sections we suppose that such distributions are available.

REMARK A particular problem is that the distribution of benefit amounts changes slightly when the cohort matures, since the exiting members have randomly differing benefit amounts. This could be taken into account by means of a database of individual insured persons, recorded with their sums insured, allowing the computer randomly to 'kill' them one by one during the simulation (see remark at section 15.2(a) above). Understandably, such an approach makes the simulations very laborious and time-consuming. A short cut is to omit this feature by assuming that the shape of the distribution of insured sums remains the same. Then the collective simulation methods of section 5.4(d) can be applied. Tests seem to suggest that the inaccuracy in the final outcomes is not great, in particular in relation to the total portfolio outcome obtained by combining all the cohorts. A compromise would be to handle the largest benefit amounts individually and the bulk of small and medium-size policies using the short-cut method.

(e) Expenditure due to the death and maturity benefits. The death strain is represented in (15.1.6) by the term $l(t - 1) \cdot q(t) \cdot [S_d(t) - v(t)^{1/2} \cdot V(t)]$. Its randomized counterpart is the sum of the individual death strains, given by the first summation term in the following expression. The second summation term introduces the corresponding expression relating to maturity payments

$$X(t) = \sum_{i=1}^{d(t)} [S_{d,i}(t) - v^{1/2} \cdot V_i(t)] + \sum_{i=1}^{e(t)} [S_{e,i}(t) - v^{1/2} \cdot V_i(t)] \quad (15.2.3)$$

where $d(t)$ is the random number of deaths of cohort members in year t (section 15.2(a)), $S_{d,i}(t)$ is the insured death benefit of the ith member to exit during year t and V is the reserve; $e(t)$, $S_{e,i}$ and V in the second summation are the corresponding quantities for maturities occurring in year t. $X(t)$ is a compound variable (3.1.1). Hence, the simulation methods dealt with in Chapter 5 are applicable. First the number of deaths is generated (section 15.2(a)), and then the matching aggregate claim $X(t)$, either by generating the S-terms in (15.2.3) one by one, or by generating their total directly using the conditional WH-generator of section 5.4(d).

Note that the maturity term in (15.2.3) may be relatively small, because the reserve builds up to the maturity sum over the period of cover. This term can, therefore, often be omitted.

(f) The expense term in (15.1.6) can be evaluated as was outlined in Chapter 11. As a rule the expenses are high in the first year or two from the inception of the insurance, after which they reach an equilibrium level. Suitable rates should be derived from the experience for the relevant environment.

An important feature is that in the long run the expenses increase according to inflation. A good first approximation may be to assume growth in proportion to the index (15.2.2).

A simple candidate for the expense term attributable to the cohort is

$$E(t) = (1 - \lambda) \cdot I(t - 1) \cdot E_1(t) \cdot I(t)/I(0) \qquad (15.2.4)$$

where the factor $1 - \lambda$ removes the estimated safety loading from the per capita premium basis $E_1(t)$, this now being calculated to correspond to the average level of the relevant benefit sums. $I(t)$ is the index of expense inflation, which may be taken as being closer to earnings inflation than price inflation (see Chapter 11 for further discussion).

(g) Lapses of policies should necessarily be incorporated into the model. The average rate of lapse $\rho(t)$ should be postulated, whereupon the number of lapses $d_{\text{laps}}(t)$ can be randomized by a similar technique to that used for the generation of deaths. Then the algorithm (15.2.1) can be completed by the recurrence realtionship.

$$l(t) = l(t - 1) - d(t) - d_{\text{laps}}(t). \qquad (15.2.5)$$

Formulae defining the surrender values $S_{\text{laps}}(t)$ payable on lapsing policies are also needed. The loss (or profit) $X_{\text{laps}}(t)$ resulting from lapses can then be simulated by the same technique as was developed for the death benefit expenditure in section 15.2(e) above. The surrender value S_{laps} replaces the death benefit sum S_d in (15.2.3).

REMARK   For convenience, the policies which become paid-up policies as a result of an early cessation of premium payment can be treated as lapses, taking the capital value of the remaining future benefits as the surrender value.

(h) Bonuses. In with-profit or participating life and pensions insurance the premium bases are assessed on particularly conservative assumptions. The business, therefore, normally generates profits. A reasonable part of this is returned to policyholders as bonuses (or dividends).

There are a great many different bonus systems. The bonus can be a specified addition to the initial sums assured to be paid in case of death or maturity. Sometimes bonuses are added during the course of the policy term and become guaranteed additions to the sum assured, although not payable until the sum assured becomes payable (**reversionary bonus**). Sometimes the bonus is only added when the policy becomes a claim, on death or maturity, giving the insurer more flexibility (**terminal bonus**). Bonuses added to the sum assured may be associated with a recalculation of the premium payable in future, corresponding to the new higher sum assured. Another possibility is for the bonus to be given in cash, in which case it may be used to reduce premiums (**premium rebate**) or placed in a separate account to accumulate until the policy becomes a claim.

Although not at the discretion of the insurer, the changes of the benefit sums of **unit-linked** and **index-linked policies** can also be interpreted as bonuses. The capital gain, and sometimes the excess total return on the investments over the deductions permitted under the contract, are allocated as increases in the sum assured. Index linkage may affect only the benefits or both the benefits and current and future premiums. The index can be external, for example the Financial Times Actuaries All-share Index of UK equities, or it may be an internal company index determined annually according to the performance of a particular fund or based on various kinds of profits available (a description and critique of a number of index linkages was given by Pentikäinen, 1968). The index linkage can be replaced by a method to amend the relevant quantities by a prefixed factor g^t $(g > 1)$.

For our purposes it is sufficient to distinguish those bonuses which are **guaranteed** once they have been awarded, and hence represent a long-term commitment, and those which are **discretionary** and are determined annually according to the resources available at the time. Figure 15.2.1 illustrates the idea.

The total amount of profit allocated to the cohort is divided between the guaranteed bonuses and the **bonus fund**, which is intended to provide the discretionary bonuses but is also available to cover potential future losses. It may consist of some specified reserves and provisions or margins inside the technical reserves and may also include understatement of asset values by reference to market values.

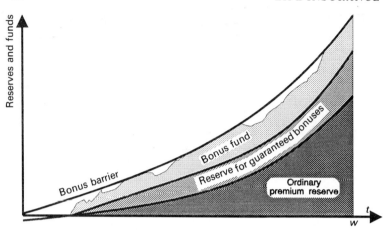

Figure 15.2.1   *The development of the total reserve consisting of the ordinary premium reserve V, the reserve for guaranteed bonuses and the bonus fund.*

Because the bonus fund functions as a buffer for the total profit and loss, its course can be expected to be irregular, as is plotted in the diagram. If the ordinary reserve is increasing, as it is in the case of endowment and pension policies, the profit gained from the investments generally grows. On the other hand, in cases where there is no mechanism for regularly increasing the premium, the expense loading may turn out to be insufficient as the cohort matures and the actual level of expenses eat into the fund.

Both the bonus systems and the relevant terminology differ greatly between countries. The guaranteed bonuses are called **reversionary bonuses** in the UK and the discretionary bonuses are known as **terminal bonuses**. The bonus fund is often termed the **investment reserve**. It supports the terminal bonuses which are paid at termination of a policy or in case of a policy becoming a claim on death.

In stochastic models a key issue is to develop rules about the apportionment of total available profit between dividends to shareholders, general strengthening of solvency, guaranteed bonuses and the bonus fund, and furthermore, how the bonuses should be allocated as between policies of different types and ages and durations, in order to achieve equity and meet the reasonable expectations of policyholders. This allocation of resources, including the size of the bonus fund, will usually be decided annually by the management, in the light of their general business strategy, as was described in

Chapter 14. For modelling it is necessary to specify a rule to mimic this process; in fact, experimenting with scenarios for alternative rules is a central aspect of the analysis of life insurance business. The rule should be constructed to reflect the traditions, practices and regulation in the country in question. In this book we can give only a simple example to indicate the type of solution which might be feasible. The overall model which we are outlining is sufficiently general that it should be able to be adapted for differing bonus systems by replacing the module determining the bonuses by another more appropriate one.

Our illustrative approach is to define a target level, referred to as the **bonus barrier**. The idea is to maintain a bonus fund which is sufficient to meet fluctuations and potential adverse events. The bonus barrier indicates the limit at which the bonus fund can be deemed to be sufficient. The remainder can be distributed as allocations of the guaranteed bonus and immediate payments of discretionary bonus. The barrier, denoted by $R(t)$, can be obtained by relating it, in one way or another, to the volume and the risk of the transactions affecting the cohort. A simple formula of the type

$$R_V(t) = b_V \cdot l(t) \cdot V(t) \tag{15.2.6}$$

would be appropriate in cases where the main risk is the volatility of return on investments. b_V is a parameter which controls the level of the bonus fund and $V(t)$ is the deterministic reserve corresponding to the average level of the relevant benefit sums S. For term insurance a formula based on capital at risk

$$R_d(t) = b_d \cdot l(t) \cdot [S_d(t) - V(t)] \tag{15.2.7}$$

might be suitable. These limits control the part of the profit which is held back as bonus fund, and hence determine the balancing amount which is available for distribution.

Another approach is to fix a target level for the guaranteed (e.g. reversionary) bonuses in order to keep them stable from year to year and let the bonus fund vary according to the available annual profit, including capital appreciation on the investments. Then the bonus barrier, if defined, may serve as an advisory measure, possibly related to notionally accruing amounts of discretionary (e.g. terminal) bonus, to indicate whether the fund is developing in a way which is sufficient to provide future bonuses which will meet the expectations of both policyholders and management.

Corresponding formulae can be constructed for index-linked policies (Beard *et al.*, 1984, section 8.3b(vi)).

The allocation of profit between shareholders and policyholders, and between different cohorts and types of policies, is a matter of equity, where each country has its own traditions and regulations. In some countries the allocation rules are laid down in the law or are subject to approval by the supervisory authority on a contract by contract basis. In the UK considerable discretion is given to insurers, but the directors are required to receive a report from the appointed actuary before deciding on any distribution of surplus, and they are obliged to have regard to the **reasonable expectations of policyholders**. Actuaries usually approach this by carrying out **notional asset share** calculations, which involve building up a retrospective reserve for different types and cohorts of policies, having regard to the actual experience of mortality, investment return, expenses, etc. In the UK the proportion of total surplus allocated to shareholders cannot be changed without going through special procedures.

The management of the bonus fund is vital for the sound financial health of the company. If too much profit is distributed in cash form as dividends and bonuses or allocated to guaranteed bonuses which require technical reserves to be established, the solvency position may be fatally weakened.

(i) The total residual amount of profit (or loss if negative) generated by the cohort in year t can now be written down by modifying (15.1.6) to allow for paid dividends and new bonus commitments.

$$Y(t) = I(t-1) \cdot [B(t) + \cdot V(t-1) - V(t) + j(t) \cdot [V(t-1) + V(t)]/2]$$
$$- X(t) - E(t) - X_{\text{laps}}(t) - D(t) - G(t). \qquad (15.2.8)$$

Here $B(t)$ and $V(t)$ are average per capita values with reference to the distribution of sums assured introduced in section 15.2(d). $X(t)$ and $X_{\text{laps}}(t)$ are now aggregate claim amounts for the cohort and $E(t)$ is the aggregate level of expenses attributable to the cohort. $D(t)$ stands for dividends to shareholders and may also be used, for example, for transfers to strengthen the general level of solvency. Finally, $G(t)$ is used for bonuses paid in cash to policyholders and for transfers to reserves in respect of guaranteed bonuses allotted. The entries in (15.2.8) refer to the midpoint of year t.

Another key variable is the **accumulated residual profit** for the cohort accrued up to the end of year t. It is obtained by the algorithm

$$Y(0,t) = (1 + j(t)) \cdot Y(0, t-1) + (1 + j(t))^{1/2} \cdot Y(t), \quad Y(0,0) = 0.$$
(15.2.9)

$Y(0,t)$ is in fact the bonus fund attributable to the cohort, as was defined above.

It is often appropriate to relate this variable to the current premium volume of the surviving cohort, for example,

$$y(0,t) = \frac{Y(0,t)}{l(t) \cdot B(t)}.$$
(15.2.10)

in cases where the premium is paid continuously during the whole term of insurance.

A useful concept is the **discounted capital value of future profits** or **net present value** at a stated rate of discount j (15.1.2)

$$Y_{\text{tot}} = \sum_{t=1}^{\infty} v^{t-\frac{1}{2}} [Y(t) + D(t) + G(t)].$$
(15.2.11)

Note that this quantity is based on the total profits emerging each year, before any allocation to policyholders or shareholders. For non-participating policies Y_{tot} represents the present value, at the time the cohort of business is written, of all the profits which will emerge from the cohort as it runs through. The discount rate can be specified according to the purpose of the calculation. It will usually be deterministic, although not necessarily the same for all parts of the business and for all time periods. In the context of assessing the profitability of a cohort of business it is usually described as the **risk rate of return** and represents the return which the providers of capital expect to obtain on their investment. An important criterion is that a cohort of policies should generate a positive expected net present value at the target risk rate of return. Otherwise the providers of capital will not consider the business worth writing.

From the point of view of the shareholders, the **discounted capital value of future shareholder dividends** deriving from the cohort of policies is also of interest. This can be seen as the value of the shareholders' interest in the cohort of business, with the shareholders' required rate of return on capital being used as the discount rate. Aggregating the discounted capital values of future shareholder dividends for the existing business of the insurer leads to the **embedded value** of the portfolio. Incorporating assumptions about the dividends as a result of future cohorts of business leads to an **appraisal value** for the insurer. Restricting the assessment to dividend payments reflects the realities of accounting and supervisory requirements as regards the timing of the emergence of profit to the shareholders. The future dividends must comply with any rules regarding the relative proportions of profit going to policyholders and shareholders. Profit can only be considered as available for distribution if statutory reserving and solvency margin requirements are met.

Even for non-participating business, in which $G(t) = 0$, a positive contribution from the cohort to $Y(0, t)$ can still often be expected towards strengthening the overall solvency position of the company. For evaluating the adequacy of the premium rates it may also sometimes be useful to employ the **discounted capital value of future residual profit**.

$$Y_0 = \sum_{t=1}^{\infty} Y(t) \cdot v^{t - \frac{1}{2}}. \qquad (15.2.12)$$

Both of these expressions may usefully be given in terms of the corresponding per capita amounts

$$\begin{aligned} y_{\text{tot}} &= Y_{\text{tot}}/l(0) \\ y_0 &= Y_0/l(0). \end{aligned} \qquad (15.2.13)$$

The discounted value of future profits is a random variable and may turn out with hindsight to have been negative, even when a satisfactory rate of return on capital and contribution to residual surplus was planned. The actual outcome on a single cohort of business may not be critical if there are offsetting excess profits on other cohorts. Balancing the profitability of the whole business in the face of random fluctuations in the experience is a key element of business strategy, for which an overall stochastic model of the business is an essential tool.

(j) Analysis of cohort behaviour. The model drafted above makes it possible to explore the profitability and risk expectations associated with the premium, reserving and bonus bases. The algorithms permit long-term simulation, which generates a distribution of outcomes. The combined effect of all the bases, including mortality, investment return, expenses, etc. can be evaluated, and the adequacy of the safety margins can be assessed, as well as exploring the resilience of different bonus strategies. Furthermore, sensitivity to potential adverse events

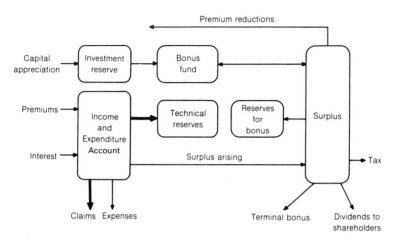

Figure 15.2.2 *The process involved in determining and allocating surplus according to the British practice. The reserve for bonus is for reversionary (guaranteed) bonuses and the bonus fund is for terminal (non-guaranteed) bonuses.*

can be tested, for example, how to cope with an increased level of inflation or an abrupt crash in equities.

Note that the model includes strongly dynamic features. Figure 15.2.2 illustrates the processes. The reserving process and the distribution of surplus can both be seen as filters in a control process. Strengthening or weakening technical reserves, or provisions, reduces or increases the surplus arising. In the distribution of surplus, dividends to shareholders, guaranteed bonuses (requiring additional technical reserves), terminal bonus and, where appropriate, premium reductions, compete for the allocation. The transfer to the bonus fund is then a balancing item, or, if this is determined by some criteria of maintaining the bonus fund at a particular size, other items, such as terminal bonus or shareholder dividends, must take the strain. The rules controlling the allocation of profits to shareholders' dividends, to guaranteed (reversionary) bonuses and to the bonus fund for discretionary (terminal) bonuses are critical. The **investment strategy** is extremely important and can be tested either in the context of the cohort alone or for the portfolio as a whole, as will be considered in section 15.3. Together with inflation, investment strategy is of key significance in the control of profitability and solvency.

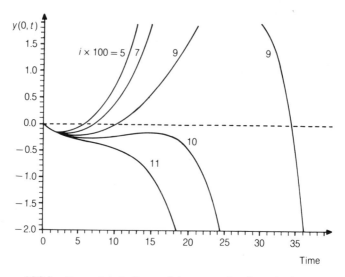

Figure 15.2.3   *Deterministic flows of the accumulated residual cohort profit* $y(0,t)$ *for different rates of inflation i.*

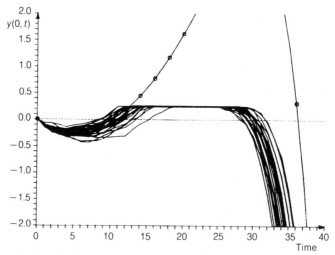

Figure 15.2.4   *Flows of the accumulated cohort profit* $y(0,t)$ *simulated by applying a bonus barrier. Profits taking* $y(0,t)$ *over the barrier are assumed to be distributed. The curve with circles is the corresponding deterministic curve with* $i = 0.09$.

(k) Examples. It is useful first to run the process deterministically, leaving aside stochastic considerations. Figure 15.2.3 illustrates an example in which the influence of the rate of inflation i is analysed. In this example all the profits are accumulated in the bonus fund, without any guaranteed bonuses being allocated. As can be seen, the postulated cohort incurs losses with high rates of inflation (in excess of the rate of return on investments).

The case $i = 0.09$ is made stochastic in Figure 15.2.4, incorporating a bonus barrier with coefficient $b = 0.2$.

REMARK   Figures 15.2.3 and 15.2.4 are taken from Beard *et al.* (1984) section 8.3, where the following bases were used: endowment policies with the term $n = 40$, mortality reserving assumption $q(t) = 0.0006 + \exp(0.115 \cdot x - 8.125)$, actual mortality $0.85 \cdot q(t)$, $l(0) = 10\,000$, expense loading $E(1) = 0.02 \cdot S$, $E(t) = 0.002$ for $t > 1$, the lapse rate $\rho(1) = 0.16$, $\rho(2) = 0.11$ and $\rho(t) = 0.01$ for $t > 2$, rate of interest $j = 0.09$, surrender value $V(t)^+ \cdot S$.

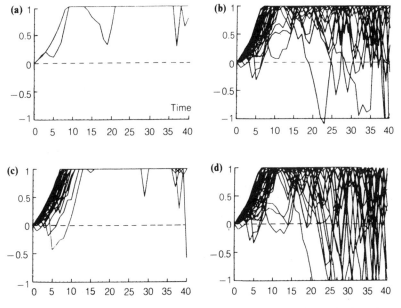

Figure 15.2.5   *Examples of the simulated development of the bonus fund, profit being truncated by a bonus barrier with $b = 1$. (a) Deterministic scenario and a single simulated realization, (b) 40 simulated realizations, $m_0 = -0.015$, $\sigma_0 = 0.026$, (c) Deterministic rate of interest, $m_0 = 0.025$, $\sigma_0 = 0.003$, (d) Average rate of inflation 5% (instead of 3%), $m_0 = -0.010$, $\sigma_0 = 0.073$.*

Clearly, the outcomes depicted in Figures 15.2.3 and 15.2.4 are not satisfactory. The next step would be to experiment with alternative scenarios until an acceptable one is found.

Pentikäinen and Pesonen (1988) demonstrated how the model could be used for **sensitivity analysis**. Figure 15.2.5 is from that paper. The standard rate of inflation was $i_0 = 0.03$, the average rate of return on investments $j_0 = 0.04$ and the bonus barrier coefficient $b = 1$; other bases are as given in the context of Figure 15.2.3.

The mean value m_0 and standard deviation σ_0 of the discounted profit y_0 (15.2.13) were also evaluated from the simulations.

As can be seen from the examples, the system was very volatile. This feature will not be discussed further here, since our purpose is only to set out tools for the investigation of cohort behaviour, and not to make the investigations (which would require a comprehensive consideration of the outcomes of many different scenarios).

15.3   Analysis of the total business

(a) Portfolio as a sum of cohorts.   The analysis of the total business of a life insurance company is based on breaking down the portfolio into cohorts of the type defined in the previous section. Each of the cohorts will be handled separately, applying the technique derived above, and the results for the whole business will be obtained by summing the cohort outcomes.

The cohorts can be specified, for example, by the following characteristics

- the type of policy;
- the year in which the policy is written;
- the sex of the insured;
- the entry age of the insured.

A sequence of cohorts with different entry years, but other characteristics the same, can be described as shown in Figure 15.3.1.

The current time is denoted by t and the time interval from the entry of the cohort up to the current year, the development time, by d. Hence, the entry year is $t - d$.

Each type of cohort gives rise to a similar sequence of consecutive cohorts as is depicted in Figure 15.3.1, for example, endowment and term insurance policies separately, specified as different cohorts according to entry age, maturity age, etc.

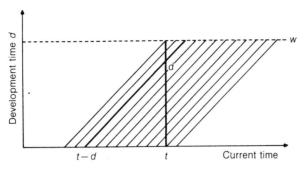

Figure 15.3.1 *A portfolio consisting of cohorts which are represented by ascending diagonal lines (Lexis diagram, cf. Figures 9.5.1 and 9.5.2).*

A practical problem is that the number of cohorts can grow quite large if the different types of policies, differing entry ages and different maturity ages, etc. are all treated as separate cohorts. However, it is likely that, without undermining the validity of the model, the number of cohorts can be kept reasonable by combining cohorts which have similar relevant characteristics. For example, the entry ages and maturity ages can be grouped into classes in fairly coarse intervals. As a short cut, all entry ages could be combined at a single average value, derived from the experience.

As time progresses, cohorts mature and exit from the portfolio, and new cohorts enter. Assumptions are required about new cohorts such as the expected size of the cohort, distribution of entry ages, etc.

(b) Bonus fund. The key variable in the stochastic analysis is the size of the bonus fund, denoted by $U_{bf}(t)$, now attributable to the total portfolio. It is simulated for $t + 1$, $t + 2$,... using an algorithm consisting of the following steps

(1) For each cohort the profit (15.2.8) is generated, denoted by $Y_j(t - d, d)$ where the subscript j specifies the cohort, $t - d$ the entry year of the cohort and t the current year for which the evaluation is made. However, Y is first calculated without the dividend and bonus terms $D(t)$ and $G(t)$ of (15.2.8). Let

$$Y_1(t) = \sum_{j,d} Y_j(t - d, d) \qquad (15.3.1)$$

be the total profit in year t from all surviving cohorts.

(2) The total profit is then apportioned as shown in Figure 15.3.2, applying rules which the user of the model must specify.

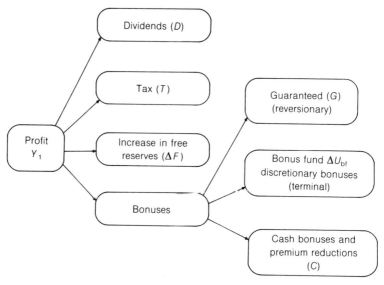

Figure 15.3.2   *Distribution of profit.*

(3) The amount of the bonus fund at the end of year t is

$$U_{bf}(t) = (1 + j(t)) \cdot U_{bf}(t - 1) + Y_1(t)$$
$$- D(t) - T(t) - \Delta F(t) - G(t) - C(t). \qquad (15.3.2)$$

REMARK   Care is required in checking that all parts of the return on investments are taken into account once and only once in the algorithm and are consistently discounted to the right time point. In order not to complicate the formulae by this kind of detail, these aspects are left to readers who are interested in applying the proposed model.

By repeating steps (1)...(3) the Monte Carlo simulation can be carried out.

(c) Examples.   Figure 15.3.3 shows examples of the use of the model outlined in the previous section. The diagrams are from the paper by Pentikäinen and Pesonen (1988), applying the same bases as in Figure 15.2.5.

Graph (a) in Figure 15.3.3 illustrates the case where no bonuses are given. The surplus grows very large, if the bases are conservative and no particular adverse event occurs. This might describe the situation if only non-participating (non-profit) policies were sold.

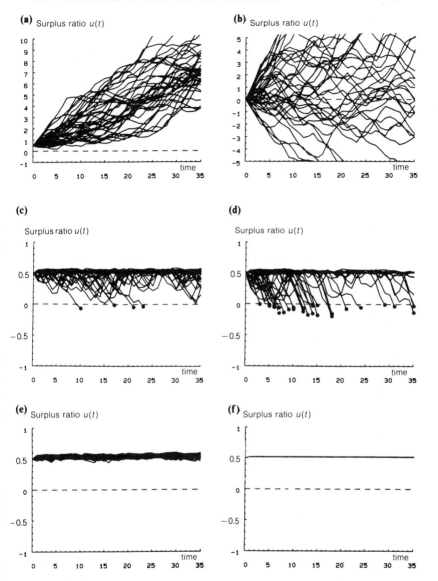

Figure 15.3.3 *Examples of the simulation of the whole portfolio. (a) All profit is accumulated in the bonus fund, no barrier, (b) No safety loadings, no ruin barrier, (c) Standard data, ruin barrier coefficient $b = 0.5$, (d) First order rate of interest $j_0 = 5\%$ (instead of standard 4%), (e) First order interest rate $j_0 = 3\%$, (f) Inflation and interest deterministic but mortality stochastic.*

The problem is to find premiums which are both competitive and still reasonably safe. Graph (b) shows the position if no safety loading is included in the premium calculation. In this situation losses are as likely to occur as profits. In practice, a configuration somewhere between the Graphs (a) and (b) should be sought to give reasonable security, in conjunction with competitive premium rates. A stochastic model can be expected to be a useful tool in testing different scenarios as an aid to decision-making.

Graphs (c), (d) and (e) demonstrate a sensitivity analysis. The primary rate of interest j_0 has the values 4%, 5% and 3% respectively.

Finally, in Graph (f) the rates of interest and inflation are made deterministic, allowing only the mortality to be stochastic. The influence of the random fluctuation is so slight that it cannot be revealed in the scale used in the diagram. This tends to confirm the view that the solvency of life insurers depends mainly on keeping under control the volatility in the return on investments and the rate of inflation.

Note that the graphs in Figure 15.3.3 relate to endowment policies. In the case of term insurance (pure death risk) the influence of the uncertainty of mortality can be expected to be more significant. Invalidity and sickness risks, which are often covered by supplementary benefits under ordinary life insurance products or even by independent policies issued by life insurers, are particularly volatile. These kinds of business are not dealt with in this book but the methods outlined can be modified to cope with them.

The examples illustrated confirm the well-known fact that the viability of life insurance depends on appropriate premium bases and investment policy, prudent reserves and bonus rules, as much as on solvency margins and capital requirements.

REMARK   Let us recall that the numerical and graphical examples set out in this book on the outcomes of the proposed models do not claim to prove any universally significant features about the phenomena which are treated. They are merely intended to illustrate the use of the methods, and their validity is limited by the particular assumptions made.

(d) Reserves and solvency tests.   The technical reserves are normally calculated deterministically using classical formulae of the type (15.1.4). In many countries the premium bases and reserving bases have traditionally been the same, in accordance with the condition (15.1.5). However, this need not be the case and no longer is so in practice. The assumptions on mortality, rate of return, etc. for the reserving basis should be amended according to the changing environment, with the proviso that they should still be sufficiently prudent. Stochastic methods can assist in evaluating whether a sufficient margin of caution has been maintained, both for individual cohorts and for the business as a whole.

The model outlined in section 15.3(b) provides an insight into various risks of stochastic nature which the classical deterministic technique cannot properly reveal. Hence, it can support the conventional actuarial analyses concerning the financial position of the insurer.

The model can also be modified to assist the solvency control which was described in section 14.6. Both the going-concern and the run-off tests can be performed (section 14.6(d)).

In the run-off case a notional discontinuation of the business in year t is assumed, so that no new cohorts are assumed to enter and bonuses continue to accrue only for the existing cohorts. In deterministic terms the condition of adequacy at the test time point t is

$$A(t) \geqslant L(t) + Y(0, t) \qquad (15.3.3)$$

where A and L are the assets and liabilities (including guaranteed bonuses but excluding the bonus fund) and $Y(0, t)$ is the expected value of the accumulated bonus fund (15.2.9). Since the discretionary bonuses are not guaranteed, the insurer can still be considered solvent if $A(t) \geqslant L(t)$. However, a normal objective of the supervisory authority might be to ensure that a reasonable level of bonus, commensurate with the level of premiums, can still be paid in the run-off situation. The test can be made stochastic by running the model of section 15.3(b), but excluding new cohorts. When all the cohorts which were in existence in year t have been exhausted, the residual amount of the sum of the free reserve and the bonus fund U_{bf} should be non-negative at an acceptable probability level. The philosophy is effectively the same as in the consideration of the run-off of outstanding claims in section 9.5; the outcome can be illustrated as in Figure 9.5.4.

There are a number of variants of the run-off or closed fund situation which was described in the previous paragraph. The full run-off process, if realized, could in practice require decades. A common procedure, therefore, in the case of new business being discontinued, is to sell or transfer the portfolio to some other insurer. This can be expected to be possible if the portfolio has a positive value, which might be the case if the condition (15.3.3) is fulfilled. In the last resort no discretionary bonuses need to be awarded, so that Y in (15.3.3) can be set to zero. If a transfer of the portfolio is not possible, and condition (15.3.3) cannot be met, even with the Y

term set to zero, then an immediate winding-up of the benefits and assets would normally be set in hand. This situation will not be dealt with in this book. It would require definition of the winding-up benefits and would depend on the forced realization of assets, expenses of liquidation, etc.

REMARK   Some possible reasons for a serious financial position of an insurer might be a lax selection of risks or an inadequate level of premiums in respect of the policies issued. Features of this kind should be taken into account in setting the basis for the reserves and when carrying out the bonus fund simulation, possibly resulting in a negative outcome of the test (15.3.3).

(e) Managerial aspects.   The model outlined above can be used to support both the premium-rating and the analysis of financial strength of a life insurer. In particular, the trade-off of risk and reward of alternative investment strategies, as well as different bonus policies, can be investigated, as can their relationship with other relevant factors.

The individual cohort approach of sections 15.1 and 15.2 is suitable for rating and the total portfolio simulation of section 15.3 both for rating and for the analysis of the current state and future prospects of the business.

The model can be developed further by incorporating dynamics to match local needs and business practices. The fact that life insurance premium rates of policies already issued cannot (as a rule) be increased can lead to difficulties in a situation where, for some reason or other, the experience deteriorates significantly, requiring, for example, strengthening of existing technical reserves. A number of strategies can be considered to retrieve the situation, for example, transfers from bonus funds and reduction of discretionary bonuses (as regards with-profit business), expense economies such as reducing staffing levels and closing branches, reducing new business strain by restricting sales of new policies, changing investment strategy, etc. The effects of alternative strategies can be tested using the model. Unfortunately, significant improvements in the financial position can usually only be achieved over time, especially if premium rates can only be increased for new policies. Analysis of financial strength, and resilience to meet future adverse scenarios is, therefore, vitally important. Stochastic modelling can support this analysis, provide early warning of problems and point to appropriate remedial action.

Another application of the model is in analysing the behaviour of components of the portfolio, for example, the profitability of different cohorts or different policy types. This is shown in Figure 15.3.4, where the profit development both for the portfolio as a whole,

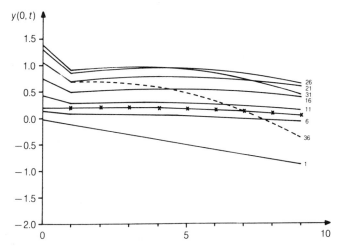

Figure 15.3.4 *Examples of the development of profit (or loss) of a supposed portfolio and eight of its cohorts with development times 1, 6, 11, ..., 36 years. The curve with asterisks represents the joint result of the portfolio. Source: Beard et al., 1984, section 8.*

and for some of the cohorts separately, is displayed. Even though much simplified, the example is realistic in that profitability resulting from inflation and other circumstances may differ considerably within the portfolio, for example, between recent and older cohorts. Features of this type are useful to know for business planning.

(f) Discussion. The techniques traditionally used in life insurance for evaluating the appropriateness and adequacy of premiums and reserves have mainly been deterministic. This gives rise to difficulties in finding a proper rationale for any particular level of safety margin or other security measures. Their purpose is, in fact, to protect the insurer from the impact of various potential adverse events, including stochastic variations. Proper analysis of the stochastic risks necessarily requires a stochastic technique, such as the one outlined above. Furthermore, it is not necessary for each element of the basis in isolation to be cautious, but only their combined effect.

Attempts to support the traditional deterministic life insurance mathematics with stochastic analysis were made at the end of the nineteenth century in the form of the individual risk theory. A summary of the ideas of the time was given by Bohlmann and others at the International Congress of Actuaries in Vienna in 1909. The idea was to calculate the capital value of future income and outgo, or rather the expected value of future profit, corresponding to the quantity y_0 in (15.2.13), as well as the related standard deviations (corresponding to σ_0 in section 15.2(k) above). Consideration was limited to the stochastic variability of mortality. The theory did not find any wide application.

More recently, the advent of electronic data processing opened up the way for more realistic stochastic analysis of life business, but actuaries were slow to see any advantage in the use of stochastic techniques, as compared to traditional deterministic models. The problem of reserving for guaranteed maturity benefits under unit-linked insurance contracts in the UK led to the development of stochastic models for investments and inflation, as described in Chapters 7 and 8. Comprehensive stochastic simulation models for life insurance have now been developed in the UK and in Finland. Analysis of financial strength by exploring the resilience of the whole portfolio to alternative adverse scenarios has become a professional requirement in North America, under the name of cash-flow testing in the USA and dynamic solvency testing in Canada. However, in neither case is any stochastic analysis required. The development of stochastic analysis of life insurance is still at an early stage, awaiting further research and the emergence of satisfactory standards for practical applications.

(g) References. A review of individual risk theory was given by Cramér (1930). A brief summary can be found, for example, in Beard *et al.*, 1984, section 8.2. Many of the formulae are presented in the US Society of Actuaries text-book *Actuarial Mathematics* (Bowers *et al.*, 1986).

A comprehensive simulation model for life insurance along the lines described in this chapter was outlined in Beard *et al.* (1984), Chapter 8 and by the Faculty of Actuaries Solvency Working Party (1986). Since then the topic has been the subject of increasing attention, both in the UK and in Finland, with papers, for example, by the Faculty of Actuaries Bonus and Valuation Group (1989), Pentikäinen and Pesonen (1988), Roff (1992) and MacDonald (1993). Brender (1988) has developed a comprehensive model for the analysis of life insurance, to support the development of dynamic solvency testing in Canada. It is deterministic but employs scenario technique to permit the investigation of alternative futures. Hardy (1991) explores the merits and demerits of the scenario technique as compared to stochastic approaches.

CHAPTER 16

Pension schemes

16.1 Pension structures and definitions

(a) Different pension structures. Pension insurance business, as traditionally offered by insurance companies, consisted of endowment assurance and deferred annuity contracts geared to the normal retirement age under the scheme. It can be considered as a special case of life insurance, although sometimes with different tax rules and restrictive conditions on surrender. However, in many countries pensions are provided by a rather different mechanism, through **pension schemes** which are established in the form of pension funds, trusts, etc. They are arranged for a particular defined group of persons, usually for the employees of a single employer or group of employers. Other funds are established for people belonging to some specified trade or profession, for example, the funds run by labour unions in the USA. These kinds of pension institutions present a different range of problems and warrant separate consideration.

REMARK The general lines of modelling to be described in what follows may also be applicable to certain types of national statutory pension schemes, for example in Finland, where a mandatory national scheme is administered by decentralized private carriers. The modelling may relate to the whole scheme or to individual carriers.

The institutional form of pension schemes varies from country to country, according to historical traditions and national regulations. The responsible body may be a legally independent fund or trust, albeit established by the employer or group of employers, or the employer may pay the pensions directly from the company's own resources. Some parts of the organization, investment management or benefit payment may be sub-contracted to an insurance company. For our present purposes the institutional structure is irrelevant. We will, therefore, simply speak about a pension **scheme**,

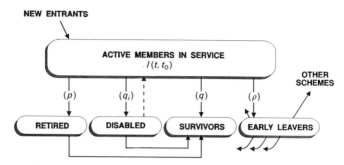

Figure 16.1.1 *Classification and movements of the members of a pension scheme.*

which is responsible for providing specified benefits to a group of **members** and receives **contributions** (corresponding to premiums) from a **sponsor** (employer) and, possibly, from the members themselves.

There is a great variety of pension schemes, differing in scope and in the definition of the benefits, as well as in the financing techniques. We can only briefly describe a few examples. The purpose is to highlight some fundamental principles. The examples which have been chosen are intended to be representative. The aim is to show how stochastic elements can be incorporated into the conventional deterministic considerations.

(b) Definitions. The benefits granted by pension schemes usually include one or more of the following:

- **old age pension** payable for the remainder of life from a specified pensionable age, x_r,
- **disability (invalidity) pension** in case of ill-health before reaching the pensionable age,
- **death benefits**, for example, survivor's pension, and
- **withdrawal** benefits for early leavers.

It is useful to classify the members and beneficiaries as is shown in Figure 16.1.1.

16.2 Pension formulae

(a) Age and time scaling. The financial position of the scheme and the benefits payable to the members are time-dependent. Both the calendar time t and the age x of the member or cohort of members

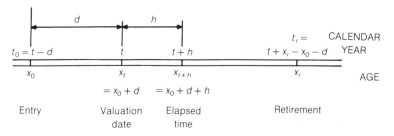

Figure 16.2.1 *Definition of time points and ages for a cohort member.*

are required. Figure 16.2.1 sets out some definitions and relations.

We assume, for the sake of brevity, that all members enter the scheme at age x_0 and retire, if still alive, at age x_r. The **current** time point t is the time at which the actuarial valuation of the scheme is performed and to which future transactions are discounted.

(b) Ageing and renewing cohorts.   All members of the same age (and, according to the above assumption, the same year of entry) constitute a cohort. Let the size of a cohort which entered in year t_0 be $l(t_0, t)$ in year t. Members exit from the cohort by retirement when they reach the pensionable age x_r, on account of ill-health at age x with frequency $q_i(x)$, on death in service at age x with frequency $q(x)$, or on leaving the scheme early at age x with frequency $\rho(x)$, as is depicted in Figure 16.1.1. Hence, the cohort, having reached the elapsed time point $t + h$ (Figure 16.2.1), ages according to the algorithm

$$l(t - d, t + h + 1) = l(t - d, t + h) \cdot [1 - q(x_{t+h}) - q_i(x_{t+h}) - \rho(x_{t+h})]$$
$$(16.2.1)$$

where $t - d$ is the year of the entry and the age of the cohort members at elapsed time $t + h$ is $x_{t+h} = x_0 + d + h$. It is assumed that the number in the cohort at the valuation date $l(t - d, t)$ is counted as an actual number from the scheme records when the valuation (analysis of the financial position) is performed in year t. This formulation is based on so-called **dependent decrements**. An alternative approach involving **independent decrements** leads to a product of terms such as $(1 - q_x(t + h))$ instead of the square-bracketed expression in (16.2.1) (Neill, 1977).

The retirement pensioners, disability pensioners, recipients of survivors' pensions and early leavers are each grouped in cohorts according to age in so far as the scheme is liable for future pensions or other benefits. Members exiting from active service give rise to a new group of additional members in one of these groups. For example, when a cohort reaches retirement age its surviving members move to become a new cohort of retired members. Similarly, the active members who are assumed to retire early on ill-health grounds in year $t + h$ move to the group of disabled at that elapsed age. The ageing of the groups of beneficiaries is dealt with by formulae analogous to (16.2.1).

For some applications an evaluation is also needed in respect of new cohorts entering after the valuation time point t (**open fund**). In this case an explicit assumption may be made about the number of new entrants in future years or an assumption may be made concerning the total size of the scheme's active membership, for example, that it will remain constant from year to year or will grow according to some specified rule. To maintain constant membership, the size of the new cohort entering in year $t + h$ must be the same as the total number of members exiting from all cohorts in the year according to (16.2.1)

$$l(t + h, t + h) = \sum_{d=1}^{\infty} [l(t + h - d, t + h - 1) - l(t + h - d, t + h)].$$

(16.2.2)

This can be further modified if a rate of growth is postulated for the overall active membership of the scheme or if allowance is made for disability pensioners to recover and to return to active service.

We will also need the probability that a member at age x_t is still an active member at age x_{t+h}. In terms of the cohort size variable (16.2.1) it is

$$p(x_t, x_{t+h}) = \frac{l(t_0, t + h)}{l(t_0, t)}.$$

(16.2.3)

Note that this expression does not depend on the entry year t_0 because the cohort size factor cancels out.

(c) **Pensionable pay.** The **salary** $S(x, t)$ of a member at age x and time t is assumed to be the basis for benefits and contributions. When considering a cohort it is the average salary of the members of the cohort at a specified age and time point. We assume that the average salary is obtained from the payroll register at time t and that it progresses thereafter as follows

$$S(x + h, t + h) = S(x, t) \cdot \frac{s(x + h)}{s(x)} \cdot \frac{I(t + h)}{I(t)}. \qquad (16.2.4)$$

Here $I(t)$ is a general earnings index (section 7.2) and $s(x)$ a **career salary scale** function which describes the increase in average salaries for the cohort, other than that which can be explained by the index. In deterministic calculations the index of general salary increases is usually replaced by an exponential multiplier $(1 + i)^h$.

The salary which is taken into account may be truncated either at the top or at the bottom or both. The latter adjustment is usually designed to integrate or partially integrate the scheme with the social security arrangements. That part of salary which is taken into account in scheme considerations is called **pensionable pay**. Appropriate modifications may need to be made to formula (16.2.4).

When new entrants are included in the analysis, a rule is required for the initial level of their salaries on entry. This needs to allow for both inflation and real growth in the intervening years.

(d) Benefits. Pensions are denoted by H modified by subscripts, and the capital values of future pensions and other benefits by X, similarly modified.

(1) **Old age pension** is defined according to the final salary principle

$$\begin{aligned} H_r(x_r, t_r) &= f_r \cdot (x_r - x_0) \cdot S(x_r, t_r) \quad \text{(initial level)} \\ H_r(x_r + h, t_r + h) &= g(h, t_r + h) \cdot H_r(x_r, t_r) \end{aligned} \qquad (16.2.5)$$

where the parameter f_r determines the pension as a fraction of pensionable pay, for example, $1/60$ (when the ages x are given in years). $S(x_r, t_r)$ is the final pay, often defined as an average of (possibly index-adjusted) pensionable pay from the latest few years (e.g. 3 or 5).

The auxiliary function $g(h, t)$ controls the development of pensions in payment from $t - h$ to t. It can introduce a direct link to an index, provide constant exponential growth, $g(h) = g^h$ (where g is a constant > 1) or it can be equal to unity if the pension amount remains constant.

In practice the actual current amounts of pensions $H_r(x_t, t)$ at the valuation date are obtained from the scheme records. Then the estimated level of pension in future is obtained from

$$H_r(x_{t+h}, t + h) = H_r(x_t, t) \cdot \frac{g(h_t + h, t + h)}{g(h_t, t)} \quad (x_t \geqslant x_r) \quad (16.2.6)$$

where h_t and $h_t + h$ are the times elapsed from the date of retirement up to t and $t + h$ respectively.

The capital value at the end of year $t \geq t_r$ of a pension already in payment is

$$X_r(x_t, t) = \sum_{h=1}^{\infty} p(x_t, x_{t+h}) \cdot v(t, t + h) \cdot H_r(x_{t+h}, t + h) \qquad (16.2.7)$$

where $p(x_t, x_{t+h})$ is the probability that a pensioner of age x_t will still be alive at age x_{t+h}.

(2) The benefit in case of early retirement owing to ill-health may be a **disability pension** H_i, which is defined in accordance with (16.2.5) but with the retirement age x_r replaced by x_i. The time span $(x_r - x_0)$ is replaced by the period from entry to early retirement $(x_i - x_0)$. However, the scheme may be designed to provide more adequate disability cover for younger members, for example by allowing extra years to count. In some schemes the benefit is based on projected service to retirement age, in which case the time from entry up to the (hypothetical) retirement age $(x_r - x_0)$ should be used, i.e. the formula (16.2.6) is applied as such. The capital value $X_i(x_t, t)$ is calculated in an analogous way to (16.2.7).

In evaluating survival probabilities for disability pensioners it should be borne in mind that the mortality experience is likely to be higher than that relating to active members or normal age retirement pensioners, particularly for an initial period. Furthermore, there is a possibility that disability pensioners may recover and return to active service. These details will not be considered here.

(3) The **death benefit**, with initial capital value X_d, may be a lump sum, or the capital value of the survivors' pensions for the widow (or widower) and children. The treatment of survivors' benefits arising on death in service is analogous to that of normal age retirement pensions and disability pensions. The death risk may be insured, in which case X_d is the (recurrent) single premium and the box for survivors in Figure 16.1.1 is empty.

Survivors' pensions may also arise on the death of a normal age retirement pensioner, disability pensioner, or early leaver with deferred rights to pension. The amount is usually directly related to the pension that was in payment to the member (or in deferment). The trigger probability is the mortality rate for the appropriate type of pensioner.

(4) Finally, the capital value X_ρ of **withdrawal benefits** for an early leaver is needed. This may correspond to the accrued liability (see section 16.3(c) below), although this will depend on whether the early leaver's rights to dynamism of deferred benefits are equivalent to the funding assumptions for accrued rights. There are many types of withdrawal benefits. If the benefit is given in the form of deferred pension, it constitutes a current liability of the scheme. If it is insured, or transferred to another scheme to which the leaver has moved, the benefit cost is the insurance premium or transfer value and thereafter there is no liability to the scheme.

For some purposes **wind-up benefits** will need to be defined for a notional discontinuation of the scheme. They might be equal to the accrued benefits, or the withdrawal benefits, if different, or have their own definition which takes into account the actual amount of assets available.

(e) Contributions. There are two principal types of scheme, **defined benefit** schemes and **defined contribution** schemes. In the former case the benefits are defined *a priori* and the contribution has to be determined to enable the benefit liabilities to be met. When the estimated value of the assets or the liabilities changes, for example as a result of inflation, investment performance, salary development, retirements, deaths and withdrawals of members, etc., a revised level of contributions can be assessed at the periodical actuarial valuation.

In a defined contribution scheme the rate of contribution is fixed and the benefits are purchased from an insurance company or directly reflect the value of the investments purchased by the contributions.

We shall consider only the defined benefit case. There are many methods used to determine the contributions. They are usually expressed as a percentage of pensionable pay

$$B(t) = c(t) \cdot S(x, t) \tag{16.2.8}$$

where the contribution rate $c(t)$ may be set individually for each member according to age or for groups of individuals. It is commonly defined as a flat percentage $c(t)$ of the pensionable payroll, hence being the same for all members in year t. It will usually remain constant during certain periods. However, it will normally be adjusted, according to the financial position of the scheme, each time that an actuarial valuation of the scheme is performed, for example, every three years in the UK, so it is appropriate to define the contribution rate $c(t)$ as a function of time t.

A specified part of the contribution, or a fixed percentage contribution rate, may be credited from the members themselves, but for our purposes only the total amount of contributions is relevant.

16.3   Deterministic methods of pension funding

(a) Evaluation of the liability of a pension scheme using deterministic methods.   The liability in respect of benefits to be paid in the future corresponds to the conventional technical reserves in life insurance. The calculation is broken down into cohorts, as is set out in Figure 16.3.1, in a way which will also be suitable for the stochastic approach. In addition to existing cohorts, future cohorts of new entrants are also depicted, because they may sometimes need to be included. Time t corresponds to the valuation date and is the time to which future incomes and expenditures are discounted.

The size $l(t + h - d, t + h)$ in year $t + h$ of a cohort which entered in year $t + h - d$ is obtained from the algorithm (16.2.1), and the initial sizes of future cohorts from (16.2.2), working on the assumption of constant total membership size. Here d is the development time, i.e. the age of the cohort counted from its date of entry.

The discounted liability $W_X(t + h, d)$ resulting from the events occurring to the cohort when it is in position $(t + h, d)$ in the figure

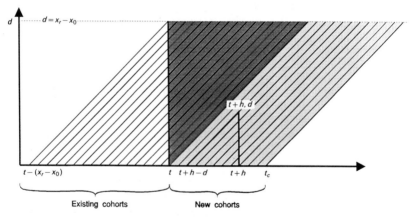

Figure 16.3.1   *The cohorts of members in service represented by ascending diagonal lines.*

(i.e. elapsed time $t + h$ and cohort development time d) is

$$W_X(t + h, d) = l(t + h - d, t + h) \cdot v(t, t + h)$$

$$\times \sum_b q_b(x_0 + d) \cdot X_b(x_0 + d, t + h) \quad (16.3.1)$$

where the subscript b refers to the type of the benefit numbered from (1) to (4) in section 16.2(d). In order to achieve a formal uniformity of the notation for the different types of benefit, the probability of normal age retirement is denoted by

$$q_r(x) = 1 \text{ for } x = x_r \text{ and otherwise} = 0. \quad (16.3.2)$$

The capital values X_b of the benefits are given by the definitions in section 16.2(d).

The total liability of the scheme is obtained from (16.3.1) by summing over all the relevant cohorts and all positive values of h and deducting the capital value of the future contributions

$$V(t) = \sum_{h,d} W_X(t + h, d) + \sum_b W_b(t) + W_E(t) - \sum_{h,d} c(t + h) \cdot W_S(t + h, d).$$

$$(16.3.3)$$

Here $W_b(t)$ is the capital value of future benefits (pensions) of type b for benefits already in payment (or in deferment) at the valuation date t. Future new beneficiaries are taken into account via the variables X_b in the expression for $W_X(t + h, d)$ (16.3.1). W_E is an estimate of the capital value of the future costs of administration and $W_S(t + h, d)$ is the discounted value of the pensionable pay of the cohort in position $(t + h, d)$.

There are differing ideas about what future commitments should be taken into account in the above summations. One approach is to limit considerations to **existing cohorts**, i.e. members in service at, or prior to, the valuation date t. Then the summation of the first and last terms of the right-hand side of (16.3.3) should cover the cohort cells in the **run-off triangle** shaded in a darker colour in Figure 16.3.1. This approach corresponds to the normal practice for defining technical reserves in both life and general insurance. Under this approach the contribution rate in respect of future entrants should be adequate to meet the liabilities generated by the new cohorts.

Another procedure is to extend the summations over the whole trapezium which is a lighter shade in the figure, hence including also the **cohorts of new entrants** with entry years $t + 1, t + 2, ..., t + t_c$. The period for which new entrants are taken into consideration, i.e. the time parameter t_c, can be the interval up to the next valuation, for example three years, or longer according to the philosophy adopted in the valuation.

The assets of the scheme are earning an overall return at rate $j(t)$ (Chapter 8) and are obtained from the algorithm

$$A(t) = (1 + j(t)) \cdot A(t - 1) + \text{contributions} - \text{expenditures}. \quad (16.3.4)$$

Pension schemes are not usually concerned about whether investment return is in the form of income or capital gains, so we will ignore the distinction, making use of the combined rate of return on investments $j(t)$. The contributions and expenditures can be obtained as by-products of the calculation of the W's in (16.3.3).

A surplus or deficit at time t can now be derived.

$$Y(t) = A(t) - V(t). \quad (16.3.5)$$

It is important that the assets and liabilities be determined in a consistent way, if spurious surpluses or deficits are to be avoided. The assets in (16.3.5) have been taken at market value, representing the retrospective accumulation of excesses of income over outgo. However, the market value of the assets currently held can also be regarded as the discounted present value, at the **market rate of discount**, of the expected future proceeds from the investments. Consistency can be achieved by discounting the liabilities at the market rate of discount, which may differ by term. Unfortunately it may be difficult to determine the corresponding market assessment of future earnings increases and future price increases (although the latter may be available in markets where there are index-linked government securities) and the key assumptions in pension fund valuations are usually the net rates of return relative to prices and earnings.

Consistency can be achieved more readily by regarding the assets as a flow of future income, arising from interest payments, dividends, rental income, maturity values of bonds and possibly the future sale value of certain assets. Income from the assets can then be discounted at the same rate of discount as is used for fixed liabilities. Equity dividends and property rental income can be assumed to grow at an annual rate consistent with the price and earnings increase assumptions.

(b) Funding rates and solvency. According to the principle of defined benefit schemes (section 16.2(e)) any shortfall is restored by amending the rate of contribution for some defined future period. Surplus may be used to increase benefits, particularly if they are not automatically increased to maintain their real value, or in certain circumstances may be returned to the sponsoring employer. However, surpluses will often be dealt with in the same way as deficits, by adjusting future contributions over a defined period. After an actuarial valuation of the financial state of the scheme the contribution rate $c(t)$ will be changed, either permanently, or with an increase or reduction spread over a certain future period. The adjustment to the contribution rate is determined so that the surplus (or deficit) Y either vanishes or attains some other target value.

This procedure results in fluctuations in the contribution rate from time to time. For example, if the rate of inflation has exceeded the level assumed in earlier valuations, this can give rise to an abrupt jump in the contributions required in schemes where the commitments are based on final salaries. If the scheme is small, individual cases of ill-health retirement or death in service can cause abrupt steps in expenditure which are considerable compared with the assets and contributions.

Solvency in respect of accrued benefits, having regard to possible fluctuations, can in principle be secured by similar methods to those employed in life insurance, namely by using cautious assumptions in valuing the liabilities and ensuring that the assets exceed the liabilities by a prudent margin.

Given that the employer may at any time cease to exist, or stop making contributions to the pension scheme, fulfilment of commitments in respect of future benefit accruals cannot be guaranteed. The main safeguards would be for an adequate level of contributions to be recommended by the actuary, for controls to be put in place to ensure that the recommended contributions are being paid into the scheme, at least so long as the employer is able to make them, and to provide a suitable structure for benefits to be run off if the scheme is discontinued.

In practice pension benefits are often not guaranteed in a contractual sense. Pension schemes are regarded as a tax-efficient mechanism for employers to pay deferred remuneration to employees. The ultimate guarantor of the benefits is the employer. In the light of this, accrued liabilities may often be assessed using less cautious assumptions than would normally be the practice in life insurance, and there is often no requirement for any solvency margin to be maintained. Any adverse developments which lead to deficits at future valuations are left to be dealt with by increased employer contributions once they have occurred and been quantified.

It may be possible for a pension scheme to reinsure certain liabilities with an insurance company, for example to cover adverse fluctuations of mortality of active members, or indeed the mortality risk associated with longevity in the case of pensions in payment.

It will be seen in section 16.4(b) that stochastic analysis can be useful in setting appropriate security levels for the accrued liabilities and for examining the overall financial dynamics of the pension scheme.

(c) Accrued liability. In the previous section the analysis was on an **ongoing basis**, i.e. the members were assumed to continue their service normally, with their contributions being paid, until exit by retirement or otherwise ((16.3.3) and (16.3.4)). However, we have already hinted that another approach is to assume a notional discontinuation of the service at the valuation date and to evaluate the accrued liability at that time point on the basis of service already completed (**past service benefits**). The idea is to ensure that the **accrued** or **earned benefits** are covered by the assets already in the fund, so as to increase the probability that these benefits at least will be able to be paid, even in the case of termination of the scheme or insolvency of the sponsoring employer. A similar approach is natural for defining the maximum level of fund which will benefit from tax-exempt status.

To evaluate the accrued liability a convention is necessary as to how the benefits are defined if the service is terminated before normal retirement age. One solution is to use the normal formula (16.2.5) but replacing the duration of service $x_r - x_0$ by $x_t - x_0$ where x_t is now the age at the (notional) termination (at the valuation date). With this change throughout in the benefit formulae, whenever the future event takes place which triggers the benefit payment, the formulae (16.3.1–16.3.3) can be applied, subject to the contribution term being dropped from (16.3.3).

Following through the benefit definition of (16.2.5), with a change only to the assumed duration of service, results in the accrued liability on the **projected unit** approach. This allows for the projected level of salary at the time the ultimate benefit comes into payment. Any guaranteed or statutorily required increases of pensions in payment would be allowed for in calculating this accrued liability. Allowance might also be made for discretionary increases if it is normal practice for them to be given.

A less generous approach to the definition of accrued liability would replace the final salary element of the benefit definitions by the salary at the valuation date t. This is the approach used in the **current unit** method of funding. It is closer to the concept of accrued rights which often applies in a winding-up situation, although this will depend on the trust deed or scheme rules. It would also correspond to typical rules for determining preserved rights for early leavers. If preserved rights have to be revalued, with a view to maintaining their real value during the years before they come into payment, or when the pensions are in payment (because of legal requirements or because of a requirement in the scheme rules), it would normally be appropriate to allow for this in assessing the value of the accrued liability.

16.4 Stochastic methods for pensions

Just as the life insurance techniques were developed in deterministic terms in section 15.1 and then transformed into stochastic form in sections 15.2 and 15.3, so the deterministic model of the pension scheme outlined in the last few sections can be elaborated into a stochastic approach.

(a) A simulation model for analysing a pension scheme can be built up with the following steps

(1) Generate the size in year t of the cohorts which entered in year t_0 using the algorithm

$$l(t_0, t) = l(t_0, t - 1) - \sum_b d_b(t). \qquad (16.4.1)$$

((16.2.1), (16.2.2), (16.3.1) and (16.3.2) give the notation). The numbers $d_b(t)$ of persons exiting from service can be simulated using the technique described in sections 15.2(a) and 15.2(g). New cohorts are obtained from (16.2.2). This gives a simulated size for the developing cohorts for all positions (t, d) of Figure 16.3.1.

(2) The existing beneficiaries, separately for each type b, are analysed in cohorts and the sizes of each age group in year t are generated as in step (1). The initial sizes of the new cohorts joining the

group of beneficiaries in year t are equal to the number of the corresponding exiting members $d_b(t)$ in (16.4.1). For example, new survivor beneficiaries are generated by deaths of members in service and by deaths of pensioners or deferred pensioners. However, in this case they are not the same individuals as the exiting members. Allowance should, therefore, be made, first for the probability that there will be a survivor to whom a benefit is due, and secondly for the likely age of the survivor.

(3) Simulate the index $I(t)$ of earnings inflation and, where appropriate for increases of pensions in payment, price inflation, by one of the methods described in Chapter 7. Then the pensionable pay $S(x, t)$ (16.2.4) and the benefits H_b, defined in section 16.2(d), can be calculated.

(4) Simulate the rate of return $j(t)$ on investments using the methods in Chapter 8. Investment strategies can be modelled in this step. The discounting factors $v(t, t + h)$ (15.1.2a) can now be calculated.

(5) Evaluate the expenditure for year t

$$X(t) = \sum_{b,x} I_b(x, t) \cdot H_b(x, t) + E(t). \tag{16.4.2}$$

The summation is over all the beneficiaries of each type b who receive benefits H in year t, together with an estimate of administrative expenses in year t.

(6) The contributions in year t can be calculated next

$$C(t) = c(t) \cdot \sum_{t_0} I(t_0, t) \cdot S(x, t). \tag{16.4.3}$$

The summation is over all the active members in year t.

(7) The simulated amount of assets at the end of year t is now given by

$$A(t) = (1 + j(t)) \cdot A(t - 1) + C(t) - X(t). \tag{16.4.4}$$

(8) The surplus or deficit for the year is the difference between the simulated assets and the amount of the then liabilities $V(t)$, evaluated by the deterministic method (16.3.3).

$$Y(t) = A(t) - V(t). \tag{16.4.5}$$

The deterministic value of liabilities $V(t)$ is obtained from the periodical actuarial valuations which can be assumed to be carried out every year for the purposes of the simulation, or at normal intervaluation intervals. If a normal intervaluation period of, say,

3 years is used, steps (1) to (7) should be repeated for the second and third years of the triennium before simulating the actuarial valuation at step (8). It can be assumed that the same technical bases will always be used, or a dynamic decision rule can be introduced to relate the assumptions in some way to the experience.

(9) It is now possible to make a new assessment of $c(t)$ for another period of years, having regard to the funding method and the results of the valuation carried out at (8), which would lead to an upwards or downwards adjustment to $c(t)$ to eradicate a deficiency or a surplus.

The desired period of projection can be achieved by performing the simulation successively for years $t + 1, t + 2, \ldots$ etc. The simulation can then be repeated as many times as required to obtain a distribution of results.

(b) Discussion. The results of the above simulation process can be used for various purposes, as the following examples show.

(1) The expenditure variable $X(t)$ (16.4.1) projects the cash flow needed to meet benefit commitments.

(2) The accumulating asset amount $A(t)$ gives guidance for the planning of investments.

(3) The profit $Y(t)$ and its variability from year to year give an indication of the volatility of the system and assist in planning how the surplus (or loss) should be spread over the years in order to avoid instability of the contribution rate, or so as best to meet the requirements of the sponsor. Then the variation range of the contribution rate $c(t)$ can be evaluated.

(4) Different investment strategies can be tested. The need to match the assets and liabilities can be taken into account, employing different definitions of matching (see section 8.6).

(5) The need to smooth the cash flow and the variability of the rate of contribution by means of insurance can be investigated, so that potentially inconvenient excessive amounts of outgo can be avoided with an acceptable probability.

(6) Sensitivity analyses can be carried out on the influence of various external and internal background phenomena.

(7) The security level for accrued benefits on notional (or actual) discontinuance of the scheme can be evaluated.

(8) The influence of the likely development in demographic structures can be investigated, for example by making the mortality and

other rates q_b depend on the calendar time t, i.e. $q_b(x) \to q_b(x, t)$. This does not lead to any overwhelming complications in the simulation technique.

Carrying out these kinds of investigation will require the model set out above to be modified to take into account the rules concerning benefits, contributions, etc. and all other relevant factors for the particular application.

The simulation procedure described in section 16.4(a) can be adapted equally well for the analysis of a number of deterministic scenarios. For some purposes this may be a helpful approach. If a fully stochastic approach is to be used, the simulation will need to be repeated a large number of times and the results will appear in the form of distributions. This may present certain practical problems with the presentation and interpretation of the results. However, the stochastic approach is essential if the effects of variability are to be explored. This will enable questions to be answered about the strength of reserving bases for accrued liabilities, for example in setting minimum funding standards or defining the maximum level of funding which will qualify for tax-exempt treatment. The stochastic modelling of assets and liabilities together may also provide helpful guidance on investment strategy, to enable the trade-off between risk and return to be properly assessed.

(c) References.   The famous mathematician Gauss (1777–1855) valued the staff pension scheme of the University of Göttingen by a technique akin to that described above. The idea was followed by other actuaries of the nineteenth century and later. The actuarial mathematics of pension plans, mainly in deterministic terms, is dealt with by Lee (1986) and in Bowers et al. (1986).

Stochastic elements were introduced among others, by a Finnish pension solvency working party (1987), Tuomikoski (1987), and by Benjamin (1989), who studied pension schemes as a control system, analysing responses to various impulses (spikes, steps, ramps, etc.), using similar ideas to those applied in section 12.3(a). Loades (1988) investigated the strength of accrued liabilities using stochastic inflation and investment models. Dufresne (1988) and Haberman (1990a, 1990b, 1992) studied the behaviour of a simplified pension scheme system subject to random investment return, as did Loades (1992) using sinusoidal input signals. Clark (1992) describes a stochastic asset/liability modelling exercise for a pension fund. Another example of an application is referred to in section 8.6(b). Some of the references of section 15.3(g) apply also to the stochastic approach to pension schemes.

The stochastic modelling of pension schemes, as with life insurance applications, is still at an early stage, notwithstanding the fact that the deterministic techniques have long traditions and are well established. Although pension provision differs a great deal between different countries, many of the same basic principles apply and simulation techniques such as those discussed above could be introduced with advantage in many markets.

Derivation of the Poisson formula

A.1 Individual and collective approaches

The Poisson formula (2.2.1) is a key primary building block in risk theory considerations. In section A.2(a) it is shown how it can be deduced from quite general assumptions about its fundamental properties; in section A.2(b) it is shown how the same formula can be obtained as a limiting expression of the binomial probability function.

(a) Individual approach. Consider a risk collective consisting of N different risk units i. In individual risk theory the total number of claims k in a risk collective during a time period, for example one year, is modelled as a sum

$$k = \sum_i k_i, \tag{A.1.1}$$

where k_i denotes the number of claims occurring in risk unit i. Each k_i is modelled separately.

Let us assume that, for each risk unit, the numbers of claims occurring in any two disjoint time intervals are mutually independent and that the probability that a claim occurs at a fixed future time point is equal to zero. Then k_i satisfies the assumptions (1)–(3) of section 2.2(a) (in the case of one risk unit the exclusion of multiple claims can in practice be taken for granted) and k_i is, therefore, Poisson(n_i) distributed with $n_i = E(k_i)$, that is

$$\text{Prob}\{k_i = k\} = e^{-n_i} \cdot \frac{n_i^k}{k!}, \tag{A.1.2}$$

as will be shown in section A.2. If, furthermore, the numbers of claims k_i of different risk units i are mutually independent, then, by additivity of Poisson variables (section 2.3(a)), the total number of claims is also

Poisson-distributed, i.e.

$$\text{Prob}\{k = k\} = e^{-n}\cdot\frac{n^k}{k!}, \quad n = \sum_i n_i. \tag{A.1.3}$$

Thus the individual approach leads to the Poisson distribution under certain idealized assumptions.

(b) Collective issue.   In collective risk theory, on the other hand, only the total number of claims is modelled and no attention is paid to individual risks. The same result (A.1.3) can be obtained by postulating that the total number of claims satisfies assumptions (1)–(3) of section 2.2(a).

Thus, under these assumptions, the individual and collective approaches lead to the same result.

(c) Poisson distribution as a limit of the binomial distribution.   The assumption of independence between the numbers of claims occurring in a risk unit in disjoint time intervals does not always hold strictly in practice. For example, if a car driver is in a serious accident, and is unable to drive his car for the rest of the year, then the risk of a claim during that period will be much reduced. In the extreme case at most one claim may occur during a year for each risk unit, i.e. when a claim has occurred, there is no further risk of a claim for that risk unit. In the special case where the probability p of a claim is the same for each risk unit, and the claims occurring to different risk units are mutually independent, the binomial distribution may be used as the claim number distribution, giving the probability

$$\text{Prob}\{k = k\} = b_k(p, N) = \binom{N}{k}\cdot p^k\cdot(1 - p)^{N-k} \tag{A.1.4}$$

that k claims will occur. This can be a satisfactory model in life insurance, for example, when at most one claim can occur during an observation period for each risk unit. This differs from the common situation in general insurance where several claim-causing events may be possible for each object within a single time period (the object is often repaired immediately after being damaged or replaced by a new one). In this case the Poisson distribution is usually a more natural choice because it allows for several claims for each individual risk unit.

It is useful to recall that the Poisson distribution is, in fact, the limit distribution of the binomial distribution as $N \to \infty$, if the expected

Table A.1.1   *Binomial(p, N) and Poisson(p · N) d.f.s; p = 0.05*

	N = 100			N = 10 000	
k	Binomial	Poisson	k	Binomial	Poisson
0	0.006	0.006	450	0.011	0.012
2	0.118	0.124	470	0.087	0.093
5	0.616	0.616	490	0.334	0.338
7	0.872	0.867	500	0.512	0.512
10	0.989	0.986	520	0.827	0.821
12	0.999	0.998	540	0.967	0.964
15	1.000	1.000	570	0.999	0.999
Mean	5	5		500	500
Standard deviation	2.18	2.24		21.8	22.4
Skewness	0.41	0.45		0.041	0.045

value $n = pN$ is kept fixed. This is seen by replacing p by n/N in (A.1.4). After some straightforward manipulations we have

$$b_k(p, N) = \frac{n^k}{k!} \cdot \left(1 - \frac{n}{N}\right)^N \cdot \frac{\left(1 - \frac{1}{N}\right)\left(1 - \frac{2}{N}\right) \cdots \left(1 - \frac{1-k}{N}\right)}{\left(1 - \frac{n}{N}\right)^k} \to \frac{n^k}{k!} e^{-n},$$

(A.1.5)

as $N \to \infty$. Hence, the binomial probability approaches the Poisson probability as the size N of the collective becomes large compared with the expected claim number n.

Table A.1.1 shows a comparison of the binomial and Poisson probabilities.

Since the difference between the Poisson and binomial distributions is usually relatively small in risk theoretical applications, the Poisson model, which is more convenient to handle, is often applied even in those cases where the binomial distribution can be regarded as a theoretically more accurate model.

A.2   Derivation of the Poisson distribution law

(a) The stationary case.   Let $k(t)$ denote the number of claims occurring in the time interval $(0, t]$, and let

$$k(u, t) = k(t) - k(u) \qquad (A.2.1)$$

denote the number of claims occurring in the interval $(u, t]$.

Suppose that the following conditions are satisfied

(1) **Independence of increments**: $k(u_1, t_1)$ and $k(u_2, t_2)$ are independent whenever $(u_1, t_1]$ and $(u_2, t_2]$ are disjoint.
(2) **Exclusion of multiple claims**: The probability that more than one claim occurs at the same time is zero.
(3)* **Stationarity of increments**: $k(t)$ and $k(u, u + t)$ are identically distributed for every $t, u > 0$.

We will show that $k(t)$ is Poisson-distributed. The more general case of section 2.2(a), where the condition (3)* is replaced by a much weaker condition (3) given in section 2.2(a), is dealt with in section A.2(b) below.

By (1) and (3)*,

$$
\begin{aligned}
p_0(s + t) &= \text{Prob}\{k(s + t) = 0\} \\
&= \text{Prob}\{k(s) = 0 \quad \text{and} \quad k(s, s + t) = 0\} \\
&= p_0(s) \cdot p_0(t). \qquad (A.2.2)
\end{aligned}
$$

Hence $p_0(t)$ is a monotonically decreasing function of t and, whenever m and n are positive integers,

$$p_0(m/n) = [p_0(1/n)]^m = [p_0(n \cdot (1/n))]^{m/n} = [p_0(1)]^{m/n}. \qquad (A.2.3)$$

Since this equation holds for every positive rational number m/n, and since $p_0(t)$ is monotonic, we have

$$p_0(t) = [p_0(1)]^t \quad \text{for every } t > 0. \qquad (A.2.4)$$

We may assume that $0 < p_0(1) < 1$. Indeed, if we had $p_0(1) = \text{Prob}\{k(1) = 0\} = 1$, then we would have the trivial case that $\text{Prob}\{k \equiv 0\} = 1$, by condition (3)*. On the other hand, if $p_0(1) = 0$, then we would have $\text{Prob}\{k(t) \geqslant 1\} = 1 - p_0(t) = 1$ for every $t > 0$, which would imply that there are infinitely many claims during any time period, since any time interval can be sub-divided into infinitely many sub-intervals. Denoting $\mu = -\ln(p_0(1))$, we have

$$p_0(t) = e^{-\mu t} = 1 - \mu \cdot t + o(t), \qquad (A.2.5)$$

where the notation $o(t)$ means that the remainder converges to zero more rapidly than t, as $t \to 0$, i.e. $o(t)/t \to 0$, as $t \to 0$.

Fix $t > 0$. Define **claim indicator variables**

$$I_0(1) = \min(1, k(0, t))$$

$$I_1(1) = \min\left[1, k\left(0, \frac{t}{2}\right)\right], \quad I_1(2) = \min\left[1, k\left(\frac{t}{2}, t\right)\right],$$

and, generally, for every integer $j \geqslant 0$,

$$I_j(v) = \min\left[1, k\left((v-1)\cdot\frac{t}{2^j}, v\cdot\frac{t}{2^j}\right)\right], \quad v = 1, 2, \ldots, 2^j,$$

and

$$k_j = \sum_{v=1}^{2^j} I_j(v). \tag{A.2.6}$$

Then, for a fixed j, the value of the indicator $I_j(v)$ is equal to one if at least one claim occurs in the vth interval $((v-1)\cdot t/2^j, v\cdot t/2^j]$, and is equal to zero otherwise. Consequently, k_j counts the number of intervals in the jth division of the interval $(0, t]$ containing at least one claim. Because the $(j+1)$th division is a sub-division of the jth division, it follows that $k_j, j = 0, 1, \ldots$, is an increasing sequence of random variables. Furthermore, for a given realization, the minimum distance, d, between two claims occurring in the time interval $(0, t]$ is positive because of exclusion of multiple claims, and therefore no sub-interval contains more than one claim from this realization as soon as j is sufficiently large that $t/2^j < d$. Therefore

$$k_j \to k(t) \quad \text{as } j \to \infty. \tag{A.2.7}$$

Since

$$\text{Prob}\{I_j(v) = 1\} = \text{Prob}\{k((v-1)\cdot t/2^j, v\cdot t/2^j) \geqslant 1\}$$
$$= 1 - p_0(t/2^j) = 1 - \exp(-\mu\cdot t/2^j),$$

and since $I_j(v)$, $v = 1, 2, 3, \ldots, 2^j$, are mutually independent by (1), it follows that the distribution of k_j is binomial $(q_j, 2^j)$ with $q_j = 1 - \exp(-\mu\cdot t/2^j) = \mu\cdot t/2^j + o(1/2^j)$. Hence

$$q_j\cdot 2^j \to \mu\cdot t \quad \text{as } j \to \infty. \tag{A.2.8}$$

Because the value of the m.g.f. of the binary variable $I_j(v)$ at s is equal to $1 - q_j + e^s\cdot q_j$, the m.g.f. M_j of the binomial $(q_j, 2^j)$ variable k_j satisfies

$$M_j(s) = (1 - q_j + e^s\cdot q_j)^{2^j}$$
$$= \left(1 + q_j\cdot 2^j\cdot\frac{(e^s - 1)}{2^j}\right)^{2^j} \to e^{\mu\cdot t\cdot(e^s - 1)}, \tag{A.2.9}$$

and, as the limit is the Poisson($\mu \cdot t$) m.g.f. (Exercise 2.3.1), it follows that the distribution of the limit variable $k(t)$ is Poisson($\mu \cdot t$).

(b) The non-stationary case. Condition (3)* of section A.2(a) will now be relaxed by allowing the process to have non-stationary increments.

Suppose that the claim number process k satisfies the conditions (1) (independence of increments) and (2) (exclusion of multiple claims) given in section A.2(a) above. The stationarity assumption (3)* is replaced by the weaker assumption (3) of section 2.2(a), which states that no time point t on the real line is in a special position, in the sense that at that specific time point there is a positive probability that a claim will occur.

The conditions (1)–(3)* imply that $p_0(t)$ is a continuous function of t. To prove this, we first note that the function $p_0(t)$ is decreasing, so that there is a discontinuity at a point u if and only if there is a downward jump of size $p_0(u-) - p_0(u) > 0$, where $p_0(u-)$ denotes the left limit $\lim_{\varepsilon \to 0+} p_0(u - \varepsilon)$. Indeed, if C_u denotes the event that a claim occurs exactly at time u, then we have

$$\mathrm{Prob}(C_u) = \lim_{\varepsilon \to 0+} \mathrm{Prob}\{k(u - \varepsilon, u) \geqslant 1\}.$$

On the other hand, because of the independence of the increments,

$$
\begin{aligned}
\mathrm{Prob}\{k(u - \varepsilon, u) \geqslant 1\} &= \mathrm{Prob}\{k(u - \varepsilon, u) \geqslant 1 \,|\, k(u - \varepsilon) = 0\} \\
&= \mathrm{Prob}\{k(u - \varepsilon) = 0 \quad \text{and} \quad k(u) > 0\}/p_0(u - \varepsilon) \\
&= [p_0(u - \varepsilon) - p_0(u)]/p_0(u - \varepsilon) \to [p_0(u-) - p_0(u)]/p_0(u-), \\
&\text{as } \varepsilon \to 0+.
\end{aligned}
\tag{A.2.10}
$$

It can be seen that $\mathrm{Prob}(C_u)$ is positive if and only if p_0 is discontinuous at the point u.

It will be shown that the condition (3)*, together with assumptions (1) and (2), are sufficient to guarantee that $k(t)$ is Poisson-distributed at every time point t.

Define the so-called **operational time** for the process $k(t)$ by setting $t^* = T(t)$, where T is the continuous increasing function

$$T(t) = -\mu \cdot \ln p_0(t), \tag{A.2.11}$$

and μ is a positive scaling factor which can be chosen freely. We define

$$k^*(t^*) = k(t). \tag{A.2.12}$$

Choosing, for example, $\mu = E(k(1))$, the expectations of both processes will be equal at time point $t = 1$.

If T is strictly decreasing there is a one-to-one correspondence between t and t^*, and it can immediately be seen that the modified stochastic process k^* also satisfies the assumptions (1) and (2).

Suppose that the function T is not strictly decreasing. Then there are one or more sub-intervals $I_j = (a_j, b_j]$ such that T is constant on each of these intervals, and is strictly decreasing outside of them. It can be seen from (A.2.10) that T is constant on each I_j, which implies that the probability of claims on these intervals is equal to zero. Thus, it may be assumed that all claims occur outside of the intervals I_j, where there is a one-to-one correspondence between t and t^*. Consequently, k^* is also well-defined when T is not strictly decreasing, for whenever $s^* = t^*$, we have $k(s) = k(t)$, even if s may not be equal to t. Furthermore, it is seen that k^* then satisfies (1) and (2).

To prove that k^* is a stationary Poisson process, it is now sufficient to prove that $\mathrm{Prob}\{k^*(t^*) = 0\}$ is equal to $e^{-\mu \cdot t^*}$, since this would imply that the process k^* satisfies (A.2.5), and the assumption (3)* was needed in section A.2(a) only for the proof of (A.2.5). We have

$$\mathrm{Prob}\{k^*(t^*) = 0\} = \mathrm{Prob}\{k(t) = 0\} = p_0(t) = e^{-T(t)} = e^{-\mu \cdot t^*},$$
(A.2.13)

which shows that k^* satisfies (A.2.5). Hence k^* is a stationary Poisson process. By (A.2.12) $k(t)$ is then Poisson-distributed.

(c) The mixed Poisson case. Note that in practice there are usually different background circumstances, such as economic conditions, which cause **random** variation in the claim propensity. Then assumption (1) regarding the independence of increments of the claim number process does not hold true. However, even then the independence assumption (1) may be satisfied **conditionally**, under the condition that the circumstances are fixed. This leads us to the so-called mixed Poisson distributions, which are introduced in section 2.4.

Pólya and Gamma distributions

(a) M.g.f. of the Gamma distribution. Let Y be a Gamma(r, a)-distributed random variable, where r and a are positive. The density of Y is given by (2.5.1). The m.g.f. M_Y of Y is, therefore,

$$
\begin{aligned}
M_Y(s) = E(e^{s \cdot Y}) &= \frac{a^r}{\Gamma(r)} \int_0^\infty e^{-(a-s) \cdot z} z^{r-1} \, dz \\
&= \frac{a^r}{\Gamma(r)} \int_0^\infty e^{-z} \left(\frac{z}{a-s} \right)^{r-1} \frac{dz}{a-s} \\
&= \left(\frac{a}{a-s} \right)^r \cdot \frac{1}{\Gamma(r)} \int_0^\infty e^{-z} z^{r-1} \, dz \\
&= \left(\frac{a}{a-s} \right)^r \cdot \frac{\Gamma(r)}{\Gamma(r)} = \left(\frac{a}{a-s} \right)^r.
\end{aligned}
\tag{B.1.1}
$$

It can be seen from (1.4.14) that the sum of independent Gamma(r_1, a) and Gamma(r_2, a)-distributed random variables is Gamma$(r_1 + r_2, a)$-distributed.

The most convenient way to calculate the basic characteristics of Y is to use the cumulant generating function $\psi_Y(s) = \ln M_Y(s) = r \cdot \ln(a) - r \cdot \ln(a-s)$ of Y (see section 1.4(c)). The jth cumulant κ_j of Y is obtained by differentiating ψ_Y as follows:

$$
\kappa_j = \psi_Y^{(j)}(0) = r \cdot \frac{(j-1)!}{(a-0)^j} = r \cdot \frac{(j-1)!}{a^j}.
\tag{B.1.2}
$$

In particular,

$$
E(Y) = \kappa_1 = \frac{r}{a}, \quad \sigma_Y^2 = \kappa_2 = \frac{r}{a^2},
$$

$$
\gamma_Y = \frac{\kappa_3}{\kappa_2^{3/2}} = \frac{2}{\sqrt{r}}, \quad \gamma_{2,Y} = \frac{\kappa_4}{\kappa_2^2} = \frac{6}{r},
\tag{B.1.3}
$$

when Y is Gamma(r, a)-distributed.

Choosing $r = a = h$ gives the main characteristics (2.5.4) for a Gamma-distributed mixing variable.

(b) Proof of (2.5.5). The proof that mixing the Poisson(n) distribution with a Gamma(h, h)-distributed mixing variable gives the Pólya distribution (2.5.3) as follows

$$
p_k = E(p_k(n \cdot q)) = \int_0^\infty e^{-nq} \frac{(nq)^k}{k!} \, dH(q)
$$

$$
= \frac{n^k \cdot h^h}{\Gamma(h) \cdot k!} \int_0^\infty e^{-(n+h)q} q^{h+k-1} \, dq
$$

$$
= \frac{n^k \cdot h^h}{\Gamma(h)} \int_0^\infty e^{-z} \cdot \left(\frac{z}{n+h} \right)^{(h+k)-1} \frac{dz}{n+h}
$$

$$
= \frac{n^k \cdot h^h}{\Gamma(h) \cdot k!} \cdot \frac{\Gamma(h+k)}{(n+h)^{h+k}} = \frac{\Gamma(h+k)}{\Gamma(h) \cdot k!} \cdot \left(\frac{h}{n+h} \right)^h \cdot \left(1 - \frac{h}{n+h} \right)^k
$$

$$
= \binom{h+k-1}{k} \cdot p^h \cdot (1-p)^k, \quad \text{where} \quad p = \frac{h}{n+h}. \tag{B.1.4}
$$

Here $p_k(n \cdot q)$ is the Poisson($n \cdot q$)-probability for k. For the binomial coefficient on the last line, see (2.5.6).

(c) Proof of (2.5.7). By (2.4.9) and by (B.1.1) the m.g.f. M of a Pólya(n, h)-distributed claim number variable k is

$$
M(s) = M_q(n \cdot e^s - n) = \left[\frac{h}{h - n \cdot e^s + n} \right]^h. \tag{B.1.5}
$$

(d) Proof of (2.5.8). Substituting (2.5.4) in (2.4.12) gives (2.5.8).

(e) Proof of (2.5.9). Since

$$
\binom{h-1}{0} = 1 \tag{B.1.6}
$$

it can be seen from (B.1.4) that

$$
p_0 = p^h, \quad p = \frac{h}{n+h}. \tag{B.1.7}
$$

If $k > 0$, then

$$p_k = \frac{\Gamma(h+k)}{\Gamma(h) \cdot k!} \left(\frac{h}{n+h} \right)^h \cdot \left(\frac{n}{n+h} \right)^k$$

$$= \frac{\Gamma(h+k-1) \cdot (h+k-1)}{\Gamma(h) \cdot (k-1)! \cdot k} \cdot \left(\frac{h}{n+h} \right)^h \cdot \left(\frac{n}{n+h} \right)^{k-1} \cdot \frac{n}{n+h}$$

$$= \frac{h+k-1}{k} \cdot \frac{n}{n+h} \cdot p_{k-1} = \left(1 + \frac{h-1}{k} \right) \cdot (1-p) \cdot p_{k-1}$$

$$= \left((1-p) + \frac{(h-1) \cdot (1-p)}{k} \right) \cdot p_{k-1}, \qquad \text{(B.1.8)}$$

where the well-known functional equality $\Gamma(z) = (z-1) \cdot \Gamma(z-1)$ is used.

Asymptotic behaviour of the compound mixed Poisson d.f.

Let X be a compound mixed Poisson variable with q as the mixing variable, and let $x = X/\mathrm{E}(X)$ be the corresponding relative variable. It will be shown that the limit d.f. of the d.f. F_x of the relative variable is the mixing d.f. H. The conditional d.f.

$$F(\cdot \mid q = q) \qquad \text{(C.1)}$$

of x is a compound Poisson($n \cdot q$) d.f., whose expectation and variance are

$$\mathrm{E}(x \mid q = q) = \frac{n \cdot q \cdot m}{n \cdot m} = q$$

$$\mathrm{Var}(x \mid q = q) = \frac{\mathrm{Var}(X \mid q = q)}{n^2 \cdot m^2} = \frac{n \cdot q \cdot a_2}{n^2 \cdot m^2} = \frac{a_2/m^2}{n} \cdot q. \qquad \text{(C.2)}$$

It follows from the latter equation that

$$\mathrm{E}[\mathrm{Var}(x \mid q)] = \mathrm{E}\left(\frac{a_2/m^2}{n} \cdot q\right) = \frac{a_2/m^2}{n} \to 0, \quad \text{as } n \to \infty. \qquad \text{(C.3)}$$

On the other hand, by the definition of conditional variance (section 1.4(g)), we have

$$\mathrm{E}[\mathrm{Var}(x \mid q)] = \mathrm{E}[\mathrm{E}((x - \mathrm{E}(x \mid q))^2 \mid q)]$$
$$= \mathrm{E}[\mathrm{E}((x - q)^2 \mid q)] = \mathrm{E}[(x - q)^2]. \qquad \text{(C.4)}$$

Together with (C.3) this implies that $\mathrm{E}[(x - q)^2]$ tends to zero or, in other words, $x \to q$ in mean square sense, as $n \to \infty$. Consequently, $F_x \to H$ as $n \to \infty$.

Numerical calculation of the normal d.f.

(a) **Numerical values of the normal d.f.** N can be obtained from standard text-books or can be programmed making use of the following expansion (Abramowitz and Stegun, 1970).

$$R = \frac{1}{\sqrt{2\pi}} e^{-x^2/2}(b_1 y + b_2 y^2 + b_3 y^3 + b_4 y^4 + b_5 y^5), \qquad \text{(D.1)}$$

where

$$y = 1/(1 + 0.2316419|x|),$$

and the values of b_i, $i = 1, 2, ..., 5$ are respectively

$$0.319381530, \ -0.356563782, \ 1.781477937,$$
$$-1.821255978, \ 1.330274429. \qquad \text{(D.2)}$$

Then

$$N(x) = \begin{cases} R & \text{for } x \leqslant 0 \\ 1 - R & \text{for } x > 0. \end{cases} \qquad \text{(D.3)}$$

(b) **Inverse formula.** When the value $N = N(x)$ is given and the matching argument x is required, first calculate

$$t = \begin{cases} \sqrt{-2\ln N} & \text{for } 0 < N \leqslant 0.5 \\ \sqrt{-2\ln(1 - N)} & \text{for } 0.5 < N < 1 \end{cases} \qquad \text{(D.4)}$$

and

$$z = t - \frac{c_0 + c_1 t + c_2 t^2}{1 + d_1 t + d_2 t^2 + d_3 t^3}, \qquad \text{(D.5)}$$

where

$$c_0 = 2.515517, \ c_1 = 0.802853, \ c_2 = 0.010328$$
$$d_1 = 1.432788, \ d_2 = 0.189269, \ d_3 = 0.001308.$$

Then

$$x = \begin{cases} -z & \text{for } 0 < N \leqslant 0.5 \\ z & \text{for } 0.5 < N < 1. \end{cases} \tag{D.6}$$

The absolute amount of the error is $< 7.5 \times 10^{-8}$ for (D.3) and $< 4.5 \times 10^{-4}$ for (D.6).

Derivation of the recursion formula for F

The recursion formula (4.1.5) for compound distributions satisfying the conditions (1) and (2) of section 4.1(a) is derived in section E(a).

In section E(b) it is shown that the mean claim size of the original claim size distribution remains unchanged when the distribution is discretized using the method described in section 4.1(d).

(a) Proof of the recursion formula. Let $X = Z_1 + Z_2 + \cdots + Z_k$ be a compound variable such that the claim number variable k satisfies the recursion formula

$$p_k = (a + b/k) \cdot p_{k-1} \quad \text{for } k = 1, 2, 3, \ldots, \tag{E.1}$$

and $s_0 + s_1 + \cdots + s_r = 1$, where $s_i = \text{Prob}\{Z = i \cdot C\}$, $i = 0, 1, 2, \ldots, r$, as in section 4.1. Without loss of generality we may assume $C = 1$.

Denote

$$s_j^{k*} = \text{Prob}\{Z_1 + \cdots + Z_k = j\} \tag{E.2}$$

for $k \geqslant 0$; especially $s_0^{0*} = 1$, and $s_j^{0*} = 0$, if $j > 0$. Then, for $k > 0$,

$$s_j^{k*} = \sum_{i=0}^{j} s_i \cdot s_{j-i}^{(k-1)*}, \tag{E.3}$$

By symmetry,

$$E\left(Z_1 \bigg| \sum_{i=1}^{k} Z_i = j\right) = \frac{1}{k} \cdot E\left(\sum_{i=1}^{k} Z_i \bigg| \sum_{i=1}^{k} Z_i = j\right) = \frac{j}{k} \tag{E.4}$$

for $k > 0$. On the other hand, if $s_j^{k*} > 0$,

$$E\left(Z_1 \bigg| \sum_{i=1}^{k} Z_i = j\right) = \sum_{i=1}^{j} i \cdot \text{Prob}\left\{Z_1 = i \bigg| \sum_{m=1}^{k} Z_m = j\right\}$$

$$= \frac{\sum_{i=1}^{j} i \cdot \text{Prob}\left\{ Z_1 = i \quad \text{and} \quad \sum_{m=2}^{k} Z_m = j - i \right\}}{\text{Prob}\left\{ \sum_{i=1}^{k} Z_i = j \right\}}$$

$$= \frac{\sum_{i=1}^{j} i \cdot s_i \cdot s_{j-i}^{(k-1)*}}{s_j^{k*}}, \tag{E.5}$$

since Z_1 and $Z_2 + \cdots + Z_k$ are independent. It follows that, for $j > 0$,

$$s_j^{k*} = \frac{k}{j} \cdot \sum_{i=1}^{j} i \cdot s_i \cdot s_{j-i}^{(k-1)*}. \tag{E.6}$$

Note that by (E.3) $s_j^{k*} = 0$ only if $s_i \cdot s_{j-i}^{(k-1)*} = 0$ for every i. Therefore, the equality (E.6) is satisfied also in the case that $s_j^{k*} = 0$.

We are now able to prove the recursion formula (4.1.8). Since

$$f_j = \text{Prob}\{X = j\} = \sum_{k=0}^{\infty} p_k \cdot s_j^{k*}, \tag{E.7}$$

it follows by (E.1), (E.3) and (E.6) that, for $j > 0$,

$$f_j = \sum_{k=1}^{\infty} (a + b/k) \cdot p_{k-1} \cdot s_j^{k*}$$

$$= \sum_{k=1}^{\infty} a \cdot p_{k-1} \cdot \sum_{i=0}^{j} s_i \cdot s_{j-i}^{(k-1)*} + \sum_{k=1}^{\infty} \frac{b}{j} p_{k-1} \cdot \sum_{i=1}^{j} i \cdot s_i \cdot s_{j-i}^{(k-1)*}$$

$$= a \cdot s_0 \cdot \sum_{k=1}^{\infty} p_{k-1} \cdot s_j^{(k-1)*} + \sum_{i=1}^{j} \left(a + i \cdot \frac{b}{j} \right) \cdot s_i \cdot \sum_{k=1}^{\infty} p_{k-1} \cdot s_{j-i}^{(k-1)*}$$

$$= a \cdot s_0 \cdot f_j + \sum_{i=1}^{j} \left(a + \frac{i \cdot b}{j} \right) \cdot s_i \cdot f_{j-i}. \tag{E.8}$$

Solving for f_j gives the recursion formula (4.1.5). The summation in the last line can be stopped at $i = r$, since $s_i = 0$ for larger values of i.

The formula (4.1.6) for initial value f_0 follows immediately from (E.7). If $s_0 = 0$ the formulae (4.1.7) are obtained directly from $f_0 = s_0$.

In order to prove (4.1.7) for positive s_0, we substitute $s = \ln s_0$ into the m.g.f. $M_k(s)$ of the corresponding claim number variable. Then, in the Poisson case, $M_k(s) = \exp(n \cdot e^s - n) = \exp(n \cdot s_0 - n)$. In the Pólya case the requested formula is obtained by substituting $h = (a + b)/a$ (see 2.5.10) in (2.5.7). For the binomial (p, N) d.f. we note first that we have (Table 4.1.1) $(a + b)/a = -N$ and $1 - a = 1/(1 - p)$. Then the last formula in (4.1.7) follows by substituting these into the binomial m.g.f. $((1 - p) + p \cdot e^s)^N$ ((A.2.9) of Appendix A).

(b) Absence of bias of the discretized distribution (4.1.13). We will show that the discretized claim size distribution $\{s_i\}_i$ defined by the equations (4.1.12), (4.1.13) and (4.1.14) has the same mean value as the original claim size d.f. S. By rescaling, if necessary, we may assume that step length C is equal to 1. Let Z be S-distributed and denote $\Delta_i = (i - 1, i]$. Then we have

$$d_i = \text{Prob}\{Z \in \Delta_i\} = l_i + r_{i-1} \tag{E.9}$$

and

$$e_i = d_i \cdot \text{E}(Z \,|\, Z \in \Delta_i) = \text{Prob}\{Z \in \Delta_i\} \cdot \text{E}(Z \,|\, Z \in \Delta_i), \tag{E.10}$$

by (4.1.14). Since $i - 1 \leqslant \text{E}(Z \,|\, Z \in \Delta_i) \leqslant i$, it is seen that $(i - 1) \cdot d_i \leqslant e_i \leqslant i \cdot d_i$, so that $0 = i \cdot d_i - i \cdot d_i \leqslant i \cdot d_i - e_i = r_{i-1} \leqslant i \cdot d_i - (i - 1) \cdot d_i = d_i$. Therefore, by (4.1.13), both l_i and r_i, and consequently s_i, are non-negative, for every i. Since every real number belongs to exactly one of the intervals Δ_i, it is clear by (4.1.12) and (E.9) that $\sum s_i = \sum d_i = 1$, showing that $\{s_i\}_i$ is a discrete probability distribution.

The mean value of Z can be written as

$$\text{E}(Z) = \sum \text{Prob}\{Z \in \Delta_i\} \cdot \text{E}(Z \,|\, Z \in \Delta_i) = \sum e_i. \tag{E.11}$$

On the other hand, the mean value of the discretized distribution $\{s_i\}_i$ is

$$\sum_i i \cdot s_i = \sum_i i \cdot r_i + i \cdot l_i = \sum_i i \cdot l_i + (i - 1) \cdot r_{i-1}$$

$$= \sum_i i \cdot (l_i + r_{i-1}) - r_{i-1} = \sum_i i \cdot d_i - r_{i-1}, \tag{E.12}$$

by (E.9) and (E.10). Since $e_i = i \cdot d_i - r_{i-1}$, by (4.1.13), the mean values (E.11) and (E.12) are equal.

Simulation

This appendix supplements Chapter 5, providing further information about simulation technique.

F.1 Uniformly distributed random numbers

(a) Generation of uniformly distributed random numbers is a basic building block in various simulation models. Random number generators have as a common feature algorithmic construction and periodicality. Any particular number is followed by another that is determined by one or more previous values. Hence, having regard to the fact that the numbers are given to a certain number d of digits, and are limited to some fixed finite interval, the generator eventually begins to repeat the same sequence. Therefore, the sequence should be long enough that it does not repeat itself in any relevant application, otherwise the outcome can be seriously impaired. Owing to the fact that these kinds of sequences, strictly speaking, are not genuine random numbers, they are often called **pseudo random numbers**. For practical purposes they are sufficient.

Computer program packages often provide generators of uniformly distributed random numbers as standard software. It is desirable to test the suitability of any such program before use, to ensure that it does not produce frequently cycling series of numbers. If the standard program is inadequate, it would be advisable to construct a new one.

The so-called **mixed congruential formula** (Rubinstein, 1981) is commonly used to generate uniformly distributed random numbers:

$$r_k = a \cdot r_{k-1} + b \ (\text{mod } 10^d) \quad (d \text{ positive integer}). \quad \text{(F.1.1)}$$

The kth random number is obtained from the expression on the right-hand side by taking the d last digits. If b is a positive integer not divisible by 2 or 5 and a is of the form $1 + 10^i$ (i being any positive integer), the formula generates a sequence of 10^d different numbers until it begins to repeat the same sequence.

It is usually best to scale the numbers so that the sequence covers the open interval (0, 1). The initial number r_0 can be chosen anywhere in this interval. It is called the **seed**. For some applications it is necessary to have the possibility of the seed being supplied by the user, in order to be able to run repeated realizations with exactly the same sequences, for example, when performing sensitivity analyses.

F.2 Normally distributed random numbers

Normally distributed random numbers can be generated by the so-called **log-and-trig formula** (Rubinstein, 1981)

$$x_1 = \sqrt{-2\ln(r_1)} \cdot \cos(2\pi r_2)$$
$$x_2 = \sqrt{-2\ln(r_1)} \cdot \sin(2\pi r_2)$$
(F.2.1)

where r_1, r_2 is a pair of uniformly distributed random numbers from the interval (0, 1) and the generated pair x_1, x_2 consists of normally distributed mutually independent random numbers having mean 0 and standard deviation 1.

REMARK Another formula, which is proposed in some textbooks, first generates a certain number of uniformly distributed random numbers and, by summing them up, derives numbers which are approximately normally distributed. This method is, unfortunately, affected by bias at the tails, and cannot therefore be recommended for risk theory applications.

F.3 Graphical presentation of outcomes

Graphical presentation of simulation outputs, which we have used frequently in this book, can provide an adequate and illustrative picture of the process and the essential features (for example, the figures in sections 5.2, 5.5, 9.4, 9.5, 9.6 and 12.4). It is important, among other things, to choose the function and the scales appropriately. It may sometimes be advisable to plot the d.f. F or $1 - F$ instead of the density, using a semi-logarithmic scale, in order to highlight the limiting properties of the tails. Figure 9.5.3 is an example. The slightly irregular flow of the curves gives an idea of the simulation inaccuracy.

F.4   Numerical outputs and their accuracy

Even though simulation is at its best when used to generate graphical diagrams, there is also a need to present the outcomes in numerical form. We provide some hints on that.

(a) A simple approach is to count the relative frequency of the events of interest. For example, if we are interested in evaluating the risk of the variable of concern, say the underwriting result, falling below a certain limit, then we can count the number of cases, c, where this occurred under a simulation. If the sample size is s, then the ratio

$$\hat{p} = c/s \qquad (F.4.1)$$

can be used as an estimate for the required probability p. Unfortunately this method is rather volatile, unless $E(c)$ is very large, as can be seen from the relative standard deviation of c:

$$\frac{\sigma_c}{E(c)} = \frac{\sqrt{p(1-p)s}}{p \cdot s} \approx \frac{1}{\sqrt{c}}. \qquad (F.4.2)$$

This expression is obtained by means of a standard binomial formula. If p is small, as it typically is in risk theory applications, this relative standard deviation turns out to be rather large. For example, if p is 0.01 and s is, say, 1000 and hence $E(c) = 10$, the 95% confidence limits for the outcome c are 4 and 16 (using the normal approximation). If a small relative error is required, the sample size has to be rather large. More exact confidence limits, for both the binomial and Poisson distributions, can be found in Johnson and Kotz, 1969, Vol. 1, pp. 60 and 67. They give the limits for the unknown 'true' value of p as a function of the simulated outcome $\hat{p}$.

A variant of the above method is to count **fractiles** from the simulated outcomes or to plot the **extreme values** appearing in the sample at each time point. Figure F.4.1 exemplifies the latter approach (the shaded area).

(b) Statistical handling of the outputs.   A much more effective method to evaluate the simulation inaccuracy and to find numerical values for the relevant quantities is to calculate the mean, standard deviation and skewness from the simulated realizations. This can easily be performed by storing the simple sums $\sum u^k$ ($k = 1, 2$ and 3) during the simulation process; from these the relevant characteristics can be calculated (Exercise F.4.1). The d.f. of the values, or rather the

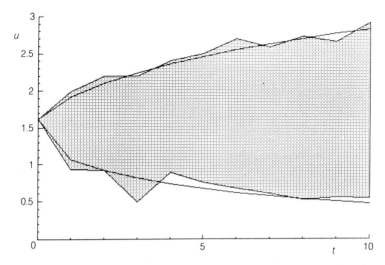

Figure F.4.1 *Simulation of the solvency ratio u(t) according to algorithm (12.3.6). The distribution is shaped by plotting the minimum and maximum value of the realizations at each time point (shaded area). For the sake of comparison the 99% confidence boundaries were also calculated using the Wilson–Hilferty formula (4.2.12) for the same basic parameters as were used in the simulation. Sample size 100.*

confidence limit $\hat{u}_\varepsilon(t)$ at each t, can be evaluated therefrom, for example by means of the WH-formula (4.2.12) or the NP-formula (4.2.6). Note that now, when the mean, standard deviation and possibly skewness are calculated, the whole set of realizations can be utilized, and not only the tail as with the formula (F.4.1). Figure F.4.2 demonstrates this method. The distribution is the same as in Figure F.4.1.

The accuracy of the confidence boundary can be enhanced by fitting a suitably defined curve $\bar{u}_\varepsilon(t)$ to smooth the confidence limits $\hat{u}_\varepsilon(t)$, which are first calculated at each t for some specified range $t_1 \leqslant t \leqslant t_2$. For example, a formula of parabolic type might be a candidate for the lower boundary:

$$\bar{u}_\varepsilon(t) = u(0) - a \cdot t^2 \tag{F.4.3}$$

where the parameter a can be determined by using the least squares method.

(c) Further methods. The simulation procedure can often be made

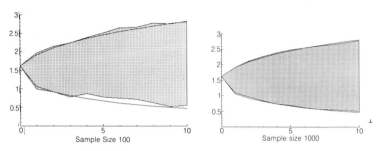

Figure F.4.2   *The area inside which the realization should remain with 99%
confidence, calculated by means of the WH-method from the characteristics
which are derived from the simulated realizations. For the sake of comparison
the confidence boundaries were also calculated directly by the WH-formula
as in Figure F.4.1. The same basic data is used as in Figure F.4.1.*

faster and the simulation inaccuracy reduced by methods which
are presented in standard study books on simulation, for example,
by variance reduction or stratifying the samples (Rubinstein,
1981). Consideration of these methods is beyond the scope of this
book.

Exercise F.4.1   Show how the characteristics $\mu(t)$, $\sigma(t)$ and $\gamma(t)$ can
be estimated if the simulation is organized according to the hints
given in section F.4(b).

F.5   Simulation of the insurance business

The simulation of various parts of the insurance process has been
dealt with in several different sections of this book. For editorial
reasons the material has been presented in numerous separate items,
beginning from the simplest cases in section 5.4, which involved only
claim numbers, and progressing towards a complete analysis of the
insurance business, where claims are stochastic in several layers, and
where inflation, investments, premiums, expenses, dynamics, etc. are
involved, as was considered in section 12.4, Chapter 13 and sections
14.2 and 14.3. For the convenience of readers the total process of
insurance simulation is now summarized, providing references to the
relevant sections and the formulae needed for the separate building

blocks. We restrict consideration to long-term modelling and to the case where there is only one line of business. This can be a particular class of insurance or several classes combined (section 3.2(d) concerning additivity).

(a) General flow chart. This is outlined in Figure F.5.1.

First, of course, the model should be specified. There are numerous alternatives to be chosen (or omitted), according to the purpose of the work, for example as follows.

- **Claims:** mixing and cycling (section 9.4), run-off, including claim reserving (section 9.5) and reinsurance (section 3.4).
- **Premiums:** premium rule (sections 10.1, 10.2 and 12.4(c)).
- **Expenses:** Chapter 11.

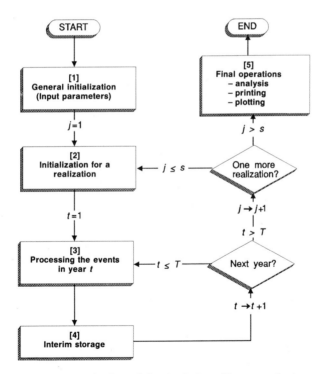

Figure F.5.1 *The flow of the simulation of insurance business.*

- **Investments:** different sub-modules for the most important investment classes (section 8.4 for the Wilkie model and section 8.5 for alternatives), portfolio selection (section 8.6) or, perhaps, the total return on all the investments taken together (section 8.5(f)).
- **Market:** section 14.3.
- **A master module** to control the flow of the simulation is obtained by suitably modifying the basic equations (14.2.1) for assets and (14.2.2) for the solvency margin. This master module is fed by auxiliary modules constructed for claims, premiums, etc.

The boxes in the chart will be commented on in subsequent sections.

(b) General initialization of parameters (Box 1). All the relevant parameters have to be given as inputs. It is convenient to construct one or more data matrices for which standard values are supplied. The user should be given the possibility of changing them and of selecting options, such as to drive some parts of the process deterministically or stochastically or to omit them altogether.

The necessary model parameters can be found from the relevant equations in the chapters and sections referred to in section F.5(a), where they appear as coefficients and various kinds of parameters. Examples are given in (6.1.11), (6.2.1), (7.3.6), in the specifications of section 8.4 (for the Wilkie model), and in (8.5.5), (8.5.9), (8.5.17) and (8.5.18). The problem of finding suitable numerical values for the model parameters is discussed in section 14.5(d).

(c) Initialization for a realization (Box 2). Some aspects, for example, the development of the index of inflation, are specific to a particular realization, but independent of the way in which the company's results develop during the realization. The rate of return on investments may fall into this category if they are handled together in bulk. It is convenient to process these items before entering the time loop of Boxes 3 and 4.

(d) Processing the events in year t (Box 3). First the primary variables – claims, premiums, expenses, return on investments, etc. – are generated (references in section F.5(a)). Then the underwriting and insurance results are obtained via the master module, together with the state of the variables at the end of year t.

(e) Interim storage (Box 4). Those quantities which are needed for the initial state of the next loop $t + 1$, and for the final operations, are stored, together perhaps with such auxiliary (cumulative) quantities as, for instance, the sums $\sum u$, $\sum u^2$, $\sum u^3$, which are used in calculating the confidence boundaries (section F.4).

(f) Final operations (Box 5) consist of calculating the required outputs of the process and elaborating them into tables, diagrams, etc., together with any analysis which is needed of the results.

Time series

G.1 Basic concepts

(a) Description of the problem. This appendix will consider the behaviour of quantities which vary by time period. Figure G.1.1 illustrates two typical examples. The curve (a) of the diagram describes a cumulative time-dependent variable $X(t)$. It is observed at discrete equidistant time points and is subject to stochastic irregularities. The indices of consumer prices and of the gross national product (GNP) are, for example, of this type.

The curve (b) portrays a variable $x(t)$ of another type. It fluctuates, with some confidence probability, between certain finite boundaries. For example, the rate of inflation or the relative increment $x(t) = [X(t) - X(t-1)]/X(t)$ of many variables of the former type may be in question. An important family of this type of variables is decribed as a **stationary stochastic process**, as will be defined more rigorously later.

Sequences of values, such as are exemplified in Figure G.1.1, are called **time series**. It is sufficient for the purposes of this book to limit consideration to **discrete equidistant series**. Unless otherwise stated, the variable t, representing time in our applications, will be scaled to integer values, corresponding to successive accounting periods, generally years.

Time series can be thought of as stochastic processes (section 1.3(b)) and vice versa those equidistant discrete stochastic processes which are required in insurance applications can be interpreted as time series. These terms are used more or less as synonyms in this book.

Chapters 2 to 6 were mainly concerned with one-year models and it was sufficient to employ cross-sectional values of the process variables $X(t)$, i.e. the values which X takes at some fixed time point t, generally at the end of the accounting year in question. When longer time periods are considered, the potential interdependence of

(a)

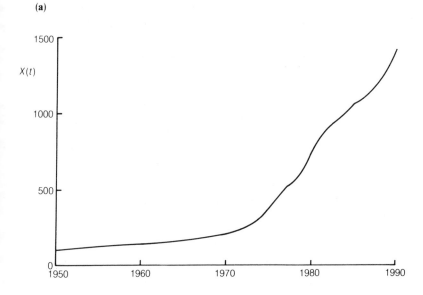

(b)

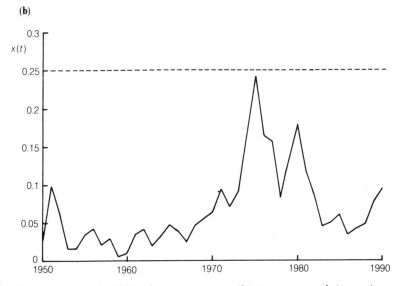

Figure G.1.1   *(a) A cumulative time series. (b) An incremental time series. The UK consumer price index and its increments.*

successive values of the variables becomes relevant. A typical feature for many economic variables is that their values in consecutive periods are not independent. This is illustrated in Figure G.1.1(b). For example, if the rate of inflation or of growth of GNP is high in year t, it is probable that it will also still be higher than normal in the following year.

In order to describe the key properties of cross-sectional variables, the basic characteristics of the distribution, in particular the mean, standard deviation and skewness, were sufficient. Now further characteristics need to be defined in order to describe the interdependence of the states of the process at different time points.

(b) Autocovariance is an important tool for characterizing time series and is defined as follows

$$\gamma(t_1, t_2) = \text{Cov}[X(t_1), X(t_2)]$$
$$= \text{E}\{[X(t_1) - \mu_X(t_1)] \cdot [X(t_2) - \mu_X(t_2)]\}. \quad (G.1.1)$$

The autocovariance expresses the extent to which the value of the time series in one time period is linearly dependent on the value in another period.

(c) Stationarity (Chatfield, 1989, section 3.2) is defined by the condition that the joint n-dimensional distribution of the vector $(X(t_1), X(t_2), ..., X(t_n))$ is the same as the joint distribution of $(X(t_1 + \tau), X(t_2 + \tau), ..., X(t_n + \tau))$ for all $t_1, t_2, ..., t_n$ and τ. Such a stochastic process is called **strictly stationary**. Heuristically, this means that the character of the process remains unchanged as time progresses and the time series curve fluctuates inside a finite 'tube', as is illustrated in Figure G.1.1(b). The stationarity condition holds for any n. In particular, taking $n = 1$, it implies that the distribution of $X(t)$ is the same for all t and that the mean $\mu_X(t)$ is independent of time t. Furthermore, the autocovariance depends only on the length of the interval τ, and not on its location on the time axis. Hence

$$\mu_X(t) = \text{E}[X(t)] = \mu$$
$$\gamma(t, t + \tau) = \gamma(\tau). \quad (G.1.2)$$

Each of these characteristics is assumed to be finite, which is a reasonable restriction in the applications with which we are concerned.

The latter condition (G.1.2) implies that the standard deviation of X, being in fact $\sqrt{\gamma(0)}$, is also constant with value $\sigma_X(t) = \sigma$.

In practice it is often sufficient to define stationarity in a less restrictive way, called **weak stationarity**, by assuming only that the conditions (G.1.2) hold, and without making any assumptions about the higher moments, nor about the distribution functions. In the cases to be dealt with in this book, the weak definition is often sufficient and will be presumed unless otherwise stated. However, if skewness is required, the third moments should be assumed to be constant in a similar way to the characteristics in (G.1.2).

(d) Autocorrelation is another key characteristic applicable to stationary time series and is defined as follows

$$\rho(\tau) = \gamma(\tau)/\sigma^2 \quad (\tau = 0, 1, 2, \dots). \tag{G.1.3}$$

It is called the **autocorrelation function** of the time series and is a straightforward variant of a normal correlation coefficient. It can assume any value within the interval $[-1, 1]$. The value $|\rho| = 1$ indicates full correlation and the value zero a perfect lack of correlation.

G.2   Autoregressive process of first order

(a) Definition.   A simple model which is sufficient to describe the observed phenomena in many applications is defined by the recurrence equation

$$x(t) - \mu = \alpha \cdot (x(t-1) - \mu) + \varepsilon(t) \quad (t \text{ an integer}), \tag{G.2.1}$$

where α is a constant, $\varepsilon(t)$ a noise term with zero mean which introduces the random impulses to the process in order to make it stochastic and μ is the time-independent mean of x.

The main properties of the process can be derived from

$$x(t) - \mu = \alpha^t \cdot (x(0) - \mu)) + \sum_{i=1}^{t} \alpha^{t-i} \varepsilon(i) \tag{G.2.2}$$

which is obtained by applying (G.2.1) repeatedly.

Furthermore, it is assumed that the noise terms $\varepsilon(t)$ are **identically distributed** and **mutually independent**, having the time-independent characteristics

$$\mu_\varepsilon = 0, \ \sigma_\varepsilon \text{ and } \gamma_\varepsilon. \tag{G.2.3}$$

$\varepsilon(t)$ can often conveniently be represented by the gamma or log-normal distribution. In the former case the Wilson–Hilferty formula (section 5.4(b)) can be used in simulations. If the noise term is not (clearly) skew, the normal distribution is appropriate and is commonly used. The conditions for applicability of these formulae should, of course, be satisfied.

(b) Properties. A necessary and sufficient condition for weak stationarity is that $|\alpha| < 1$. Then the first term on the right-hand side of (G.2.2) tends rapidly to zero and the distribution of the cross-sectional $x(t)$ approaches an equilibrium distribution.

REMARK   Rigorously taken the stationarity concept applies to a time series which is already in an equilibrium state so that the effect of the initial state $x(0)$ has disappeared (which theoretically requires an infinitely distant initial time point). Because, however, the effect of the initial state usually becomes insignificant in a fairly short time, we speak about stationarity irrespective of this theoretical point.

The basic characteristics of the equilibrium distribution are now (Exercise G.1)

$$\mu_x = \mu$$

$$\sigma_x^2 = \frac{\sigma_\varepsilon^2}{1 - \alpha^2}$$

$$\gamma_x = \gamma_\varepsilon \frac{(1 - \alpha^2)^{3/2}}{1 - \alpha^3} \qquad \text{(G.2.4)}$$

$$\rho(\tau) = \alpha^\tau.$$

A typical realization of the first order process is illustrated in Figure G.2.1. Note the skewness upwards, which is generated by a skew noise term. It may be useful in modelling processes of the character of Figure G.1.1(b).

It is helpful to look at the special case where $\varepsilon(t) \equiv 0$ for all $t > 0$,

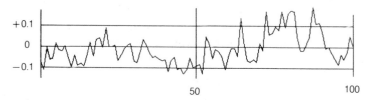

Figure G.2.1   *A realization of a time series process (G.2.1) with $\mu = 0$, $\alpha = 0.6$, $\sigma_\varepsilon = 0.05$ and $\gamma_\varepsilon = 1$.*

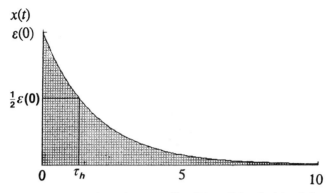

Figure G.2.2   *An impulse and its run-off; $x(0) = \varepsilon(0)$ and $\varepsilon(t) = 0$ for $t > 0$, hence $x(t) - u = \varepsilon(0) \cdot \alpha^t$; $\mu = 0$, $\alpha = 0.6$, $\tau_h = 1.36$.*

as shown in Figure G.2.2. This can be interpreted as an impulse given to the process variable $x(t)$ at $t = 0$. Its effect runs off at a speed which is determined by α. The half-life time τ_h is plotted in the diagram.

To understand the character of the process (G.2.1) it is useful to note that it is a **superimposition** of a sequence of impulses such as that illustrated in Figure G.2.2. More generally, impulses of other types, such as ramps, trends, etc., also appear, as is demonstrated in Figure 12.3.1.

(c) Determining the coefficients.   Text-books, for example Chatfield, 1989, or Box and Jenkins, 1976, propose standard statistical methods for deriving the required coefficients $\alpha, \sigma_\varepsilon$ and γ_ε from actual data (Exercise G.2). Unfortunately the methods are only satisfactory if the available data sequence is rather long. As Chatfield advises, it is best first to plot the material in diagrams (correlograms) for a visual examination. An estimate can usually be obtained from such a diagram. Coefficients which have been derived from live data can be tested by simulating the resulting process and comparing the outcomes with the original data.

G.3   Autoregressive process of second order

(a) Definition.   Some of the time series observed in insurance appli-cations have periodical, cyclical deviations from their mean which are longer, broader and more regular than can be generated by the

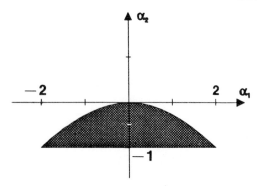

Figure G.3.1   *The area from which α_1 and α_2 are to be chosen in order to obtain a stationary cyclical process.*

first order model of section G.2. A better fit can often be achieved by extending the feedback time lag from one year to two years, and in some cases longer. For this purpose a further term will be added to (G.2.1)

$$x(t) - \mu = \alpha_1 \cdot (x(t-1) - \mu) + \alpha_2 \cdot (x(t-2) - \mu) + \varepsilon(t). \qquad (G.3.1)$$

Depending on the choice of the coefficients α_1 and α_2, this algorithm defines a great variety of processes. For our purposes, the sub-family of such processes which are stationary, permitting the simulation of cyclical time series, is of particular interest. This is achieved by the constraints

$$-1 < \alpha_2 < -\tfrac{1}{4}\alpha_1^2, \qquad (G.3.2)$$

which limit considerations to the pairs (α_1, α_2) located in the shaded area of Figure G.3.1. The proof can be found in Chatfield, 1989, §3.4.4(b) or in Cox and Miller, 1965, §7.2; see also Exercise G.3.

Note the conceptual meaning of the term cycle as explained in the Remark in section 12.2(b).

(b) Properties.   As in section G.2 the character of the second order process is best understood by putting the noise term $\varepsilon(t) \equiv 0$. Then the solution of the difference equation (G.3.1) proves to be the damped oscillation function (Exercise G.3)

$$x(t) - \mu = (-\alpha_2)^{t/2} \cdot a \cdot \cos(\omega t + \phi), \qquad (G.3.3)$$

where the amplitude factor a and the phase ϕ are parameters, which can be determined from the given initial positions $x(-1)$ and $x(0)$.

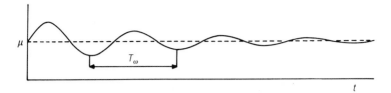

Figure G.3.2   *An example of the damped oscillation function.*

The coefficient ω, called (angular) frequency, is obtained from

$$\tan \omega = \sqrt{-1 - 4\frac{\alpha_2}{\alpha_1^2}}. \qquad (G.3.4)$$

The wavelength is

$$T_\omega = \frac{2\pi}{\omega}. \qquad (G.3.5)$$

Figure G.3.2 displays an example of the damped oscillation function (G.3.3).

In the general case the noise terms $\varepsilon(t)$ are random variables such as were specified in section G.2. Each of them gives rise to a damping oscillation and the total outcome is a superimposition of them.

Figure G.3.3 illustrates a realization of a typical process (G.3.1). Note that in the borderline case when (α_1, α_2) approaches the upper boundary of the shaded area in Figure G.3.1, i.e. when the right-hand

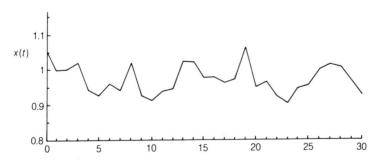

Figure G.3.3   *An example of a time series of second order.* $\alpha_1 = 0.6$, $\alpha_2 = -0.3$, $\mu = 0$, $\sigma_\varepsilon = 0.05$ *and* $\gamma_\varepsilon = 0$.

side inequality in (G.3.2) is replaced by equality, $\omega \to 0$ and $T_\omega \to \infty$. The outcome is still stationary but does not cycle (see (G.3.3)).

(c) **Characteristics.** As a time series of type (G.3.1) which fulfils the stationarity conditions (G.3.2) develops onwards from an initial state, the distribution of the cross-sectional variable $x(t)$ approaches a limit (equilibrium) distribution. The standard deviations of the noise term $\varepsilon(t)$ and of the limit distribution are linked as follows:

$$\sigma_x^2 = \frac{1 - \alpha_2}{(1 + \alpha_2)[(1 - \alpha_2)^2 - \alpha_1^2]} \cdot \sigma_\varepsilon^2 \qquad (G.3.6)$$

and the autocorrelations of the limit distribution $x(t)$ are

$$\rho(\tau) = (-\alpha_2)^{\tau/2} \cdot \frac{\cos(\omega\tau + v)}{\cos v} \qquad (G.3.7)$$

where v is determined so that

$$\rho(1) = \frac{\alpha_1}{1 - \alpha_2}. \qquad (G.3.8)$$

In practice one usually has at least a general idea of the wavelength and the amplitude of the cycles, as well as of $\rho(1)$. Then the relevant coefficients can be found by solving them from the above set of equations, as will be demonstrated in Exercise G.2. More advanced methods for the estimation of the parameters are given, for example, in Chatfield, 1989, or Box and Jenkins, 1976.

G.4   Generalizations and variants

(a) **Time series of higher order** can be created by adding further terms to (G.3.1). Some of the models also propose using several $\varepsilon(t - \tau)$ terms, for example, Cox and Miller, 1965, or Box and Jenkins, 1976. This permits better matching with the empirical data owing to increased flexibility. On the other hand, the models will only be valid if there is a sizeable amount of data from which the parameters can be estimated, for example, the observed time series should be long and sufficiently homogeneous over the whole period (Exercise G.2). It is often difficult to satisfy this latter requirement. There is also a risk of overparametrization, since by adding further terms and parameters to the model it can be made to fit the observed past data better and better, but without necessarily improving the model as a forecasting tool.

Higher order time series are not dealt with in this book, partly because of the need to avoid overcomplication of the text and partly because of uncertainty as to their usefulness in practical insurance applications.

(b) Mixed series. There are some phenomena which appear to affect each other to a significant extent. This suggests the need to find links between the models which might otherwise be constructed to describe each of them separately. For example, inflation is certainly a factor affecting the return which is available on investments. This suggests an approach where inflation is simulated first and the resulting process is used in generating the time series for investment returns. This approach, which has been described as a **cascade** system of autoregressive processes, is applied in Chapters 7, 8 and 12.

(c) Logarithmic increments. In handling positive cumulative series $X(t)$ of the type illustrated in Figure G.1.1(a), it is sometimes useful to work with the logarithm of the increment ratio

$$r_X(t) = X(t)/X(t-1) = 1 + i_X(t) \tag{G.4.1}$$

where $i_X = [X(t) - X(t-1)]/X(t)$.

For example, the first order times series algorithm (G.2.1) can be formulated as follows

$$\ln r_X(t) - \mu_r = \alpha \cdot [\ln r_X(t-1) - \mu_r] + \varepsilon(t) \tag{G.4.2}$$

where $\mu_r = E(\ln r_X)$ can be derived, just as μ in (G.2.1), from past logarithmic increments. This approach somewhat changes the properties of the series, making it skew upwards and preventing r_X from taking negative values (see Exercise G.5). It can be appropriate in cases where X is exponentially increasing and μ_r (approximately) constant. Examples are equations (8.4.2) and others in Wilkie's model of section 8.4, as well as (8.5.3). If $\bar{\imath}$ is small, the numerical differences between (G.2.1) and (G.4.2) seem to be minimal.

Exercise G.1 Derive the formulae (G.2.4).

Exercise G.2 Evaluate the standard deviation σ_X and the half-time from Figure G.1.1(b) and calculate the model parameters α and σ_ε

Table G.4.1

t	x	t	x	t	x	t	x
1950	0.027	1960	0.010	1970	0.064	1980	0.180
1951	0.099	1961	0.035	1971	0.095	1981	0.119
1952	0.063	1962	0.042	1972	0.071	1982	0.086
1953	0.016	1963	0.020	1973	0.092	1983	0.046
1954	0.017	1964	0.033	1974	0.159	1984	0.051
1955	0.035	1965	0.048	1975	0.243	1985	0.061
1956	0.043	1966	0.039	1976	0.166	1986	0.034
1957	0.021	1967	0.025	1977	0.158	1987	0.042
1958	0.030	1968	0.047	1978	0.083	1988	0.048
1959	0.006	1969	0.055	1979	0.134	1989	0.078
						1990	0.095

therefrom. Do this separately for the period 1950–1970 and 1970–1990 and for the whole period 1950–1990 (noticing the obvious difference in the course of the process for the two separate 20-year periods). Calculate the model parameters directly from Table G.4.1 using the estimate (Chatfield, 1989, p. 25)

$$r_\tau = \frac{\displaystyle\sum_{t=1}^{N-\tau} (x_t - \bar{x})(x_{t+\tau} - \bar{x})}{\dfrac{N-\tau}{N} \cdot \displaystyle\sum_{t=1}^{N} (x_t - \bar{x})^2} \tag{G.4.3}$$

for the autocorrelation coefficient at lag τ. Compare σ_ε and r_1 with the values which were evaluated directly. Simulate the resulting time series for a lengthy period.

Exercise G.3   Solve the difference equation (G.3.1) in the deterministic case $\varepsilon(t) \equiv 0$. When is the solution an oscillating function of t? (Hint: the solution is of the form $f(t) = x(t) - \mu = A \cdot b^t$, A and b being constants; furthermore if $f_1(t)$ and $f_2(t)$ are solutions, $C_1 f_1(t) + C_2 f_2(t)$ is as well.)

Exercise G.4   Study the behaviour of the time series defined by the difference equation (G.3.1) where $x(-1) = 0.1$, $x(0) = 0.05$, $\alpha_1 = 0$, $\alpha_2 = -1$, $\mu = 0$ and the ε-terms are normally distributed having mean $= 0$ and $\sigma = 0.03$. Plot $x(t)$ for $t = 1, 2, \ldots, 25$, first putting $\sigma = 0$

and then for three (or more) realizations with different values of σ. Verify that the wavelength is the same as that obtained from (G.3.5).

Exercise G.5 Construct a parallel time series to that in Exercise G.2 using the logarithmic increment (G.4.2). Simulate and plot both series on the same diagram, using the same sequence of random numbers, in order to demonstrate the difference between the linear and logarithmic approaches.

Portfolio selection

Section 8.6 introduced the portfolio selection model which was developed by Wise (1984). In this appendix we explore some of the results developed by Wilkie (1985) and Wise (1987) in subsequent papers.

Following the notation introduced by Wilkie (1985) and Wise (1987) we suppose that there are n assets S_i with cash flows represented by vectors $\mathbf{q}_i$ ($i = 1, 2, \ldots, n$), where the element q_{ij} is the cash flow in year j from unit holding of asset i ($j = 1, 2, \ldots, m$). We define $\mathbf{Q}$ as the $m \times n$ matrix $\{q_{ij}\}$.

Note that in this appendix vectors and matrices are typeset in bold Roman, for instance, $\mathbf{x}$ and $\mathbf{A}$.

Let $\mathbf{x}'$ be the vector $\{x_1, x_2, \ldots, x_n\}$ where x_i is the number of units of asset i held in the portfolio. The total cash flow can then be represented by vector $\mathbf{a}$, where $\mathbf{a} = \mathbf{x}' \cdot \mathbf{Q}$.

It is assumed that the excess cash flow available in any year can be rolled up by reinvestment, or by holding the money on deposit, so that the ultimate proceeds (at time T, say) from investing in a unit of asset S_i at time zero is a random variable R_i, with expected value E_i and variance V_i ($= \sigma_i^2$).

We can consider the liabilities L as a negative asset, with corresponding outflows in years 1 to m, which can also be rolled up (accumulated) to time T. The accumulated amount of the liability is a random variable R_L, with expected value E_L and variance V_L ($= \sigma_L^2$).

Denote the returns per unit of asset S_i ($i = 1, 2, \ldots, n$) by the row vector $\mathbf{r}' = (R_1, R_2, \ldots, R_n)$. The expected value of the vector $\mathbf{r}'$ is denoted by $\mathbf{e}' = (E_1, E_2, \ldots, E_n)$.

The variance/covariance matrix is denoted by

$$\mathbf{V} = \begin{bmatrix} V_1 & C_{12} & \cdots & C_{1n} \\ C_{21} & V_2 & \cdots & C_{2n} \\ \cdots & \cdots & \cdots & \cdots \\ \cdots & \cdots & \cdots & \cdots \\ C_{n1} & C_{n2} & \cdots & V_n \end{bmatrix}$$

where $C_{ij} = \rho_{ij}\sigma_i\sigma_j$ is the covariance between assets i and j and ρ_{ij} is the corresponding correlation coefficient. Since $C_{ij} = C_{ji}$ the matrix is symmetric. We define $\mathbf{c} = (C_{L1}, C_{L2}, \dots, C_{Ln})$ as the covariance vector of the liabilities with the assets; $C_{Li} = \mathrm{Cov}(L, S_i)$.

We can now define our investment portfolio as made up of x_i units of asset S_i ($i = 1, 2, \dots, n$). Then the portfolio security vector is $\mathbf{x}' = (x_1, x_2, \dots, x_n)$.

The ultimate surplus (at time T) is given by

$$R = \mathbf{x}' \cdot \mathbf{r} - R_L = \mathbf{r}' \cdot \mathbf{x} - R_L \qquad (\mathrm{H.1})$$

and the corresponding mean ultimate surplus and variance are given by

$$E = \mathbf{x}' \cdot \mathbf{e} - E_L = \mathbf{e}' \cdot \mathbf{x} - E_L; \quad V = \mathbf{x}' \cdot \mathbf{V} \cdot \mathbf{x} - 2\mathbf{x}' \cdot \mathbf{c} + V_L. \quad (\mathrm{H.2})$$

Finally, we denote the mean square ultimate surplus (at time T) by G, where

$$G = E^2 + V \qquad (\mathrm{H.3})$$

where V is the variance of the ultimate surplus.

With this notation we can see that a match, in the sense introduced by Wise (section 8.6(i)), minimizes G. An unbiased match minimizes G subject to $E = 0$, and a positive unbiased match introduces the additional conditions $x_i \geqslant 0$ ($i = 1, 2, \dots, n$). A further element of the portfolio selection problem is the current market value of the assets, since this defines the basis on which one asset can be exchanged for another in the search for an optimal portfolio. We define the row vector of prices as $\mathbf{p}' = (P_1, P_2, \dots, P_n)$, where P_i is the market price per unit of asset S_i. The total market value, or price, of the portfolio represented by the portfolio security vector $\mathbf{x}'$ is P, where

$$P = \mathbf{x}' \cdot \mathbf{p} = \mathbf{p}' \cdot \mathbf{x}. \qquad (\mathrm{H.4})$$

The characteristics of the portfolio in terms of riskiness relative to the liabilities can be examined in terms of the three quantities P, E and V

P is the aggregate market value or price of all the assets in the portfolio;

E is the expected value of the ultimate surplus at time T;

V is the variance of the ultimate surplus.

In conventional portfolio theory (section 8.6(d)) efficient portfolios are defined as those not dominated by other portfolios within the two-dimensional (r, σ) space. Wilkie (1985) generalized the concept to the three-dimensional (P, E, V) space, using exactly the same principles to define an efficient portfolio. In the (P, E, V) space a portfolio is dominated by another if

- for the same P and E, the other has a lower V;
- for the same P and V, the other has a higher E;
- for the same E and V, the other has a lower P;
- for the same P, the other has a higher E and lower V;
- for the same E, the other has a lower P and lower V;
- for the same V, the other has a lower P and higher E;
- the other has a lower P, a higher E and a lower V.

According to this definition, Wise's optimal portfolios, which minimize the mean square ultimate surplus, are not efficient, although in practical examples it seems that they are close to being so. Wise (1987) showed that his approach and that of Wilkie (1985) were two alternative ways of defining optimality in the (P, E, V) space.

The problem facing the investment manager is to use a given market value of assets to achieve a satisfactory return at acceptable risk. Within the theoretical three-dimensional (P, E, V) space he might, therefore, seek to define a two-dimensional set of efficient portfolios by requiring minimum V for any combination of P and E. He then identifies a unique portfolio by specifying the price P and the trade-off between risk and return in terms of

$$\frac{\partial V}{\partial E} = \frac{1}{\mu}. \tag{H.5}$$

The actuary's problem is usually rather different. Given a set of liabilities, the actuary is concerned to establish an appropriate value of assets (P) so as to ensure that some criterion is met regarding E, for example $E = 0$ or $E = E_T > 0$, subject to acceptably low V.

An efficient portfolio from the actuary's point of view is one which, for given E, minimizes V subject to P, and minimizes P subject to V. The actuary thus will seek to define a two-dimensional set of efficient portfolios which minimizes V for any combination of E and P. He then identifies a unique portfolio by specifying the expected ultimate surplus E and the trade-off between risk and price in terms of the degree of risk v, where

$$\frac{\partial V}{\partial P} = -v. \tag{H.6}$$

Positive values of v represent the marginal trade-off between risk (expressed in terms of the variance of ultimate surplus, V) and market value (or price, P). It measures the extent of departure from the portfolio at which V attains its minimum for given E (where $\partial V/\partial P = 0$).

The portfolio specified by $E = 0$ and $v = 0$ is the unbiased match defined earlier. This may not be an efficient portfolio, since the points on the efficient frontier which correspond to $E = 0$ may have a minimum value of v greater than zero. However, for realistic price differentials between different securities, the loss of efficiency (i.e. the value of v for the corresponding efficient portfolio) is likely to be very small if the unbiased match is chosen.

We have

$$P = \mathbf{x}' \cdot \mathbf{p}$$
$$E = \mathbf{x}' \cdot \mathbf{e} - E_L \tag{H.7}$$
$$V = \mathbf{x}' \cdot \mathbf{V} \cdot \mathbf{x} - 2\mathbf{x}' \cdot \mathbf{c} + V_L$$

and a solution is required for the portfolio holding vector $\mathbf{x}'$ which minimizes the variance of the ultimate surplus (V) for given current portfolio (P) and expected ultimate surplus (E). The parameters E and $v\,(= -\partial V/\partial P)$ can then be specified to obtain a unique portfolio.

Exercise 8.6.1   Show that, if E and v are given, the portfolio which minimizes V is given by

$$\mathbf{x}' = \mathbf{c}' \cdot \mathbf{V}^{-1} - \tfrac{1}{2}v\mathbf{p}' \cdot \mathbf{V}^{-1} + \frac{1}{\alpha}(E + E_L - \varepsilon + \tfrac{1}{2}v\gamma)\mathbf{e}' \cdot \mathbf{V}^{-1} \tag{H.8}$$

and that for the unbiased match

$$\mathbf{x}' = \mathbf{c}' \cdot \mathbf{V}^{-1} + \frac{\varepsilon}{\alpha}\mathbf{e}' \cdot \mathbf{V}^{-1}. \tag{H.9}$$

with suitably chosen α, ε and γ.

Exercise 8.6.2   Show that, if P and μ are given, the portfolio which minimizes V is given by

$$\mathbf{x}' = \mathbf{c}' \cdot \mathbf{V}^{-1} - \left[\frac{\pi - P}{\beta} + \frac{\gamma}{2\mu\beta}\right] \cdot \mathbf{p}' \cdot \mathbf{V}^{-1} + \frac{1}{2\mu}\mathbf{e}' \cdot \mathbf{V}^{-1}. \tag{H.10}$$

APPENDIX I

Solutions to exercises

1.4.1. $\alpha_1 = a^{-1} \cdot (1 - e^{-aM})$, by (1.4.9). Since $\sigma^2 = \alpha_2 - \alpha_1^2$, integration by parts gives:

$$\sigma^2 = \int_0^M Z^2 \cdot a \cdot e^{-aZ} \, dZ + M^2 \cdot e^{-aM} - \alpha_1^2 = \frac{1}{a^2} - \frac{2M}{a} e^{-aM} - \frac{1}{a^2} e^{-2aM}.$$

1.4.2. Substitute $e^{tX} = 1 + tX + \cdots + t^2 \cdot X^2/2! + \cdots$ into (1.4.11). Integration term by term gives (1.4.13). (1.4.12) follows from (1.4.13). Formulae (1.4.14) and (1.4.15) are deduced by substituting $X = X_1 + X_2$ and $Y = aX + b$ in (1.4.11).

1.4.3.
$$M(s) = \int_{-\infty}^{\infty} e^{sx} \cdot \frac{e^{-((x-\mu)/\sigma)^2/2}}{\sigma\sqrt{2\pi}} \, dx$$

$$= e^{\frac{1}{2}s^2\sigma^2 + \mu s} \cdot \int_{-\infty}^{\infty} \frac{e^{-((x-\mu-s\sigma^2)/\sigma)^2/2}}{\sigma\sqrt{2\pi}} \, dx,$$

where the latter equality is obtained by manipulating the exponents. The last integral is equal to 1 since the integrand is the $N(\mu + s\sigma^2, \sigma)$-density. The result, $\psi(s) = \mu s + \frac{1}{2}\sigma^2 s^2$, is obtained by taking logarithms.

1.4.4. (a) Straightforward calculation gives $\alpha_j = j!/a^j$. Substitution in (1.4.22) gives $\kappa_1 = 1/a$, $\kappa_2 = 1/a^2$, $\kappa_3 = 2/a^3$ and $\kappa_4 = 6/a^4$. (b) The derivatives of $\psi(s) = \ln M(s) = \ln(a/(a-s))$ at $s = 0$ are $\kappa_j = \psi^{(j)}(0) = (j-1)!/a^j$.

1.4.5. Denoting $m_j(s) = M^{(j)}(s)/M(s)$ and taking successive derivatives of $\psi(s) = \ln M(s)$, we get $\psi^{(1)} = m_1$, $\psi^{(2)} = m_2 - m_1^2$, $\psi^{(3)} = m_3 - 3 \cdot m_1 \cdot m_2 + 2m_1^3$ and $\psi^{(4)} = m_4 - 4m_1 \cdot m_3 - 3m_2^2 + 12m_1^2 \cdot m_2 - 6m_1^4$.

Since $\kappa_j = \psi^{(j)}(0)$ and $m_j(0) = \alpha_j/M(0) = \alpha_j$, the formulae (1.4.22) follow.

1.4.6. We have $\mathrm{Var}[\mathrm{E}(X\,|\,Y)] = \mathrm{E}[\mathrm{E}(X\,|\,Y)^2] - [\mathrm{E}[\mathrm{E}(X\,|\,Y)]]^2 = \mathrm{E}[\mathrm{E}(X\,|\,Y)^2] - \mathrm{E}(X)^2$, by (1.4.32), and $\mathrm{Var}(X\,|\,Y) = \mathrm{E}[(X - \mathrm{E}(X\,|\,Y))^2\,|\,Y] = \mathrm{E}(X^2\,|\,Y) - 2\cdot\mathrm{E}[X\cdot\mathrm{E}(X\,|\,Y)\,|\,Y] + \mathrm{E}[\mathrm{E}(X\,|\,Y)^2\,|\,Y] = \mathrm{E}(X^2\,|\,Y) - 2\cdot\mathrm{E}(X\,|\,Y)\cdot\mathrm{E}(X\,|\,Y) + \mathrm{E}(X\,|\,Y)^2 = \mathrm{E}(X^2\,|\,Y) - \mathrm{E}(X\,|\,Y)^2$, by (1.4.36) and (1.4.37). Consequently, $\mathrm{Var}[\mathrm{E}(X\,|\,Y)] + \mathrm{E}[\mathrm{Var}(X\,|\,Y)] = \mathrm{E}[\mathrm{E}(X\,|\,Y)^2] - \mathrm{E}(X)^2 + \mathrm{E}[\mathrm{E}(X^2\,|\,Y) - \mathrm{E}(X\,|\,Y)^2] = -\mathrm{E}(X)^2 + \mathrm{E}(X^2) = \mathrm{Var}(X)$.

1.4.7. $\mathrm{Cov}[\mathrm{E}(X\,|\,Z), \mathrm{E}(Y\,|\,Z)] = \mathrm{E}[\mathrm{E}(X\,|\,Z)\cdot\mathrm{E}(Y\,|\,Z)] - \mathrm{E}[\mathrm{E}(X\,|\,Z)]\cdot \mathrm{E}[\mathrm{E}(Y\,|\,Z)] = \mathrm{E}[\mathrm{E}(X\,|\,Z)\cdot\mathrm{E}(Y\,|\,Z)] - \mathrm{E}(X)\cdot\mathrm{E}(Y)$ and $\mathrm{E}[\mathrm{Cov}(X, Y\,|\,Z)] = \mathrm{E}[\mathrm{E}(X\cdot Y\,|\,Z) - \mathrm{E}(X\,|\,Z)\cdot\mathrm{E}(Y\,|\,Z)] = \mathrm{E}(X\cdot Y) - \mathrm{E}[\mathrm{E}(X\,|\,Z)\cdot \mathrm{E}(Y\,|\,Z)]$. The sum of these two terms is $-\mathrm{E}(X)\cdot\mathrm{E}(Y) + \mathrm{E}(X\cdot Y) = \mathrm{Cov}(X, Y)$.

1.4.8. Let I_A be a random variable, such that $I_A = 1$ in A, and $I_A = 0$ outside A (I_A is called the indicator of the event A). We have $\mathrm{Prob}(A) = \mathrm{E}(I_A) = \mathrm{E}[\mathrm{E}(I_A\,|\,Y)] = \mathrm{E}[\mathrm{Prob}(A\,|\,Y)]$, where the second identity follows from iterativity (1.4.32). The second row of (1.4.39) follows, if the first row is applied for the event $A = \mathrm{Prob}\{X \leqslant x\}$. The third row of (1.4.39) follows directly from (1.4.32): $M(s) = \mathrm{E}[\mathrm{E}(e^{sX}\,|\,Y)] = \mathrm{E}[M(s\,|\,Y)]$.

1.4.9. Taking logarithms in (1.4.15) gives $\psi_{aX+b}(s) = \psi_X(as) + bs$. Differentiation gives $\kappa_1(aX + b) = a\cdot\kappa_1(X) + b$, and further differentiation gives $\kappa_j(aX + b) = a^j\cdot\kappa_j(X)$ for $j > 1$. By (1.4.25) $\sigma_{aX+b} = |a|\cdot\sigma_X$ and $\gamma_{aX+b} = \gamma_X$, if $a > 0$, and $= -\gamma_X$, if $a < 0$.

1.4.10. See the solution of the previous exercise.

1.4.11. For the sum $X + Y$ of independent random variables $\gamma_{X+Y}\cdot \sigma_{X+Y}^3 = \kappa_3(X + Y) = \kappa_3(X) + \kappa_3(Y) = \gamma_X\cdot\sigma_X^3 + \gamma_Y\cdot\sigma_Y^3$. The corresponding kurtosis formula is obtained similarly just by replacing 3 by 4.

1.4.12. It is seen from the proof of formula (1.4.26) that a corresponding formula holds true in the case of several independent summands. Hence the skewness of k independent and identically distributed random variables is simply $(k\cdot\gamma\cdot\sigma^3)/(\sigma\cdot\sqrt{k})^3 = \gamma\cdot k^{-1/2}$, where γ

and σ denote the skewness and standard deviation of the identically distributed summands. Similarly the kurtosis is equal to γ_2/k.

2.3.1. $M(s) = \sum_{k=0}^{\infty} e^{sk} \cdot e^{-n} \frac{n^k}{k!} = e^{-n} \sum_{k=0}^{\infty} \frac{(ne^s)^k}{k!} = e^{-n} \cdot e^{ne^s} = \exp(n \cdot (e^s - 1)).$

2.3.2. By (2.3.3) $\psi^{(j)}(s) = n \cdot e^s$ for every $j > 0$. Putting $s = 0$ gives $\kappa_j = n$.

2.3.3. $p_k = (n^k/k!) \cdot e^{-n} = (n/k) \cdot (n^{(k-1)}/(k-1)!) \cdot e^{-n} = (n/k) \cdot p_{k-1}.$

2.3.4. It is seen from (2.3.7) that p_k increases as a function of k for values $k < n$ and decreases when $k > n$. The mode is therefore the largest integer $\leqslant n$. For integer n the distribution is bimodal, since $p_{n-1} = p_n$.

2.3.5.
$$\sum_{k=0}^{\infty} k \cdot (k-1) \cdot \cdots \cdot (k-i) \cdot e^{-n} \frac{n^k}{k!}$$
$$= n^{i+1} \cdot e^{-n} \sum_{k=i+1}^{\infty} \frac{n^{k-i-1}}{(k-i-1)!} = n^{i+1}.$$

Applying this for $i = 0$, 1 and 2, we obtain $E(k) = \alpha_1 = n$, $E(k \cdot (k-1)) = \alpha_2 - \alpha_1 = n^2$ and $E(k \cdot (k-1) \cdot (k-2)) = \alpha_3 - 3\alpha_2 + 2\alpha_1 = n^3$. Solving alphas and substituting into (1.4.22) gives $\kappa_2 = \kappa_3 = n$. By (1.4.25) we have $\gamma = \kappa_3/\kappa_2^{3/2} = 1/\sqrt{n}$.

2.3.6. Suppose that the number of claims is Poisson(n). Since $(10\,131 - 9025)/2 = 553$, at least one of the two given values differs more than 500 from n. The best estimate for n is the average 9578. But the standard deviation σ of Poisson(9578) is $\sqrt{9578} \approx 98$. By using the normal approximation we get $\mathrm{Prob}\{|k - n| > 500 \approx 5 \cdot \sigma\} \approx 2 \cdot (1 - N(5)) < 10^{-6}$. It is therefore very unlikely that the number of claims is Poisson(n) distributed for a fixed n.

2.4.1. For Poisson($n \cdot q_i$) d.f. we have $\alpha_2(n \cdot q_i) = n \cdot q_i + n^2 \cdot q_i^2$. Then $\alpha_2 = 14\,275$, by (2.4.7). Since $\alpha_1 = 100$, we have $\sigma^2 = \alpha_2 - \alpha_1^2 = 4275 \approx (65.4)^2$. Another, shorter solution: first calculate the variance of q and then use (2.4.12).

2.4.2. Solving σ_q from $\sigma^2 = n + n^2 \cdot \sigma_q^2$ gives $\sigma_q = (\sigma^2 - n)^{1/2}/n \approx 0.10$.

2.4.3. The derivatives of the c.g.f. (2.4.10) are $\psi'(s) = \psi'_q(\varphi(s)) \cdot \varphi'(s)$, $\psi'' = \psi''_q(\varphi) \cdot \varphi'^2 + \psi'_q(\varphi) \cdot \varphi''$ and $\psi^{(3)} = \psi^{(3)}_q(\varphi) \cdot \varphi'^3 + \psi''_q(\varphi) \cdot 2 \cdot \varphi' \cdot \varphi'' + \psi''_q(\varphi) \cdot \varphi' \cdot \varphi'' + \psi'_q(\varphi) \cdot \varphi^{(3)}$. Putting $s = 0$ and noting that $\varphi(0) = 0$, $\varphi^{(j)}(0) = n$ and $\psi'_q(0) = E(q) = 1$, we get $\kappa_1 = n$, $\kappa_2 = \kappa_2(q) \cdot n^2 + n$ and $\kappa_3 = \kappa_3(q) \cdot n^3 + 3 \cdot \kappa_2(q) \cdot n^2 + n$.

2.4.4. The formulae (2.4.12) followed by (1.4.25) from the result of Exercise 2.4.3.

2.4.5. (a) The variance of the mixed Poisson variable is $n + n^2 \cdot \text{Var}(q) > n = \text{Var}(\text{Poisson}(n))$. (b) Choose such a mixing variable that its skewness is negative. Then the numerator in the skewness formula (2.4.12) is polynomial of n of degree 3 and has a negative third order coefficient $\kappa_3(q)$. Therefore the skewness of the corresponding mixed Poisson d.f. has negative skewness when n is chosen large enough.

2.4.6. The kurtosis is obtained as in the solutions of Exercises 2.4.3 and 2.4.4 by derivating the cumulant generating function once more and evaluating the derivative at $s = 0$.

2.5.2. $p_k / p_{k-1} = ((h + k - 1)/k) \cdot (n/(n + h)) \geqslant 1$ if and only if $k \leqslant n - n/h$.

2.5.3. The Pólya m.g.f. $M_{h,n}$ given by (2.5.7) satisfies $(M_{h,n}(s))^2 = (h/(h + n - n \cdot e^s))^{2h} = (2h/(2h + 2n - 2n \cdot e^s))^{2h} = M_{2h,2n}(s)$.

2.5.4. Since $E(q) = 1$, we have $H(q) = 1 - e^{-q}$. By (2.5.1) q is then, in fact, Gamma$(1, 1)$-distributed. Then $b = 0$ in (2.5.10). By (2.5.9) k has geometric d.f.

3.2.1. By (2.3.3), $\psi_k(s) = \psi_q(n \cdot e^s - n)$. By (3.2.5) $\psi_X = \psi_k(\psi_Z(s)) = \psi_q(n \cdot M_Z(s) - n)$.

3.2.2. Substitute the Pólya m.g.f. M_k (see (2.5.7)) into the equation $M_X(s) = M_k(\ln M_Z(s))$.

3.2.3. By (3.2.5) and (2.4.10) the c.g.f. ψ_X of a compound mixed Poisson variable X is $\psi_X(s) = \psi_q(\psi(s))$, where $\psi(s) = \varphi(\psi_Z(s))$ is the corresponding (non-mixed) compound Poisson c.g.f. having cumulants (3.2.12), i.e. $\psi^{(j)}(0) = n \cdot a_j$. As in the solution of Exercise 2.4.3,

we obtain $\kappa_2(X) = \kappa_2(q) \cdot (n \cdot a_1)^2 + \kappa_1(q) \cdot (n \cdot a_2) = \sigma_q^2 \cdot n^2 \cdot m^2 + n \cdot a_2$
and $\kappa_3(X) = \kappa_3(q) \cdot n^3 \cdot m^3 + 3 \cdot \kappa_2(q) \cdot m \cdot n^2 \cdot a_2 + n \cdot a_3$. The formulae
(3.2.14) then follow by (1.4.25).

3.2.4. The fourth derivative of the c.g.f. $\psi_X(s) = \psi_q(\psi(s))$ is (see the
solution of Exercise 2.4.3):

$$\psi_X^{(4)} = \psi_q^{(4)}(\psi) \cdot \psi'^4 + \psi_q^{(3)}(\psi) \cdot 3 \cdot \psi'^2 \cdot \psi'' + 3 \cdot \psi_q^{(3)}(\psi) \cdot \psi'^2 \cdot \psi''$$
$$+ 3\psi_q''(\psi) \cdot [\psi''^2 + \psi' \cdot \psi^{(3)}] + \psi_q''(\psi) \cdot \psi' \cdot \psi^{(3)} + \psi_q'(\psi) \cdot \psi^{(4)}.$$

It is seen that $\gamma_2 = \kappa_4(X)/\sigma_X^4$,

$$\kappa_4(X) = \kappa_4(q) \cdot n^4 \cdot m^4 + 6\kappa_3(q) \cdot n^3 \cdot m^2 \cdot a_2$$
$$+ \kappa_2(q) \cdot n^2 \cdot [3a_2^2 + 4n \cdot a_3] + n^4 \cdot a_4.$$

3.2.5. $\text{Var}(X) = \text{Var}[\text{E}(X|k)] + \text{E}[\text{Var}(X|k)] = \text{Var}(k \cdot m) + \text{E}(k \cdot \text{Var}(Z)) = m^2 \cdot \text{Var}(k) + n \cdot \text{Var}(Z)$.

3.2.7. $r_2 - 1 = (a_2 - m^2)/m^2 = \text{Var}(Z)/m^2 \geqslant 0$. On the other hand,
if $Z \geqslant 0$, we have $a_2^2 = [\text{E}(Z^2)]^2 = [\text{E}(\sqrt{Z} \cdot \sqrt{Z^3})]^2 \leqslant \text{E}(\sqrt{Z^2}) \cdot \text{E}(\sqrt{Z^6}) = m \cdot a_3$. So

$$r_2^2 = \frac{a_2^2}{m^4} \leqslant \frac{m \cdot a_3}{m^4} = r_3.$$

3.2.8. The proof of (2.3.11) applies also for the compound Poisson d.f.

3.2.9. Cf. section 6.6.2(b).

3.2.10. Since $x^{2/3}$ is a concave function for $x > 0$, we have $a_2 = \text{E}(Z^2) = \text{E}(Y^{2/3}) \leqslant [\text{E}(Y)]^{2/3} = a_3^{2/3}$.

3.2.11. By Exercise 3.2.10 we have $\gamma_k = n^{-1/2} \leqslant n^{-1/2} \cdot a_3/a_2^{3/2} = \gamma_X$.

3.2.12. Replace ψ_q and φ in the solution of Exercise 2.4.3 by the
c.g.f.s ψ_k and ψ_Z, respectively.

3.2.13. Since $\text{Cov}(k, X|k) = 0$ (the first variable being conditionally
constant), we have, by (1.4.40), $\text{Cov}(k, X) = \text{Cov}(\text{E}(k|k), \text{E}(X|k)) = \text{Cov}(k, m \cdot k) = m \cdot \text{Cov}(k, k) = m \cdot \text{Var}(k)$. The correlation coefficient
ρ is equal to $m \cdot \sigma_k/\sigma_X$. Especially, in the compound mixed Poisson

case

$$\rho = \sqrt{\frac{1 + n \cdot \sigma_q^2}{1 + \frac{\sigma_Z^2}{m^2} + n \cdot \sigma_q^2}}$$

The correlation is never negative, it is close to zero when the number of claims is small and claim sizes have large variance. The larger n, the closer to 1 the correlation coefficient (if $\sigma_q > 0$).

3.2.14. Since X_1 and X_2 are conditionally independent, given q_1 and q_2, we have $\mathrm{Cov}(X_1, X_2 | q_1, q_2) = 0$. Therefore, by (1.4.40), we have $\mathrm{Cov}(X_1, X_2) = \mathrm{Cov}[E(X_1 | q_1, q_2), E(X_2 | q_1, q_2)] = \mathrm{Cov}[E(X_1 | q_1),$ $E(X_2 | q_2)] = \mathrm{Cov}(m_1 \cdot n_1 \cdot q_1, m_2 \cdot n_2 \cdot q_2) = m_1 \cdot n_1 \cdot m_2 \cdot n_2 \cdot \mathrm{Cov}(q_1, q_2) = 250 \cdot \rho$. Then $\sigma_X = (6025 + 250 \cdot \rho)^{1/2} = 76.0, 77.6, 79.2$ for $\rho = -1$, 0, 1.

3.2.15. Clearly $\mu_X = nm = Nn_j m = 1840$ in both cases. In case (i)

$$\sigma_X = \sqrt{na_2 + n^2 m^2 \sigma_q^2} = 383,$$

and in case (ii) each risk unit forms a section, and therefore

$$\sigma_X = \sqrt{\sum_j (n_j a_2 + n_j^2 m^2 \sigma_{q_j}^2)} = 105.$$

3.3.1. Denote the unknown rate of death by q. Then the following values for the step function S follow from (3.5.1)

$$S(100) = \frac{5000q}{5000q + 1000q + 2 \cdot 2000q} = 0.5;$$

$$S(250) = 0.6 \text{ and } S(500) = 1.$$

3.3.2. Then $S_i(Z) = 0$, if $Z < Z_i$, and $S_i(Z) = 1$, if $Z \geqslant Z_i$.

3.3.4. $\mathrm{Prob}\{k_i = 1 | k = 1\} = \dfrac{\mathrm{Prob}\{k_i = 1 \text{ and } k_j = 0 \text{ for every } j \neq i\}}{\mathrm{Prob}\{k = 1\}}$

$$= \frac{n_i \cdot e^{-n_i} \cdot \prod_{j \neq i} e^{-n_j}}{n \cdot e^{-n}} = \frac{n_i}{n}.$$

3.3.5. The sum term in (3.3.4) is equal to 4.381. The integral term (note truncation at $M = 100\,000$) is equal to 2.804. Then $a_1 = 7.185$.

3.3.6. Suppose that $Z_{i-1} < Z_i$ and that $S(Z) = S(Z_{i-1})$ for $Z_{i-1} \leqslant Z < Z_i$. Then the integral term and $S(M)$ are both constants if M is restricted to the interval $[Z_{i-1}, Z_i)$. Hence, L is linear in that interval.

3.3.7. Since the sample d.f. S gives the same probability $1/n$ for every observation Z_i, we have

$$L(M) = \mathrm{E}(\min(M, Z)) = \frac{1}{n} \sum_{i=1}^{n} \min(M, Z_i),$$

where Z is distributed according to the sample d.f. S.

3.3.8. $L_{a \cdot Z + b}(M) = \mathrm{E}(\min(a \cdot Z + b, M)) = \mathrm{E}(\min(a \cdot Z, M - b) + b)$

$$= \mathrm{E}\left(a \cdot \min\left(Z, \frac{M - b}{a}\right) + b\right)$$

$$= a \cdot L_Z\left(\frac{M - b}{a}\right) + b.$$

3.3.9. Assume $M_1 < M_2$. Since $0 \leqslant \min(M_2, Z) - \min(M_1, Z) \leqslant M_2 - M_1$, we have $0 \leqslant L(M_2) - L(M_1) \leqslant M_2 - M_1$. Therefore L is increasing and continuous.

3.3.10. The derivative of L, at every point M where S is differentiable, is

$$L'(M) = M \cdot S'(M) + (1 - S(M)) - M \cdot S'(M) = 1 - S(M).$$

Since $1 - S(M)$ is a decreasing function of M, it is seen that L is concave.

Another proof: Let $\lambda \in (0, 1)$. Since the function $M \to \min(M, Z)$ is obviously concave, we have $\lambda \cdot \min(x, Z) + (1 - \lambda) \cdot \min(y, Z) \leqslant \min(\lambda \cdot x + (1 - \lambda) \cdot y, Z)$, if $0 < \lambda < 1$. Therefore,

$$\lambda \cdot L(x) + (1 - \lambda) \cdot L(y) \leqslant L(\lambda \cdot x + (1 - \lambda) \cdot y)$$

whenever $0 < \lambda < 1$. It is well known that this implies that L is concave.

3.3.11. $\min(M, Z)$ increases to Z as $M \to +\infty$.

3.3.12. (a) If S is differentiable at a point M, then $L'(M) = M \cdot S'(M) + 1 - S(M) - M \cdot S'(M) = 1 - S(M)$, by (3.3.5). Since this holds true at almost every point M, and since S is right continuous, we have $L'(M+) = 1 - S(M)$ for every M. On the other hand, S has a left limit $S(M-)$ at every point M, so that $L'(M-) = 1 - S(M-)$. (b) Since $L'(M) = 1 - S(M)$ at almost every point M, and since L is continuous, the requested integral formula holds true.

3.3.13. (a) If $S(M-) = 0$, we have $L(M) = E(\min(M, \mathbf{Z})) = M$. (b) By Exercise 3.3.12(b) $L(x) - L(y) = \int_y^x 1 - S(Z)\,dZ \leqslant \int_y^x 1\,dZ = x - y$.

3.3.14. By Exercise 3.3.12(b) we have

$$L_T(M) - L_T(T) = \int_T^M 1 - \frac{S(Z) - S(T)}{1 - S(T)}\,dZ = \int_T^M \frac{1 - S(Z)}{1 - S(T)}\,dZ$$

$$= \frac{L(M) - L(T)}{1 - S(T)}, \quad M \geqslant T.$$

3.3.15. Clearly L is continuous, $L(M) = L_1(M)$ for $M \leqslant A$ and $L'(M+) = L_2'(M+) = 1 - S_2(M)$ for $M > A$.

3.3.16. This follows from (3.2.30).

3.3.18. Substitute $r = 1$ in (2.5.1).

3.3.19. The proof of section 2.3(e) of asymptotic normality of Poisson distribution can be applied, since gamma distributions have the additivity property described in section 3.3.5(c).

3.3.20. Since $S(Z) = N(\ln(Z - d) - \mu)/s$ and since the derivative of $\ln(Z - d)$ is $1/(Z - d)$, we have $S'(Z) = N'(\ln(Z - d) - \mu)/s \cdot ((Z - d) \cdot s)^{-1}$.

3.3.21. (a) Since $d = 0$, we have $a_k = E(\mathbf{Z}^k) = E(e^{k \cdot Y}) = M(k)$. (b) The mean m is equal to $d + M(1)$. For the derivation of the higher characteristics we may assume $d = 0$. The variance σ_Z^2 is equal to $M(2) - (M(1))^2 = e^{2 \cdot \mu} \cdot e^{s^2} \cdot (e^{s^2} - 1)$. By (1.4.22) the third cumulant of $\mathbf{Z}$ is equal to $M(3) - 3 \cdot M(2) \cdot M(1) + 2 \cdot (M(1))^3 = e^{3 \cdot \mu} \cdot e^{(3/2) \cdot s^2} \cdot (e^{s^2} - 1)^2 \cdot (e^{s^2} + 2)$. (c) The only real root of the equation (3.3.11) is $\eta = (e^{s^2} - 1)^{1/2}$.

Thus $s^2 = \ln(1 + \eta^2)$, and μ can be solved from $m - d = e^{\mu + s^2/2}$ by noting that $m - d = \sigma_Z/\eta$.

3.3.22. $\operatorname{Prob}\{Z - c > Z\} = \operatorname{Prob}\{Z > Z + c\} = \left(\dfrac{D + \beta}{Z + \beta}\right)^{\alpha}$

$$= \left(\frac{(D - c) + (\beta + c)}{Z + (\beta + c)}\right)^{\alpha}.$$

3.3.23. Let $Z > T$. Then $\operatorname{Prob}\{Z > Z \mid Z > T\} = \dfrac{\operatorname{Prob}\{Z > Z\}}{\operatorname{Prob}\{Z > T\}}$

$$= \left(\frac{T + \beta}{Z + \beta}\right)^{\alpha}.$$

3.3.24. $\displaystyle\int_{D}^{\infty} (Z + \beta)^k \cdot \mathrm{d}S(Z) = \int_{D}^{\infty} \alpha \cdot (D + \beta)^{\alpha} \cdot (Z + \beta)^{k - \alpha - 1}\, \mathrm{d}Z.$

3.3.25. The mean m is obtained by substituting $k = 1$ in the formula of Exercise 3.3.24. Similarly, by making use of the fact that $\sigma_Z^2 = \sigma_{Z+\beta}^2$, the variance formula is obtained.

3.3.26. By (1.4.22) $\kappa_3(Z + \beta) = \mathrm{E}((Z + \beta)^3) - 3\mathrm{E}(Z + \beta) \cdot \mathrm{E}((Z + \beta)^2) + 2(\mathrm{E}(Z + \beta))^3 = 2 \cdot \alpha \cdot (\alpha + 1) \cdot (\alpha - 1)^{-3} \cdot (\alpha - 2)^{-1} \cdot (\alpha - 3)^{-1}$. The skewness in (3.3.15) is obtained by noting that

$$\gamma_Z = \gamma_{Z+\beta} = \frac{\kappa_3(Z + \beta)}{\sigma_Z^3}, \quad \text{since } \sigma_Z = \sigma_{Z+\beta}.$$

3.3.27. (a) $L(M) = L(D) + \displaystyle\int_{D}^{M} 1 - S(Z)\,\mathrm{d}Z = D + \int_{D}^{M} \left(\frac{D + \beta}{Z + \beta}\right)^{\alpha} \mathrm{d}Z$

$$= D + \frac{D + \beta}{\alpha - 1} - \frac{M + \beta}{\alpha - 1}[1 - S(M)]$$

$$= \frac{\alpha D + \beta - (M + \beta) \cdot [1 - S(M)]}{\alpha - 1}.$$

(b) $|L_\alpha(M) - L_1(M)| \leqslant \displaystyle\int_{D}^{M} |S_1(Z) - S_\alpha(Z)|\,\mathrm{d}Z$

$$\leqslant |M - D| \cdot \max_{Z} |S_1(Z) - S_\alpha(Z)| \to 0, \text{ as } \alpha \to 1.$$

(c) For $\alpha = 1$, we have $L(M) = D + \int_D^M \dfrac{D + \beta}{Z + \beta} dZ = D + (D + \beta) \cdot$
$[\ln(M + \beta) - \ln(D + \beta)]$ for $M \geqslant D$.

3.3.28. Denote $M = Z_{\max}$. We have

$$E((Z_{tr} + \beta)^i) = \int_D^\infty (Z + \beta)^i dS(Z)$$

$$= \int_D^M \alpha \cdot (D + \beta)^\alpha \cdot (Z + \beta)^{i - \alpha - 1} dZ + (M + \beta)^i \cdot [1 - S(M)],$$

and

$$\alpha_k(M) = E(Z_{tr}^k) = E[((Z_{tr} + \beta) - \beta)^k]$$

$$= \sum_{i=0}^k \binom{k}{i} \cdot (-\beta)^{k-i} \cdot E((Z_{tr} + \beta)^i).$$

3.3.29. $\alpha_k = \displaystyle\int_{-\infty}^C Z^k \dfrac{dS(Z)}{S(C)}$

$$= \frac{1}{S(C)} \cdot \left[\int_{-\infty}^C Z^k dS(Z) + (1 - 1) \cdot C^k \cdot (1 - S(C)) \right]$$

$$= \frac{1}{S(C)} \cdot [a_k(C) - C^k (1 - S(C))].$$

3.3.30. Let $M \leqslant C$. Then

$$L_C(M) = \frac{1}{S(C)} \cdot \int_{-\infty}^M Z \, dS(Z) = \frac{1}{S(C)} \cdot [L(M) - M \cdot (1 - S(M))].$$

3.3.31. $\ln(1 - S(Z)) = \ln(D^\alpha \cdot Z^{-\alpha}) = \ln(D^\alpha) - \alpha \cdot \ln(Z).$

3.3.32. Use (3.3.30) and the properties of L.

3.3.33. $c_{\text{franch}}(D) = 1 - \dfrac{L(D) - D \cdot (1 - S(D))}{m}.$

3.3.34. $P = 0.01 \cdot \displaystyle\int_2^{10} 1 - S(Z) \, dZ = 0.0024.$

3.4.1.
$$S_{re}(Z) = \text{Prob}\{Z - M \leqslant Z | Z > M\}$$
$$= \frac{S(Z + M) - S(M)}{1 - S(M)} = 1 - e^{-cZ}.$$

Assuming that the aggregate claim amount is a compound mixed Poisson variable, we have

$$\text{Var}(X_{re}) = n_M \cdot a_2(M) + n_M^2 \cdot m_M^2 \cdot \sigma_q^2 = \frac{n \cdot e^{-cM}}{c^2} \cdot (2 + n \cdot e^{-cM} \cdot \sigma_q^2).$$

3.4.2. Answer to the last question: 33.1% of the reinsurance risk premium.

3.4.3. By (3.3.11) and (3.3.12) $a = 5$, $s = 1.683$ and $\mu = 1.802$. By (3.3.13) we get $P_{re} = 1200$, 810, 500 and 40 for the given values of M. The total risk premium is 3000.

3.4.4. (a, b) By (3.4.8) we have

$$a_k(Z_{re}) = \int_M^\infty (Z - M)^k dS(Z) = \sum_{i=0}^k \binom{k}{i} \cdot (-M)^{k-i} \cdot \int_M^\infty Z^i dS(Z)$$

$$= \sum_{i=0}^k \binom{k}{i} \cdot (-M)^{k-i} \cdot [a_i - a_i(M) + M^i \cdot (1 - S(M))]$$

$$= \sum_{i=1}^k \binom{k}{i} \cdot (-M)^{k-i} \cdot [a_i - a_i(M)].$$

(c) Apply the above formula to the claim size d.f. S truncated at $M + A$.

3.4.5. We have $\text{Var}(X_{re}) = n \cdot a_2(Z_{re}) + n^2 \cdot (a_1(Z_{re}))^2 \cdot \text{Var}(q)$. The a_k-moments are obtained by applying (3.4.9) to the d.f. S truncated at $M + A$.

3.4.6. For $Z < 102.4$ the probability mass was assumed to be concentrated to the class averages $\bar{Z}_i$ given in Table 3.3.1. For $\bar{Z}_i \leqslant M \leqslant \min(102.4, \bar{Z}_{i+1})$:

$$a_k(M) = \sum_{j \leqslant i} \bar{Z}_j^k \cdot (S(Z_j) - S(Z_{j-1})) + M^k \cdot (1 - S(Z_i)).$$

For $T \leqslant Z < 100000$ we have $S(Z) = 1 - c \cdot Z^{-\alpha}$, where $T = 102.4$,

$\alpha = 1.3938$ and $c = 7.3208$. Hence, for $T \leqslant M < 100\,000$,

$$a_k(M) = a_k(T) - T^k \cdot (1 - S(T))$$
$$+ \int_T^M Z^k \, dS(Z) + M^k \cdot (1 - S(M)).$$

The integral term is equal to $\alpha \cdot c \cdot (M^{k-\alpha} - T^{k-\alpha})/(k - \alpha)$.

3.4.7. $P_{re}(r) = n \cdot r \cdot (L(\infty) - L(M/r)) = c \cdot r^\alpha \cdot (M + r \cdot \beta)^{1-\alpha}$, where c does not depend on M.

3.4.8. Check the cases $\{Z_{tot} \leqslant A\}$, $\{A < Z_{tot} \leqslant M + A\}$ and $\{Z_{tot} > M + A\}$.

3.4.9. Take the derivative of (3.4.8).

3.4.10. Recall that $L' = 1 - S$.

3.4.11. Make use of the equation $P = P_{tot} - P_{re}$.

3.4.12. (a) Both the expected number of claims exceeding the retention limit, and the size of the reinsurer's share of a claim remain at least the same.
(b) In the case that $S(M/r) = S((M + A) -)$.
(c) $P_{re}(r) = n \cdot (L(\infty) - L(M/r)) = c \cdot r^\alpha$, where c does not depend on M.

3.4.13. The conditional claim size d.f. of Z_{re} under the condition $\{Z_{re} > 0\}$ is $S_{re}(Z) = [S(M + Z) - S(M)]/[1 - S(M)]$, $M \leqslant Z \leqslant M + A$, $S_{re}(M + A) = 1$. By the formula given in Exercise 3.4.9, we have

$$a_k(Z_{re} | Z_{re} > 0) = \int_0^A k \cdot Z^{k-1} \cdot \left(1 - \frac{S(M + Z) - S(M)}{1 - S(M)}\right) dZ,$$

from which the asked formula follows by noting the equality (3.4.9).

3.4.14. $m_{ced} = r \cdot m$, $\sigma_{ced} = r \cdot \sigma$, $\gamma_{ced} = \gamma$.

3.4.15. $S_{ced}(Z) = S\left(\dfrac{Z}{1 - r}\right)$ for $Z < M$ and $S_{ced}(Z) = 1$ otherwise.

3.4.16. $n \cdot m_M = n \cdot \sum_j E(r_M \cdot Z_{tot} | Q = Q_j) \cdot \dfrac{n_j}{n} = \sum_j r_M(Q_j) \cdot n_j \cdot m(Q_j)$.

Similarly, $E((r_M \cdot Z_{tot})^2 | Q = Q_j) = (r_M(Q_j))^2 \cdot a_{2,j}$.

3.4.17. Replace $r_M(Q_j)$ by $(1 - r_M(Q_j))$.

3.4.18. (b) $Z_{re,sur} > 0$ implies $Q > M$. Then $Z_{re,sur} = \left(1 - \dfrac{M}{Q}\right) Z_{tot} \geqslant Z_{tot} - M$, provided that $Q \geqslant Z_{tot}$.

3.4.19. (a) $\dfrac{d}{dM}\left[M \cdot (1 - N(M)) - \dfrac{1}{\sqrt{2\pi}} e^{-M^2/2} \right] = 1 - N(M)$.

(b, c) By Exercise 3.3.8 $P_{re} = \sigma \cdot L_N\left(\dfrac{M - P_{tot}}{\sigma}\right)$.

3.4.20. By (3.4.27) (a) $P_{re} \approx 0.2931$; (b) $P_{re} \approx 2.336$.

4.1.1. $a = \dfrac{1}{6}, b = \dfrac{3}{2}, f_0 = p_0 = \left(\dfrac{5}{6}\right)^{10} \approx 0.1615, f_1 = \left(a + \dfrac{b}{1}\right) \cdot s_1 \cdot f_0 = \dfrac{10}{6} \cdot 0.2 \cdot \left(\dfrac{5}{6}\right)^{10} \approx 0.0538, f_2 = \left(a + \dfrac{b}{2}\right) \cdot s_1 \cdot f_1 + \left(a + \dfrac{2b}{2}\right) \cdot s_2 \cdot f_0 \approx 0.2252$, etc.

4.1.3. $p_k = \binom{N}{k} p^k \cdot (1 - p)^{N-k} = \dfrac{(N - k + 1) \cdot N!}{k \cdot (k-1)! \cdot (N - k + 1)!} \cdot \dfrac{p}{1-p} \cdot p^{k-1}$.

$(1 - p)^{N-k+1} = (N - k + 1)/k \cdot p/(1 - p)) \cdot p_{k-1} = (a + b/k) \cdot p_{k-1}$, where $a = -p/(1 - p)$ and $b = a \cdot (N + 1)$.

4.1.5. (a) When $R = S$, and when $T = S$, respectively. (b) It is easy to see that whenever F and G are d.f.s such that $F \leqslant G$, the convolution satisfies the inequality $F*F \leqslant F*G \leqslant G*G$. Since $R \leqslant S \leqslant L$, it follows that $R^{k*} \leqslant S^{k*} \leqslant L^{k*}$, for every positive integer k. By (3.1.4) we then have $F_R \leqslant F_S \leqslant F_L$. It is easy to see that $R \leqslant U \leqslant L$, and that $R \leqslant T \leqslant L$ in the case that $S(r \cdot C) - S(0) = 1$, so that the remaining inequalities also hold true.

4.1.6. We have

$$\int_m^M (X - m)^2 \, dF_X(X) \leqslant (M - m) \cdot \int_m^M (X - m) \, dF_X(X)$$
$$= (M - m) \cdot (\mu - m)$$

where $\mu = E(X)$. On the other hand, since $E(Y) = \mu$, we have $\text{Prob}\{Y = M\} = (\mu - m)/(M - m) = 1 - \text{Prob}\{Y = m\}$, and therefore

$$\int_m^M (Y - m)^2 \, dF_Y(Y) = 0 + (M - m)^2 \cdot \frac{\mu - m}{M - m} = (M - m) \cdot (\mu - m).$$

Since expected values of both variables are equal, it follows that $\text{Var}(X) \leqslant \text{Var}(Y)$.

4.1.7. Let us use the notation of (4.1.12–14), and denote $I_i = \{(i-1) \cdot C < Z \leqslant i \cdot C\}$. Let Y_i be a random variable such that

$$Y_i = \begin{cases} 0 & \text{in } \{Z \notin I_i\} \\ (i-1) \cdot C & \text{by probability } r_{i-1} \\ i \cdot C & \text{by probability } l_i \end{cases}$$

Z being an S-distributed random variable; note that $\text{Prob}\{Z \notin I_i\} + r_{i-1} + l_i = (1 - d_i) + d_i = 1$. Then the sum $Z^* = \Sigma Y_i$ is distributed according to the equidistant distribution (4.1.12). By Exercise 4.1.6 we have $\text{Var}(Z^* | Z \in I_i) = \text{Var}(Y_i | Z \in I_i) \geqslant \text{Var}(Z | Z \in I_i)$. Hence $\text{Var}(Z^*) = \Sigma d_i \cdot [E(Z^* | Z \in I_i) - E(Z^*)]^2 + \Sigma d_i \cdot \text{Var}(Z^* | Z \in I_i) \geqslant \Sigma d_i \cdot [E(Z | Z \in I_i) - E(Z)]^2 + \Sigma d_i \cdot \text{Var}(Z | Z \in I_i)$, by (1.4.38).

4.1.8. We will find out for which pairs (a, b) of real numbers (4.1.1) defines a probability d.f. We exclude the case that $p_0 = 1$ (binomial with $p = 0$). Clearly, we must have $p_0 > 0$, since otherwise all p_k's are zero.

(1) Case $a = 1$. Then $b < 0$, since otherwise $p_k > p_0 > 0$ for all k. On the other hand, $p_1 \leqslant (1 + b) \cdot p_0$ implies $b \geqslant -1$. Therefore, $p_2 = (1 + b/2) \cdot p_1 \geqslant p_1/2, p_3 = (1 + b/3) \cdot p_2 \geqslant 2/3 \cdot p_2 \geqslant p_1/3$ and, generally, $p_k \geqslant p_1/k$. But $\Sigma p_k \geqslant \Sigma 1/k = \infty$ is a contradiction. So $a \neq 1$.

(2) Case $a > 1$. Then $a + b/k \to a$, as $k \to \infty$, and therefore we must have $p_k = 0$ for some k. Let $m = \min\{k : p_k = 0\}$. Then $p_{m-1} > 0$ and $a + b/m = 0$, or $b = k \cdot m$. Since $p_1 = a \cdot (1 - m) \cdot p_0 \geqslant 0$, we have $m = 1$, and therefore $p_0 = 1$, which case has already been excluded.

(3) Case $0 < a < 1$. Since $p_1 = (a + b) \cdot p_0$, we have $b > -a$. Choosing $p = 1 - a$ and $h = b/a + 1$ (then $n = h/p - h > 0$), we obtain the Pólya (n, h) distribution.

(4) Case $a = 0$. $b > 0$ and we have the Poisson(b) distribution.

(5) Case $a < 0$. Then $a + b/k < 0$ when k is large enough. Therefore, there exists $N \geqslant 2$ such that $p_N = 0$ and $p_{N-1} > 0$. Since $(a + b/N) = 0$, we have $b = -N \cdot a$. The Binomial(p, N) distribution with $p = a/(a - 1)$ is then the case.

4.2.1. Differentiating the r.h.s. of (3.4.27) gives

$$P'_{re}(M) = \sigma \cdot \left[\frac{\gamma}{6} y'(M) \cdot N'(y(M)) \right.$$
$$\left. + \left(1 + \frac{\gamma}{6} \cdot y(M) \right) \cdot N'(y(M)) \cdot (- y(M)) \cdot y'(M) \right]$$
$$- (1 - N(y(M))) + (M - P_{tot}) \cdot N'(y(M)) \cdot y'(M)$$
$$= \sigma \cdot N'(y(M)) \cdot y'(M) \cdot \left[\frac{\gamma}{6} - \left(y(M) + \frac{\gamma}{6} (y(M))^2 \right) + \frac{M - P_{tot}}{\sigma} \right]$$
$$- (1 - N(y(M))) = - (1 - N(y(M))),$$

since the [ ]-term is equal to zero, by (3.4.28). Therefore, P_{re} satisfies (3.4.26) for $F(X) = N(y(X))$.

4.2.2. Use (B.1.3) given in Appendix B.

5.2.1. Since $F^{-1}(r) = \min\{X \mid F(X) \geq r\}$, we have $\text{Prob}\{X \leq X\} = \text{Prob}\{F^{-1}(r) \leq X\} = \text{Prob}\{r \leq F(X)\} = F(X)$.

5.2.2. $F^{-1}(r) = - \ln(1 - r)/a$ and $1 - r$ is uniform$(0, 1)$ distributed if and only if r is.

5.2.3. (a) Generate an $N(0, 1)$-random number r by using the log-and-trig formula, set $Y = \mu + s^2 \cdot r$ (see (3.3.12)) and use formula (3.3.9). (b) If r is uniform$(0, 1)$, then $Z = F^{-1}(1 - r) = (D + \beta)/r^{1/\alpha}$ is Pareto(α, β, D) distributed.

6.2.1. By (6.1.10) $U_r = y\sqrt{n \cdot a_2} - \lambda \cdot n \cdot m$ and $U'_r = y \cdot \sqrt{n' \cdot a_2} - \lambda' \cdot n' \cdot m$, where $n' = 0.9 \cdot n$, $\lambda \cdot n \cdot m = 5$ and $\lambda' \cdot n' \cdot m = (1 + \lambda) \cdot n \cdot m - n' \cdot m = 15$. Hence $U_r - U'_r \approx 11.3$ (£million).

6.2.2. Let $0 \leq p \leq 1$ ($100 \cdot p\%$ chooses the sum 1 [in £100]). Then $U_r(p) = y\sqrt{(4 - 3p)n} - \lambda(2 - p)n$, by (6.1.10). The required maximum is obtained for $p \approx 0.52$.

6.2.3. Putting $U_r = 0$ gives $\lambda \approx 2.33\sqrt{2/1000 + 0.04^2} \approx 0.14$.

6.2.4. Use (3.2.27) to calculate the characteristics of the merged company. Verify that the skewness of the aggregate claim amount of both C_1 and the merged company are smaller than unity (otherwise (6.1.10), based on the NP-approximation, cannot be used). Solving y from (6.1.10) gives $y = 2.816$. The corresponding U is £512 000.

6.2.5. We have $\sigma \leqslant \sqrt{\sigma_1^2 + \sigma_2^2 + 2 \cdot \sigma_1 \cdot \sigma_2} = \sigma_1 + \sigma_2$, where equality holds true if and only if $X_1 = a \cdot X_2 + b$, where $a > 0$. As in (6.2.7), it is seen that $U_r \leqslant U_{r,1} + U_{r,2}$.

6.2.6. Under the condition $q_A = q$, the assumptions of section 6.2(e) are in force, so that (6.2.8) holds true for every possible value q of the mixing variable q_A. Therefore, the capital at risk $U_{r,1} + U_{r,2}$ for the merged company would lead to improved security level. In other words, for the merged company, the minimum capital at risk U_r corresponding to the original security level $1 - \varepsilon$ is smaller than $U_{r,1} + U_{r,2}$.

6.2.7. Apply (6.1.7). Now $X_{\varepsilon,i} = 0$ and hence $U_{r,i} = -\lambda_i \cdot P_i$, while $U_r \geqslant 1 - \lambda_1 \cdot P_1 - \lambda_2 \cdot P_2 > U_{r,1} + U_{r,2}$, since $\text{Prob}\{X_1 + X_2 \geqslant 1\} = 1 - 0.9^2 = 0.19 > 0.11 = \varepsilon$.

6.3.1. By Exercise 3.4.9 (or by (3.4.8)) we have $a_k'(M) = k \cdot M^{k-1} \cdot (1 - S(M))$. Substitute these into $U_r'(M) = \frac{1}{2} y_\varepsilon a_2'(M) \sqrt{n/a_2(M)} - \lambda \cdot n \cdot a_1'(M)$.

6.3.2. Since $S(0) = 0$, $a_2'(M)/M^2 \to 1$, as $M \to 0+$. It follows that the limit of $U_r'(M)$ as $M \to 0+$ is negative if and only if $\lambda > y_\varepsilon/\sqrt{n}$. (Note that it was assumed that the safety loading coefficient does not depend on the retention limit M!).

6.3.3. Let M_0 denote the smallest value M such that $S(M) = 1$; M_0 may be ∞. For $M < M_0$ the bracket expression in

$$U_r'(M) = n \cdot (1 - S(M)) \cdot \frac{M}{\sqrt{a_2(M)}} \cdot \left[\frac{y_\varepsilon}{\sqrt{n}} - \lambda \cdot \sqrt{\frac{a_s(M)}{M^2}} \right],$$

determines the sign of $U_r'(M)$. Since $a_2'(M) = 2 \cdot M \cdot (1 - S(M)) < 2 \cdot M = dM^2/dM$, if $S(M) > 0$, $a_2(M)/M^2$ is a strictly decreasing and positive function of M except for values M such that $S(M) = 0$, for which $a_2(M)/M^2 = 1$. Hence, if $U_r'(0+) > 0$, then $U_r(M)$ is a strictly increasing function of M for $M < M_0$. For $M \geqslant M_0$, the value of $U_r(M)$ remains, of course, constant. The case $U_r'(0+) = 0$ is otherwise similar, but

$U_r(M)$ first remains constant until $S(M)$ becomes positive. If $U_r'(0+) < 0$ (i.e., if $\lambda \geqslant y_\varepsilon/\sqrt{n}$, by Exercise 6.3.2), then either $U_r(M)$ is a strictly decreasing function of M for all $M < M_0$ or the sign of the bracket expression changes at a point $p < M_0$, in which case $U_r(M)$ is strictly decreasing for values $0 < M < p$ and strictly increasing for $p < M < M_0$, and attains its minimum at the point p.

6.3.4.   The values $M = 1.8$ and $P = 68$ can be read from Figure 6.3.1. Substituting into (6.3.1) and (6.3.6a) approximate values 0.7 and 0.6 for M are obtained. From Figure 3.3.6, by iteration value, $K = 0.43$ can be evaluated. Substituting into (6.3.1) and (6.3.6a) revised values $M = 1.8$ and 1.6 ensue.

6.3.5. $\Delta U_r = U_r'(M)\Delta M$ and $\Delta P = P'(M)\Delta M$ and the return $R = \lambda \Delta P$. The required rate is

$$i(M) = R/\Delta U_r = \lambda P'(M)/U_r'(M).$$

$U_r'(M)$ was calculated for Exercise 6.3.1 and

$$P'(M) = na_1'(M) = n(1 - S(M)).$$

Then after some reductions and using data (6.2.1), we obtain

$$i = \frac{\lambda}{yM/\sqrt{n \cdot a_2(M)} - \lambda} = 0.07$$

6.3.6.   The verification is performed by substituting M and related $a_2(M)$ values to the expression of the rate i of the solution of Exercise 6.3.5. The required limit is obtained by interpolation between the lines of Table 3.4.1, $M \approx 1.7$ £million.

6.3.8.   According to the second line in the formulae (3.3.18), we have $a_i(M) = (\alpha - i \cdot M^{i-\alpha})/(\alpha - i)$. By (6.1.9) $R_y(M) = (2.33^2 - 1)/6 \cdot a_3(M)/a_2(M)$. By trial and error it is found that $U_r(4.8) < 30$ and $U_r(4.9) > 30$. Since the skewness γ_X satisfies $\gamma_X \approx 0.15 < 1$ for these values of M, formula (6.1.10) based on the NP-approximation can be applied. Hence $M = 4.8$.

6.6.1.   Since σ_{X_i} is the same for all companies, the optimal solution is $Y_i = c \cdot (X_1 + \cdots + X_N)$, $c = 1/N$. Hence $r = \sigma_{Y_i}/\sigma_{X_i} = 1/\sqrt{N}$.

6.6.2.   Substituting into (6.6.19) and (6.6.20):

$$G_1 = 1 - \exp[-0.01 \cdot (100 - B)] \text{ and}$$
$$G_{II} = 0.99 \cdot [1 - \exp(-0.01 \cdot 100)]$$
$$+ 0.01 \cdot [1 - \exp(-0.01 \cdot (100 - 75))] = 0.628.$$

Equating G_1 and G_{II}, $B = 1.11$ follows.

6.6.3. $E(G(U)) > E(G(V))$ if and only if $E(\bar{G}(U)) = a \cdot E(G(U)) + b > a \cdot E(G(V)) + b = E(\bar{G}(V))$.

8.5.1. First simulate the yield $j_k(t)$ of the interest as proposed in section 8.5(f). Then

$$A(t) = \sum_{\tau = t}^{T} a(\tau) \cdot [1 + j(\tau)]^{-(\tau - t)}$$

where t is the present time and T the termination time.

8.5.2. Provide (8.5.2) with the factor (8.5.19).
(1) The deterministic case: give t_0, d_0 and b_0 manually to fit the application of concern.
(2) Stochastic case: define ϕ in (8.5.19) to have the values either 0 or 1 by probabilities $p_0(n(t))$ and $1 - p_0(n(t))$ according to the Poisson law (2.2.1), where

$$n(t) = \alpha \cdot t \cdot [I_e(t) - \bar{I}_e(t)]^+.$$

Here α is a parameter, the index $I_e(t)$ is as in (8.5.2) and $\bar{I}_e(t)$ the reference index (i.e. the former without the factor $(1 + \delta(t))$). Simulate ϕ for $t = 1, 2, \ldots$ until it first has the value 1 and then put $t_c = t$ and continue as described in the text following (8.5.19). Variable $n(t)$ describes the accumulating pressure for 'the bubble bursting' induced when the yield has exceeded the reference level for a lengthy time. Parameter α controls the frequency of the postulated crashes and can after some trial and error be determined so that the average density of the appearance of the events is as intended.

8.6.1. Minimize V subject to $E = E^*$ and $P = P^*$ by defining the Lagrangian

$$L = V - 2g_1(E - E^*) - 2g_2(P - P^*)$$
$$= \mathbf{x}' \cdot \mathbf{V} \cdot \mathbf{x} - 2\mathbf{x}' \cdot \mathbf{c} + V_L - 2g_1 \cdot (\mathbf{e}' \cdot \mathbf{x} - E_L - E^*) - 2g_2(\mathbf{p}' \cdot \mathbf{x} - P^*)$$

and setting the partial derivatives of L, with respect to x, g_1 and g_2, equal to zero. The portfolio which minimizes V is given by

$$\mathbf{x}' = \mathbf{c}' \cdot \mathbf{V}^{-1} + \frac{\gamma(P - \pi) - \beta(E + E_1 - \varepsilon)}{\gamma^2 - \alpha\beta} \mathbf{e}' \cdot \mathbf{V}^{-1}$$

$$+ \frac{\gamma(E + E_1 - \varepsilon) - \alpha(P^* - \pi)}{\gamma^2 - \alpha\beta} \mathbf{p}' \cdot \mathbf{V}^{-1}$$

where

$$\alpha = \mathbf{e}' \cdot \mathbf{V}^{-1} \cdot \mathbf{e}; \quad \beta = \mathbf{p}' \cdot \mathbf{V}^{-1} \cdot \mathbf{p}; \quad \gamma = \mathbf{e}' \cdot \mathbf{V}^{-1} \cdot \mathbf{p} = \mathbf{p}' \cdot \mathbf{V}^{-1} \cdot \mathbf{e};$$

$$\varepsilon = \mathbf{c}' \cdot \mathbf{V}^{-1} \cdot \mathbf{e}; \quad \pi = \mathbf{c}' \cdot \mathbf{V}^{-1} \cdot \mathbf{p}; \quad \varphi = \mathbf{c}' \cdot \mathbf{V}^{-1} \cdot \mathbf{c}$$

Then use the expression

$$\frac{\partial P}{\partial x_i} = \frac{\partial V}{\partial P} \cdot \frac{\partial P}{\partial x_i} + \frac{\partial V}{\partial E} \cdot \frac{\partial E}{\partial x_i}$$

for $i = 1, \ldots n$ to obtain

$$\mathbf{x}' = \mathbf{c}' \cdot \mathbf{V}^{-1} - \tfrac{1}{2}v\mathbf{p}' \cdot \mathbf{V}^{-1} + \frac{1}{2\mu} \mathbf{e}' \cdot \mathbf{V}^{-1}$$

and substitute for μ using

$$E = \mathbf{x}' \cdot \mathbf{e} = \varepsilon - \tfrac{1}{2}v \cdot \gamma + \frac{\alpha}{2\mu} - E_L$$

8.6.2. The solution is as for Exercise 8.6.1, until the final stage, where you should substitute for v using

$$P = \mathbf{x}' \cdot \mathbf{p} = \pi - \tfrac{1}{2}v\beta + \frac{\gamma}{2\mu}$$

9.5.1. The cohort reserve is

$$C_c(t - d, t) = \sum_{\tau = 0}^{d} X_p(t - d, t - d + \tau) \cdot \sum_{l=i}^{d_{\max} - d} \left[\prod_{k=0}^{l-2} a(d + k) \right]$$

$$\times [a(d + i - 1) - 1] \cdot v^{l - \frac{1}{2}}$$

where the first sum stands as in (9.5.8) for the payments up to the end of year t and the terms of the second sum for the payments in each of the later years.

9.5.2. A straightforward method is to fill the boxes belonging to a cohort (issued in year u in Figure 9.5.1) with the expected values of the claim payments $E(X(u, u + d))]$ (which are available from the simulation procedure described in section 9.5.3(c) and Exercise 9.5.4). Their sum X_u is the expected value of the total aggregate claim attributable to this cohort. Let X_d be that part of this sum where the development time exceeds a given d. Then $c_p(d) = X_d/X_u$.

Another approach is to base the calculation on the observed data.

Replace the parallelogram of Figure 9.5.1 so far left that it totally covers only already settled claims, i.e. right-hand end of its base should be $\leqslant t - d_{max}$. Then the claims of that part of the parallelogram which is over the level d in ratio of all claims is an estimate for $c_p(d)$. Because owing to inflation and real growth of the portfolio the latest cohorts usually have the greatest weight, it might be appropriate to unify the initial monetary value and the volume of each cohort belonging to the parallelogram.

A variant employing the latest available settled claims is suggested by Pentikäinen and Rantala (1992), p. 200, based on a rectangular instead of a parallelogram configuration.

9.5.3. $\rho(d) = [c(d-1) - c(d)]/[c(d-1) \cdot c(d)]$.

9.5.4. The steps in the requested flow chart are as follows: (1) generate the index $I(t)$ of inflation for period $t - t_{max}, t + t_{max}$ (see section 7.3(c, d)); (2) generate the rate of interest $j(t)$ (section 8.5(f)); (3) generate the mixing variable $q(t)$ (sections 9.2(c, d, e)); (4) calculate $n(u)$ (9.2.9); (5) calculate $n(u, t)$ (9.5.13); (6) Calculate μ_X, σ_X and γ_X for each cell (u, t) (9.3.5); (7) generate $X(u, t)$ (section 5.4(c)), see also section F.5 of Appendix F.

9.5.5. (1) Generate $X(u, t)$ for all the cells of Figure 9.5.1 (see Exercise 9.5.4); (2) Calculate $C(t)$ (according to (9.5.2), (9.5.3) and some of the rules of section 9.5.2); (3) Calculate $R(t)$ from (9.5.14).

9.5.6. This formula can be obtained by direct reasoning from the claim transactions and credited interest, or by noting that claims $X(u, t + \tau)$, where $u < t$, constitute in reserve $C(t-1)$ a term $X(u, t + \tau) \cdot v^{\tau + \frac{1}{2}}$ and in $C(t)$ a term $X(u, t + \tau) \cdot v^{\tau - \frac{1}{2}}$, hence they contribute in (9.5.16) to the term $C(t) - C(t-1)$ by

$$X(u, t + \tau) \cdot v^{\tau - \frac{1}{2}} \cdot (1 - v) \simeq X(u, t + \tau) \cdot v^{\tau - \frac{1}{2}} \cdot j$$

or if $\tau = 0$, by $X(u, t + \tau) \cdot \frac{1}{2} j$. The correction term in (9.5.16) consists of these terms inherent from all of the relevant cohorts. It corresponds to a similar increase in incomes in the yield gained for the reserve.

It depends on the application and definitions whether it is appropriate to write down this term explicitly or to let it be embedded in the claim and income terms. For example, the simulated X's might not give a mean which is an unbiased estimate of the postulated μ_X, if this term is omitted.

Note that $1 - v$ was approximated by j. A strict expression would be $1 - v = j \cdot v$.

10.3.1. $100 = \alpha_m \cdot 5 \cdot 10$ implies $\alpha_m = 2$; total initial cost is $5 \cdot 10 + 30 = 80$, it is amortized in 9 years.

12.3.1. Since X_t and P_t are independent, we have

$$E(X_t/P_t) = E(X_t) \cdot E(1/P_t) > E(X_t)/E(P_t),$$

by Jensen's inequality, since $1/x$ is concave for $x > 0$ (see Exercise 3.2.9).

14.6.1. $U_{req} = y_\varepsilon \cdot P \cdot \sqrt{r_2/n + \sigma_q^2} - \lambda \cdot P = 4.97$ in £ million.
(a) $T = 2$; n is now $2 \cdot 10\,000$ and $U_{req} = 7.18$, increase being 44%.
(b) $y_{0.001} = 3.09$; $U_{req} = 7.25$, increase 46%.

F.4.1. From the estimates of the moments about zero

$$\hat{\alpha}_k = \frac{1}{s} \sum_{i=1}^{s} u_i^k$$

the required characteristics can be obtained by means of (1.4.22) and (1.4.25).

G.1. Having regard to $\kappa_j(a \cdot x) = a^j \cdot \kappa_j(x)$ (see Exercise 1.4.9), the additivity of the cumulants and that for the equilibrium distribution $t = \infty$ we have

$$\kappa_j(x) = \sum_{i=0}^{\infty} \kappa_j(\varepsilon) \cdot \alpha^{ji} = \frac{\kappa_j(\varepsilon)}{1 - \alpha^j}.$$

The required formulae of μ_X, σ_X and γ_X are obtained by substituting into (1.4.25). The formula for $\rho(\tau)$ is derived similarly.

G.2. Solution is left to the readers.

G.3. Denote $x(t) - \mu = f(t) = Ab^t$. Then

$$f(t) - \alpha_1 f(t-1) - \alpha_2 f(t-2) = Ab^{t-2}[b^2 - \alpha_1 b - \alpha_2] = 0.$$

Equating the expression in brackets to zero, two values are obtained

$$b = \tfrac{1}{2}\alpha_1 \pm \tfrac{1}{2}\sqrt{\alpha_1^2 + 4\alpha_2},$$

and the general solution is

$$f(t) = A_1 b_1^t + A_2 b_2^t.$$

The coefficients A_1 and A_2 are determined by initial conditions.
The solution is oscillating if b_1 and b_2 are complex, i.e. when

$$\alpha_1^2 + 4\alpha_2 < 0.$$

Moving into polar coordinates

$$b_{1,2} = re^{\pm i\omega},$$

with

$$r = \sqrt{-\alpha_2} \quad \text{and} \quad \omega = \arctan\left[\frac{\sqrt{-\alpha_1^2 - 4\alpha_2}}{\alpha_1}\right] \quad (0 < \omega < \pi).$$

The general solution can be written in the form

$$f(t) = A_1 r^t e^{i\omega t} + A_2 r^t e^{-i\omega t},$$

or, since $e^{i\omega t} = \cos(\omega t) + i\sin(\omega t)$ and by some straightforward transformations

$$f(t) = Ar^t \sin(\omega t + v),$$

where A and v are other coefficients. The oscillation is damped if $r < 1$. The wavelength is directly seen from condition $2\pi = \omega T$, i.e. $T = 2\pi/\omega$.

G.4.

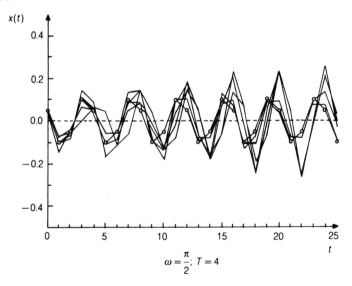

$$\omega = \frac{\pi}{2};\ T = 4$$

G.5. The solution is straightforward. The differences between the linear and logarithmic approaches prove insignificant (providing that μ, resp. μ_r are derived separately from the observed i_x- and the logarithmic increment series).

Bibliography

AB = ASTIN Bulletin
CA = Transactions of the International Congress of Actuaries
AF = Proceedings of the AFIR Colloquium
JIA = Journal of the Institute of Actuaries
$JIASS$ = Journal of the Institute of Actuaries Students' Society
$JRSS$ = Journal of the Royal Statistical Society
MS = Mitteilungen der Vereinigung Schweizerischer Versicherungs-
mathematiker
$PCAS$ = Proceedings of the Casualty Actuarial Society
SA = Scandinavian Actuarial Journal
TFA = Transactions of the Faculty of Actuaries
TSA = Transactions of the Society of Actuaries

Abramowitz, M. and Stegun, I. (1970) *Handbook of mathematical functions*, Dover Publications, London.
Adelson, R.M. (1966) Compound Poisson distributions. *Operational Research Quarterly*, **17**, 73–5.
Ajne, B. and Sandström, A. (1991) New standard regulations regarding allocation of the safety reserve in Sweden. *Transactions of the ASTIN Colloquium* 1991, Stockholm, 3–28.
Akhurst, R.B., Abbott, W.M., Cooper, S.M., *et al.* (1988) Corporate planning in general insurance, Report presented to the General Insurance Study Group in Harrogate.
Ammeter, H. (1948) A generalization of the collective theory of risk in regard to fluctuating basic probabilities. *SA*, **31**, 171–98.
Anderson, J.C.H. (1959) Gross premium calculations and profit measurement for non-participating insurance. *TSA*, **11**, 357–420.
Anscombe, F.J. (1960) Notes on sequential sampling plans. *JRSS Series A*, **123**, 297–306.
Ashe, F. (1986) An essay at measuring the variance of estimates of outstanding claim payments. *AB*, **16**, S99–S113.
Balzer, L.A. and Benjamin, S. (1980) Dynamic response of insurance

systems with delayed profit–loss sharing feedback to isolated un-predicted claims. *JIA*, **107**, 513–28.

Beard, R.E. (1956) Some statistical aspects of non-life insurance. *JIASS*, **13**, 139–57.

Beard, R.E., Pentikäinen, T. and Pesonen, E. (1984) *Risk theory*, 3rd edition, Chapman & Hall, London.

Becker, F. (1981) Analyse und Prognose von wirtschaftlichen Zeitreihen der Deutschen Schaden- und Unfallsversicherung, Inaugural-Dissertation, Universität Mannheim. *Veröffentlichungen des Instituts für Versicherungswissenschaft der Universität Mannheim*, **19**.

Benjamin, S. (1989) Driving the pension fund. *JIA*, **116**, 717–35.

Benktander, G. (1984) An approach to credibility in calculating IBNR for casualty excess reinsurance. *GIRO Bulletin*, **36**, 19–21.

Bernoulli, D. (1738) Specimen theoriae novae de mensura sortis. *Commentarii Academicae Scieantiarum Imperialis Petropolitae* V, 175–92.

Bertram, J. (1981) Numerischen Berechnung von Gesamtschaden-verteilung. *Blätter der Deutschen Gesellschaft für Versicherungsmathematik,* **15**, 175–94.

Bohlmann, G. (1909) Die Theorie des mittleren Risikos in der Lebensversicherung. *CA*, **1.1**, 593–683.

Bohman, H. and Esscher, F. (1964) Studies in risk theory with numerical illustrations concerning distribution functions and stop loss premiums. *SA*, **46**, 173–225 and **47**, 1–40.

Bonsdorff, H. (1990) On changing the parameter of exponential smoothing in experience rating. *AB*, **20**, 191–200.

Bonsdorff, H. (1991) A model for investment return: asymptotic behaviour. *AF*, **4**, 27–37.

Borch, K. (1960) Reciprocal reinsurance treaties. *AB*, **1**, 170–91.

Borch, K. (1961) The utility concept applied to the theory of insurance. *AB*, **1**, 245–55.

Borch, K. (1962) Application of game theory to some problems in automobile insurance. *AB*, **2**, 208–21.

Borch, K. (1963) Recent developments in economic theory and their application to insurance. *AB*, **2**, 322–41.

Borch, K. (1974) *The mathematical theory of insurance*, Lexington Books, Lexington.

Bornhuetter, R.L. and Ferguson, R.E. (1972) The actuary and IBNR. *PCAS*, **59**, 181–95.

Borregaard, J. Dengsoe, C., Hertig, J., et al. (1991) Equalization reserves: Reflections by a Danish working party. *Transactions of the ASTIN Colloquium* 1991, Stockholm, 61–70.

Bowers, N., Gerber, H., Hickman, J., *et al.* (1986) *Actuarial mathematics*, Society of Actuaries, Schaumburg.

Box, G.E.P. and Cox, D.R. (1964) An analysis of transformations. *JRSS Series B*, **26**, 211–52.

Box, G.E.P. and Jenkins, G.M. (1976) *Time-series analysis*, Holden Day, New York.

Boyle, P.P. (1978) Immunization under stochastic models of the term structure. *JIA*, **105**, 177–87.

Brender, A. (1988) Solvency requirements for life insurers in Canada. *CA*, **1**, 49–59.

Brennan, M.J. and Schwartz, S.E. (1977) Savings bonds, retractable bonds and callable bonds. *Journal of Financial Economics*, **5**, 67–88.

Bühlmann, H. (1970) *Mathematical methods of risk theory*, Springer Verlag, Heidelberg.

Butsic, R.B. (1992) Solvency measurement for property-liability risk-based capital applications. *Insurer Financial Solvency*, Vol. I, 311–54, Casualty Actuarial Society, Arlington.

Campagne, C. (1961) Minimum standards of solvency of insurance firms–Report of the ad hoc Working Party on Minimum Standards of Solvency, OEEC, TP/AS(61)1.

Campagne, C. (1975) Minimum standards of solvency for insurance firms, OEEC, TFD/PC/565.

Canadian Institute of Actuaries (1992) *Standard of practice on dynamic solvency testing*, Ottawa.

Casualty Actuarial Society (1990) *Foundations of Casualty Actuarial Science*, Arlington.

Chatfield, C. (1989) *The analysis of time series*, 4th edition, Chapman & Hall, London.

Clark, G. (1992) Asset and liability modelling – the way ahead? Paper presented to the Staple Inn Actuarial Society, London.

Clarkson, R.S. (1991) A non-linear stochastic model for inflation. *AF*, **3**, 233–53.

Cornish, E.A. and Fisher, R.A. (1937) Moments and cumulants in the specification of distributions. *Rev. Int. Statist. Inst.*, **5**, 307–20.

Coutts, S.M. and Clark, G.J. (1991) A stochastic approach to asset allocation within a general insurance company. *AF*, **4**, 95–112.

Coutts, S.M. and Devitt, E.R.F. (1988) Simulation models and the management of a reinsurance company. *CA*, **1**, 109–17.

Coutts, S.M. Devitt, E.R.F. and Ross, G.A. (1984) A probabilistic approach to assessing the financial strength of a general insurance company. *CA*, **3**, 129–36.

Cox, D.R. and Miller, H.D. (1965) *The theory of stochastic processes*, Chapman & Hall, London.

Cox, J.C., Ingersoll, J.E. and Ross, S.A. (1977) A theory of the term structure of interest rates and the valuation of interest-dependent claims, Paper presented at the Western Finance Association meeting, Anaheim.

Cox, J.C., Ingersoll, J.E. and Ross, S.A. (1985) A theory of the term structure of interest rates. *Econometrica*, **53**, 385–407.

Cramér, H. (1926) Review of F. Lundberg, *SA*, **9**, 223–45.

Cramér, H. (1930) On the mathematical theory of risk. *Skandia Jubilee Volume*, Stockholm.

Cummins, J.D. and Derrig, R.A. (1988) *Classical insurance solvency theory*, Kluwer Academic Publishers, Dordrecht.

Cummins, J.D. and Derrig, R.A. (1989) *Financial models of insurance solvency*, Kluwer Academic Publishers, Dordrecht.

Cummins, J.D. and Derrig, R.A. (1991) *Managing the insolvency risk of insurance companies*, Kluwer Academic Publishers, Dordrecht.

Cummins, J.D. and Harrington, S.A. (1987) *Fair rate of return in property-liability insurance*, Kluwer-Nijhoff Publishing, Dordrecht.

D'Arcy, S.P. and Doherty, N.A. (1988) *The financial theory of pricing property-liability insurance contracts*, Huebner Foundation, Monograph 15, Philadelphia.

Daykin, C.D. (1976) Long-term rates of interest in the valuation of a pension fund. *JIASS*, **21**, 286–340.

Daykin, C.D. (1984) The development of concepts of adequacy and solvency in non-life insurance in the EEC. *CA*, **3**, 299–309.

Daykin, C.D. (1987) Further thoughts on long-term rates of interest in the valuation of a pension fund. *JIASS*, **30**, 117–40.

Daykin, C.D., Ballantine, D.G. and Anderson, W.D.B. (1993) Modelling the assets and liabilities of a pension fund. *AF*, **2**, 525–37.

Daykin, C.D., Bernstein, G.D., Coutts, S.M., *et al.* (1987) Assessing the solvency and financial strength of a general insurance company. *JIA*, **114**, 227–310.

Daykin, C.D., Bernstein, G.D., Coutts, S.M., *et al.* (1987a) The solvency of a general insurance company in terms of emerging costs. *AB*, **17**, 85–132 and in Cummins and Derrig (1989).

Daykin, C.D., Devitt, E.R., Khan, M.R. and McCaughan, J.P. (1984) The solvency of general insurance companies, *JIA*, **111**, 279–336.

Daykin, C.D. and Hey, G.B. (1990) Managing uncertainty in a general insurance company, *JIA*, **117**, 173–277.

De Finetti, B. (1957) Su un'impostazione alternativa della teoria collettiva del rischio. *CA*, **2**, 433–43.

De Moivre, A. (1725) *Annuities on lives.*

De Wit, J. (1671) *The value of life annuities in proportion to redeemable rents.* English translation reprinted in *JIA*, **3**, 93–120.

De Wit, G. and Kastelijn, W. (1980) The solvency margin in non-life companies. *AB*, **11**, 136–44.

D'Hooge, L. and Goovaerts, M.J. (1976) Numerical treatment of the determination of the structure function of a tariff class. *CA*, **1**, 53–65.

Dingell, J.D. (1990) Failed promises: Insurance company insolvencies. *Report of the Subcommittee on Oversight and Investigations of the Committee on Energy and Commerce, US House of Representatives,* February 1990.

Driessen, T. (1988) *Cooperative games,* Kluwer, Dordrecht.

Dufresne, D. (1988) Movements of pension contributions and fund levels when rates of return are random. *JIA*, **115**, 535–44.

van Eeghen, J. (1981) *Loss reserving methods.* Surveys of Actuarial Studies 1, Nationale-Nederlanden, N.V.

van Eeghen, J., Greup, E.K. and Nijssen, J.A. (1983) *Rate making.* Surveys of Actuarial Studies 2, Nationale-Nederlanden, N.V.

Eggenberger, F. and Pólya, G. (1923) Über die Statistik der vergetteter Vorgänge. *Zeitschrift für angewandte Mathematik und Mechanik,* **3**, 279–89.

Elton, E.J. and Gruber, M.J. (1987) *Modern portfolio theory and investment analysis,* Wiley, New York.

Engle, R.F. (1982) Autoregressive conditional heteroscedasticity with estimates of the variance of the United Kingdom inflation. *Econometrica,* **50**, 987–1007.

Faculty of Actuaries' Solvency Working Party (1986) The solvency of life assurance companies. *TFA*, **39**, 251–340.

Faculty of Actuaries' Valuation Reserve Group (1989) Bonus rates, valuation and solvency during transition between higher and lower investment returns. *TFA*, **40**, 490–585.

Feigin, P.D. and Tweedie, R.L. (1985) Random coefficient autoregressive processes, a Markov chain analysis of stochasticity and finiteness of moments. *Journal of Time Series Analysis,* **6**, 1–14.

FIMAG (1992) *see* Geoghegan *et al.* (1992).

Fine, A.E.M., Headon, C.P., Hewitson, T.W., *et al.* (1988) Proposals for the statutory basis of valuation of the liabilities of linked long-term insurance business. *JIA*, **115**, 555–630.

Finnish Working Party, Solvency of general insurers (1982), *see* Pentikäinen and Rantala (1982).

Finnish Working Party, Solvency of pension insurers (1987), *see*

Pentikäinen *et al.* (1987).

Finnish Working Party, Solvency of life insurers (1992), *see* Rantala *et al.* (1992).

Friedman, J.W. (1977) *Oligopoly and the theory of games*, North-Holland, Amsterdam.

Galbraith, J. (1974) *Economics and the public purpose*, Deutsch, London.

Geoghegan, T.J., Clarkson, R.S., Feldman, K.S., *et al.* (1992) Report on the Wilkie stochastic model. *JIA*, **119**, 173–228.

Gerber, H. (1979) *An introduction to mathematical risk theory*, S.S. Huebner Foundation, University of Pennsylvania, Philadelphia.

Godolphin, E.J. (1980) Specifying univariate models for the de Zoete equity index, Maturity Guarantees Working Party. *JIA*, **107**, 116–33.

Goovaerts, M.J. and Hoogstad, W.J. (1987) *Credibility theory*, Nationale-Nederlanden, N.V.

Goovaerts, M.J., DeVylder, F. and Haezendonck, J. (1984) *Insurance premiums*, North-Holland, Amsterdam.

Gossiaux, A.-M. and Lemaire, J. (1981) Méthodes d'ajustement de distributions de sinistres. *MS*, 1981, 87–94.

Gragnola, J.B. (1985) *Corporate modeling at All State Insurance Company, strategic planning and modeling in property-liability insurance*, Kluwer-Nijhoff, Boston.

Grote, J.D. (1975) *The theory and application of differential games*, D. Reidel Publishing Company, Dordrecht.

Haberman, S. (1990a) Variability of pension contributions and fund levels with random and autoregressive rates of return. *ARCH*, 1990, **1**, 141–71.

Haberman, S. (1990b) Stochastic approach to pension funding methods. *AF*, **4**, 93–112.

Haberman, S. (1992) Pension funding methods and autoregressive interest rates. *CA*, **2**, 83–98.

Haldane, J.B.S. (1938) The approximate normalization of a class of frequency distributions. *Biometrika*, **29**, 392–404.

Halley, E. (1693) An estimate of the degrees of the mortality of mankind. Reprinted in *JIA*, **112**, 283–301.

Hardy, M. (1991) *Aspects of the assessment of life office solvency.* Paper presented to Third International Conference on Insurance Finance and Solvency, Rotterdam.

Hart, D.G., Buchanan, R.A. and Howe, B.A. (1987) *Actuarial practice of general insurance*, Institute of Actuaries of Australia, Sydney.

Heiskanen, J. (1982) Degree of loss and surplus reinsurance of fire insurance. Unpublished research work, Helsinki.

Hertig, J. (1992) *How to use internal equalisation funds to replace or complement reinsurance*, International Co-operative Reinsurance Bureau.

Hewitt, C.C. (1986) Discussion contribution on claims reserves, *Insurance: Mathematics & Economics*, **5**, 31–3.

Hogg, R.V. and Klugman, S.A. (1984) *Loss distributions*, Wiley, New York.

Huang, C.-F. and Litzenberger, R.H. (1988) *Foundations for financial economics*, Elsevier Science Publishing Co., Amsterdam.

Ibbotson, R.G. and Sinquefield, R.A. (1977) *Stocks, bonds, bills, and inflation: the past (1926–1976) and future (1977–2000)*, Financial Analysts Research Foundation, Charlottesville.

Ibbotson, R.G. and Sinquefield, R.A. (1982) *Stocks, bonds, bills, and inflation: the past and the future*, Financial Analysts Research Foundation, Charlottesville.

Institute of Actuaries (1991) *Claim reserving manual*, London.

Jewell, W. (1980) Models in insurance. *CA*, S, 87–141.

Jewell, W. and Sundt, B. (1981) Improved approximations for the distribution of a heterogeneous risk portfolio. *MS*, 1981, 221–39.

Johnson, N.L. and Kotz, S. (1969) *Distributions in statistics. Discrete distributions*, Wiley, New York.

Johnson, P.D. and Hey, G.B. (1971) Statistical studies in motor insurance. *JIA*, **97**, 199–249.

Kastelijn, W.M. and Remmerswaal, J.C.M. (1986) *Solvency*. Surveys of Actuarial Studies 3, Nationale-Nederlanden, N.V.

Kaufman, A.M. and Liebers, E.C. (1992) NAIC risk-based capital efforts in 1990–91, *Insurer Financial Solvency*, Vol. I, 123–78, Casualty Actuarial Society, Arlington.

Kauppi, L. and Ojantakanen, P. (1969) Approximations of the generalised Poisson function. *AB*, **5**, 213–26.

Kendall, M.G. and Stuart, A. (1977) *The advanced theory of statistics*, Charles Griffin & Co., London.

Kotler, P. (1975) *Marketing decision making, a model building approach*, Holt, Rinehart and Winston, New York.

Landin, D. (1980) Risk accumulation. A lecture paper.

Lee, E.M. (1986) *Introduction to pension schemes*, Institute of Actuaries, London.

Lee, R.E. (1985) A prophet of profits. *JIASS*, **28**, 1–42.

Lemaire, J. (1977) Echange de risques entre assureurs et des jeux. *AB*, **9**, 155–80.

Lemaire, J. (1984) An application of game theory: cost allocation. *AB*, **14**, 61–81.

Levin, A.M. (1986) Tough lessons. *Business Insurance*, 24 November 1986.

Loades, D.H. (1988) Assessing the security of pension fund valuation bases using a stochastic investment model. *CA*, **5**, 105–23.

Loades, D.H. (1992) Instability in pension funding. *CA*, **2**, 137–54.

Loimaranta, K., Jacobsson, J. and Lonka, H. (1980) On the use of mixture models in clustering multivariate frequency data. *CA*, **2**, 147–61.

Lundberg, F. (1909) *Über die Theorie der Rückversicherung*. *CA*, **1.2**, 877–955.

Lundberg, F. (1919) *Teorin för riskmassor*, Försäkringsinpektionen, Stockholm.

MacDonald, A.S. (1993) What is the value of a valuation? *AF*, **2**, 725–43.

Mariathasan, J.W.E. and Rains, P.F. (1993) Strategic financial management in a general insurance company. *AF*, **2**, 745–69.

Maturity Guarantees Working Party (1980) Report. *JIA*, **107**, 103–231.

Meyers, G. and Schenker, R. (1983) Parameter uncertainty in the collective risk model. *PCAS*, **70**, 111–43.

Molenaar, W. (1974) Approximations to the Poisson, binomial and hypergeometric distribution functions, *Math. Centre Transactions*, Amsterdam.

Mooney, S. (1986) The cyclicality of property and casualty insurance: is federal regulation the answer? *National Underwriter*, 28 March 1986.

Morgan, K., Akhurst, R.B. Orros, G.C., *et al.* (1992) *Equalisation reserves on a European basis*, General Insurance Study Group, Bournemouth.

Neill, A. (1977) *Life contingencies*, Heinemann, London.

von Neumann, J. and Morgenstern, O. (1947) *Theory of games and economic behaviour*, 2nd edition, Princeton University Press, Princeton, reprinted by Wiley, 1964.

Norberg, R. (1986) A contribution to the modelling of IBNR claims. *SA*, 1986, 155–203.

Oppenheimer, P. (1991) The cycle of opportunity. *The Economist*, June.

Owen, G. (1969) *Game theory*, W.B. Saunders, London.

Panjer, H. (1981) Recursive evaluation of a family of compound distributions. *AB*, **12**, 22–6.

Panjer, H.H. and Willmot, G.E. (1992) *Insurance risk models*, Society of Actuaries, Schaumburg.

Patrik, G. (1980) Estimating casualty insurance loss amount distributions. *PCAS*, **67**, 57–109.

Pentikäinen, T. (1952) On the net retention and solvency of insurance companies. *SA*, **35**, 71–92.

Pentikäinen, T. (1968) Linking life and private pension insurance to the price index. *CA*, **1**, 847–59.

Pentikäinen, T. (1977) On the approximation of the total amount of claims. *AB*, **9**, 281–9.

Pentikäinen, T. (1979) Dynamic programming, an approach for analysing competition strategies. *AB*, **10**, 183–94.

Pentikäinen, T. (1987) Approximative evaluation of the distribution function of aggregate claims. *AB*, **17**, 15–39.

Pentikäinen, T., *et al.* (1987) Equalization reserves for pension insurance, Report presented to the Ministry of Social Affairs and Health, available in Finnish only.

Pentikäinen, T. and Pesonen, M. (1988) Stochastic dynamic analysis of life insurance. *CA*, **1**, 421–37.

Pentikäinen, T. and Rantala, J. (1982) *Solvency of insurers and equalization reserves, Vols. I and II*, The Insurance Publishing Company, Helsinki.

Pentikäinen, T. and Rantala, J. (1986) On the run-off risk. *AB*, **16**, 113–47.

Pentikäinen, T. and Rantala, J. (1992) A simulation procedure for comparing different claims reserving methods. *AB*, **22**, 191–216.

Pentikäinen, T., Bonsdorff, H., Pesonen, M., *et al.* (1989) *Insurance solvency and financial strength*, Insurance Publishing Company, Helsinki.

Pentikäinen, T., Bonsdorff, H., Pesonen, M., *et al.* (1992) Capital needs of an insurer. *CA*, **2**, 213–31.

Pesonen, M. (1984) Optimal reinsurance. *SA*, 1984, 65–90.

Pesonen, M. (1989) Random number generators for compound distributions. *SA*, 1989, 47–60.

Pitkänen, P. (1975) Tariff theory. *AB*, **8**, 204–28.

Press, W.H., *et al.* (1986) *Numerical recipes,* Cambridge University Press, Cambridge.

Priestley, M.B. (1980) State dependent models: a general approach to non-linear time series analysis. *Journal of Time Series Analysis*, **1**, 47–71.

Priestley, M.B. (1981) *Spectral analysis and time series*, Academic Press, London.

Purchase, D.E., Fine, A.E.M., Headdon, C.P., *et al.* (1989) Reflections on resilience: some considerations of mismatching tests, with particular reference to non-linked long-term insurance business. *JIA*, **116**, 347–452.

Ramlau-Hansen, H. (1988) A solvency study in non-life insurance, Part I: Analysis of fire, windstorm and glass claims. *SA*, 1988, 3–34.

Rantala, J. (1984) An application of stochastic control theory to insurance business, Ph.D. Thesis, University of Tampere, Finland.

Rantala, J. (1988) Frequency response functions – a technique to evaluate the variation ranges in certain insurance indicators. *CA*, **1**, 453–68.

Redington, F.M. (1952) A review of the principles of life office valuation. *JIA*, **78**, 286–340.

Reid, D.H. (1984) Solvency: the expression of the relationship between capital and insurance markets? *CA*, **3**, 375–90.

Renshaw, A.E. (1989) Chain ladder and interactive modelling (claims reserving and GLIM). *JIA*, **116**, 559–87.

Richard, S.E. (1976) *An analytical model of the term structure of interest rates*, Working paper, Graduate School of Industrial Administration, Carnegie-Mellon University, Pittsburgh, Pennsylvania.

Roff, T. (1992) Asset and liability studies on a with-profit fund. Paper presented to the Staple Inn Actuarial Society, October 1992, London.

Roth, R.J. Jr (1992) California earthquake zoning and probable maximum loss evaluation program, California Department of Insurance, Los Angeles.

Rubinstein, R. (1981) *Simulation and the Monte Carlo method*, Wiley, New York.

Sandström, A. (1991) On moment corrections when data are grouped into non-equidistanced intervals. *Transactions of the ASTIN Colloquium*, Stockholm, 333–52.

Seal, H. (1971) Numerical calculation of the Bohman–Esscher family of convolution-mixed negative binomial distribution functions. *MS*, **71**, 71–94.

Seal, H. (1977) Approximations to risk theory's $F(x, t)$ by means of the gamma distribution. *AB*, **9**, 213–18.

Smart, I.C. (1977) Pricing and profitability in a life office. *JIA*, **104**, 125–58.

Smith, M. (1981) The underwriting cycle in property and casualty insurance, University of Utah, Salt Lake City, Utah.

Stanard, J.N. (1986) A simulation test of prediction errors of loss reserve estimation techniques. *PCAS*, **72**, 124–53.

Straub, E. (1984) Actuarial remarks on planning and controlling in reinsurance. *AB*, **14**, 183–91.

Sundt, B. (1990) On excess of loss reinsurance with reinstatements. *Transactions of the ASTIN Colloquium*, Montreux, 2.83–2.93.

Taylor, G.C. (1986) *Reserving in non-life insurance*, North-Holland, Amsterdam.

Taylor, G.C. (1986a) Underwriting strategy in a competitive insurance environment. *Insurance Mathematics and Economics*, **5**, 59–77.

Taylor, G.C. (1988) An analysis of underwriting cycles in relation to insurance pricing and solvency. Published in Cummins and Derrig (1991).

Taylor, S. (1986) *Modelling financial time series*, Wiley, New York.

Thompson, J. (1979) *Non-life insurance business game*, GRE, London.

Tiller, M.W. (1990) Individual risk rating. In *Casualty Actuarial Society*, Chapter 3.

Tuomikoski, J. (1987) Simulation of pension insurance companies. An appendix to committee report 1987:26, Ministry of Social Affairs and Health, Helsinki (available in Finnish only).

Venezian, E. (1985) Ratemaking methods and profit cycles in property and liability insurance. *Journal of Risk and Insurance*, **52**, 477–500.

Verrall, R.J. (1989) A state space representation of the chain ladder linear model. *JIA*, **116**, 589–609.

Verrall, R.J. (1990) Bayes and empirical Bayes estimation for the chain ladder model. *AB*, **20**, 217–43.

Wilkie, A.D. (1984) Steps towards a comprehensive stochastic investment model. *Occasional Actuarial Research Discussion Paper No. 36*. Institute of Actuaries, London.

Wilkie, A.D. (1985) Portfolio selection in the presence of fixed liabilities: a comment on 'The matching of assets to liabilities'. *JIA*, **112**, 229–77.

Wilkie, A.D. (1986) A stochastic investment model for actuarial use. *TFA*, **39**, 341–403.

Wilkie, A.D. (1986a) Some applications of stochastic investment models. *JIASS*, **29**, 25–51.

Wilkie, A.D. (1990) Modern portfolio theory – some actuarial problems. *AF*, **1**, 199–215.

Wilkie, A.D. (1991) Introductory review of the papers for the 2nd

AFIR International Colloquium. *AF*, **1**, 1–22.

Willmot, G. (1986) A class of counting distributions with insurance applications. Thesis presented to the University of Waterloo, Waterloo.

Wilson, E.B. and Hilferty, M. (1931) The distribution of chi-square. *Proceedings of the National Academy of Science, USA*, **17**, 684–8.

Wise, A.J. (1984) The matching of assets to liabilities. *JIA*, **111**, 445–501.

Wise, A.J. (1987) Matching and portfolio selection. *JIA*, **114**, 113–33, 551–68.

Subject index

Note:

Pages marked in bold represent the reference where the word is defined or introduced.

Author index

There are numerous references throughout the book to *Risk Theory* by Beard *et al.*,*Insurance Solvency and Financial Strength* by Pentikäinen *et al.*, *Managing Uncertainty in a General Insurance Company* by Daykin and Hey and the papers of the British Solvency Working Party (Daykin, Coutts, Bernstein, Devitt, Hey, Reynolds and Smith). These references are listed separately under the group of authors and are not included under the names of individual authors.